Smart Nanomaterials Technology

Series Editors

Azamal Husen, Wolaita Sodo University, Wolaita, Ethiopia

Mohammad Jawaid, Laboratory of Biocomposite Technology, Universiti Putra Malaysia, INTROP, Serdang, Selangor, Malaysia

Nanotechnology is a rapidly growing scientific field and has attracted a great interest over the last few years because of its abundant applications in different fields like biology, physics and chemistry. This science deals with the production of minute particles called nanomaterials having dimensions between 1 and 100 nm which may serve as building blocks for various physical and biological systems. On the other hand, there is the class of smart materials where the material that can stimuli by external factors and results a new kind of functional properties. The combination of these two classes forms a new class of smart nanomaterials, which produces unique functional material properties and a great opportunity to larger span of application. Smart nanomaterials have been employed by researchers to use it effectively in agricultural production, soil improvement, disease management, energy and environment, medical science, pharmaceuticals, engineering, food, animal husbandry and forestry sectors.

This book series in Smart Nanomaterials Technology aims to comprehensively cover topics in the fabrication, synthesis and application of these materials for applications in the following fields:

- Energy Systems—Renewable energy, energy storage (supercapacitors and electrochemical cells), hydrogen storage, photocatalytic water splitting for hydrogen production
- Biomedical—controlled release of drugs, treatment of various diseases, biosensors,
- Agricultural—agricultural production, soil improvement, disease management, animal feed, egg, milk and meat production/processing,
- Forestry—wood preservation, protection, disease management
- Environment—wastewater treatment, separation of hazardous contaminants from wastewater, indoor air filters

Azamal Husen · Rakesh Kumar Bachheti ·
Archana Bachheti
Editors

Current Trends in Green Nano-emulsions

Food, Agriculture and Biomedical Sectors

Springer

Editors
Azamal Husen
Wolaita Sodo University
Wolaita, Ethiopia

Rakesh Kumar Bachheti
Department of Industrial Chemistry
Addis Ababa Science and Technology University
Addis Ababa, Ethiopia

Archana Bachheti
Department of Environment Science
Graphic Era University
Dehradun, Uttarakhand, India

ISSN 3004-8273 ISSN 3004-8281 (electronic)
Smart Nanomaterials Technology
ISBN 978-981-99-5397-4 ISBN 978-981-99-5398-1 (eBook)
https://doi.org/10.1007/978-981-99-5398-1

This Springer imprint is published by the registered company Springer Nature Singapore Pte Ltd.
The registered company address is: 152 Beach Road, #21-01/04 Gateway East, Singapore 189721, Singapore

Preface

Recent developments in nanotechnology have opened up new opportunities for several applications in a range of sectors, including food, cosmetics, drugs, biomedicine, environment, and agricultural sectors. Nanoemulsions are one of the promising formulations based on nanotechnology that have drawn more and more attention because of their unique properties and potential applications. Nanoemulsions are emulsions with droplet size on the order of 100 nm. Nanoemulsion contains oil, water, and an emulsifier. The addition of an emulsifier is critical for the creation of small-sized droplets. Emulsifier plays an important role in stabilizing the nanoemulsions. Mainly, emulsifier is used as a surfactant, but proteins and lipids have also been effective in the preparation of nanoemulsions. Nanoemulsions have a large surface area for interaction and better bioavailability due to their small droplet sizes, transparency, and heat stability.

The chapters in this book, written by experts in the area, cover a range of subjects pertaining to green nanoemulsions. The book is divided into 16 chapters that discuss the introduction, properties, and applications of green nanoemulsions in a range of fields, including agriculture, food production, cosmetics, pharmaceuticals, and biomedicine. The first chapter introduces readers to the concept of green nanoemulsions and summarizes its features and benefits over conventional emulsions. The ensuing chapters go over the creation and characterization of citrus fruit, seed, and essential oil-based nanoemulsions as well as their potential applications in food and agricultural preservation. The use of green nanoemulsions in the treatment of stomach, breast, liver, and lung cancers as well as their antibacterial and anticancer capabilities are also covered in the book. The usage of green nanoemulsions in the medical, cosmetic, and wound-healing fields is also covered. Along with the safety criteria that apply to nanoemulsions in the food, agricultural, and biomedical sectors, the book also discusses the difficulties and potential uses of green nanoemulsions as a drug delivery mechanism.

Overall, this book aims to contribute to the recent understanding and in-depth details on the creation, evaluation, and use of green nanoemulsions. The book in

Basic of Nanoemulsions

Introduction to Green Nanoemulsions and Their Properties

Aashna Sinha, Manjari Bhatia, Pranchal Rajput, Kundan Kumar Chaubey, Manideep Sood, Atreyi Pramanik, Anis Kumar Pal, Sujata Jayaraman, Shalini Jha, Km. Bhawna, Rakesh Kumar Bachheti, and Archana Bachheti

Abstract The creation of effective and novel organic, nontoxic, and eco-friendly supplements is currently gaining popularity. The need for further greener products will increase exponentially over time. Healthy, green, and sustainable colloidal systems that satisfy these requirements and standards include green nanoemulsions. As the name implies, green nanoemulsions are nano-sized emulsions that are safer, more functionally effective, and kinetically stable than standard emulsions. Oil and water are often distributed into one another to form nanoemulsions, with one of the components acting as a medium (continuous or external phase). The other, however, is suspended within the medium (discontinuous or internal phase). The synthesis of biological nanoemulsion is called 'Green Chemistry', as it is the more advanced field in science and technology for enhancing biomedicine and plant based or agricultural studies. The size of nanoparticles in the nanoemulsions are extremely small that ranges between 1 and 100 nm. Phase inversion, micro fluidization, solvent dispersion etc. are some of the method for the production of nanoemulsions. Different microscopy and electronic instruments are used for the characterization of nanoemulsions like zeta potential, transmission electron microscopy etc. The usage of emulsion are in cosmetics, lubricants, nano-medicine, pain, catalyst and many more. The recent

A. Sinha · P. Rajput · K. K. Chaubey (✉) · M. Sood · A. Pramanik · A. K. Pal · S. Jayaraman · S. Jha · Km. Bhawna
Division of Research and Innovation, Uttaranchal University, Arcadia Grant, P.O. Chandanwari, Premnagar, Dehradun, Uttarakhand 248007, India
e-mail: kundan2006chaubey@gmail.com

M. Bhatia
Department of Bio-Technology, GLA University, Post-Chaumuhan, Dist. Mathura, Uttar Pradesh 281406, India

K. K. Chaubey
School of Basic and Applied Sciences, Sanskriti University, Mathura, Uttar Pradesh 281401, India

R. K. Bachheti
Centre of Excellence in Nanotechnology, Department of Industrial Chemistry, Addis Ababa Sciences and Technology University, P.O. Box-16417, Addis Ababa, Ethiopia

A. Bachheti
Department of Environment Science, Graphic Era University, Dehradun, Uttarakhand 248002, India

A. Husen et al. (eds.), *Current Trends in Green Nano-emulsions*, Smart Nanomaterials Technology, https://doi.org/10.1007/978-981-99-5398-1_1

advancements in science have come up with many new branches and catapulted the existing fields to new heights and one such field is nanotechnology that has led to advancements in many areas. Hence, the purpose of this chapter is based on the different methods for the production of nanoemulsions with their properties.

Keywords Nano-emulsion · Advantages · Emergent · Green chemistry · Properties · Sustainable · Green revolution

1 Introduction

Nanoemulsions are tiny size emulsions that are synthesized for enhancing the transfer of pharmaceutical active ingredients. The process in which two immiscible aqueous solution mix together with surfactant (Emulsifying Agent) to form one resultant phase that contains thermodynamically stable compounds are nanoemulsions. The emulsifying substance is sometimes referred to as interphase or intermediate. Miniemulsions, which are fine oil/water or water/oil dispersions stabilised by an interfacial coating of surfactant molecules with droplet sizes ranging from 20 to 600 nm, are also referred to as nanoemulsions. The transparency of nanoemulsions is due to their tiny size. There are three different forms of nanoemulsions that may be created: (a) water in oil nanoemulsions (water droplets are scattered in the continuous oil phase), (b) oil in water nanoemulsions (oil is disseminated in the continuous aqueous phase), and (c) bi-continuous nanoemulsions. The most important ingredients are surfactant and co-surfactant because they give nanoemulsions stability [34]. The next green revolution in India will be built on the paradigm of green nanotechnology. India, a developing nation with an agricultural economy, is important in defining the objectives of sustainable development (SDGs). With the intention of removing hunger and providing food-security for basic nutrition with the goal of sustainable agriculture. Promoting sustainable agriculture has been designated as goal 2 of the 17 sustainable development goals. In India, numerous initiatives have been made to eradicate hunger. Nanotechnology is a benefit for humanity since it is a powerful instrument for improving human wellbeing [40].

A growing number of economic sectors around the world are using nanoparticles, which is generating interest in their creation for environmental safety and biological synthesis [30–32, 37]. Physiochemical methods are frequently used to produce nanoparticles, which can be expensive and eventually result in environmental harm that is incurable [43]. Wide-ranging biological research is urgently required as an alternative, successful, inexpensive, and ecologically acceptable strategy for the creation of nanoparticles with particular features. The phytoconstituents (e.g. caffeine, saponins, vitamins, digoxin, peptides flavonoids etc.) found in different plant extracts were used for the formation of green nano compounds. These biomolecules act as a very good reducing agents that are important in the formation of nano-particulates [28, 29, 33, 56, 57, 73]. Similar to this, microorganisms have a larger number of biochemically active compounds that can interchange organic metallic ions into metallic nanoparticles, this has been discovered very recently and also the most understudied area of analysis [5, 6, 56, 60–65]. It is possible for some plants, fungi, bacteria, actinomycetes, and agricultural waste to biosynthesize nanoparticles. The usage of nano emulsions in several sectors, including the food, hygiene products, and pharmaceutical industries, is expanding due to its potential advantages over regular emulsions. Different names for nano-emulsion include submicron, tiny, and ultrafine. Different types of nano-emulsion surfactant used are ionic, anionic, cationic and zwitter ion surfactant rapidly [48]. Due to their comparatively small droplet size, they show higher amount of stability to droplet aggregation and creaming [46]. Small enough droplets in nano-emulsions (d50 nm) scatter light waves very weakly, making them useful for variety of different use. Low droplet concentrations can produce nano emulsions having high viscosity and gel like properties because the microscopic droplets' surrounding polymer or electrostatic layers overlap [8]. The stability, rheology, and appearance of nano-emulsions are highlighted in this chapter's summary of their physicochemical characteristics. The evolution of nano-emulsion-based output with desirable photo physical qualities and functional impute is discussed along with the bond between the properties of nano emulsions. By altering the rheological characteristics of macro emulsion and closely linked colloidal dispel, nano-emulsion is a latest, more enhanced physiological build of emulsion [44]. A dispel solution of oil and water is called an emulsion, and nano emulsions with a size of between 10 and 200 nm are stoichiometric fluctuating but kinetically organized [46]. It appears optically transparent or murky due to its small size. Hydrophilic components usually cover hydrophobic ones in a nano-emulsion. A three-part system called a nano-emulsion consists of liquid phase, a surfactant, and oil.

Here we discuss about the different methods for production of the nano-emulsions and their particular properties.

2 Green Nanoemulsion

Healthy, green, and sustainable colloidal systems that meet these requirements and standards include green nanoemulsions. As the name implies, green nanoemulsions are nanosized emulsions that are safer, more functionally effective, and kinetically stable than standard emulsions. Oil and water are typically dispersed into one another to form nanoemulsions, with one of the components acting as a medium (continuous or external phase). However, the other is suspended within the medium (discontinuous or internal phase) [51].

3 Different Methods for the Production of Green Nano-Emulsions

Two immiscible liquids are combined to form nanoemulsions, which are made up of 20–500 nm-sized droplets of one liquid phase (the dispersed phase) floating in the other liquid phase (the continuous phase). In the case of O/W nanoemulsions, (Fig. 1) in which oil droplets are suspended in water, oil is referred to as the dispersed phase and water as the continuous phase, and vice versa for W/O nanoemulsions. It is nearly always necessary to use a surfactant (or emulsifier) to stabilise nanoemulsion droplets. Sodium dodecyl sulphate (SDS, anionic) and cetyltrimethylammonium bromide (CTAB, cationic) are two typical ionic surfactants, whereas Tween, Span, and Brij are typical non-ionic surfactants. Although synthetic surfactants have been used to create the majority of nanoemulsions, certain research have also used natural biopolymers to create nanoemulsions. Nanoemulsion preparation can be divided into two groups of methods: high-energy methods and low-energy methods. High-energy methods are energy intensive methods and typically require power density input of 10^7–10^9 W/kg^1 [18]. Ultrasonic emulsification, high-pressure homogenization and micro fluidization, are all components of the high-energy emulsification process [71]. In contrast, low-energy methods only require power density input of 10^3–10^5 W/kg^1 [18]. Phase inversion temperature and spontaneous emulsification are all components of the low-energy emulsification technique [15, 36]. Reverse nanoemulsion may be created in a very viscous solution using a combination technique that combines high-energy and low-energy emulsification. The different methods for the process of nanoemulsions gel are given below.

4 High-Energy Methods Emulsification Processes

1. **High Pressure Homogenization**

High-pressure homogenization is necessary for the creation of nanoemulsion. Utilizing a high-pressure homogenizer or piston homogenizer, this method creates nanoemulsion with very small particle sizes (up to 1 nm) [17]. A combination of oil, water, and surfactant(s) is forced through a tiny gap of 5–10 mm in high-pressure homogenizers [11, 12, 19]. Due to the relatively tiny gap size, pressure droplets can reach up to a few thousand bars and are deformed by extremely high shear force values, which cause the drop to shatter into smaller pieces [22, 23]. The mixture must typically pass through the homogenizer 15–20 times before the droplet size stabilises [19, 21, 44].

2. **Ultrasonic Emulsification**

It involves the initial mixing of the emulsions which has the frequency of 20 khz that converts to nano-droplets size. High-pressure homogenization and ultrasonication have a similar mechanism. Electrical input in an ultrasonicator is transformed into

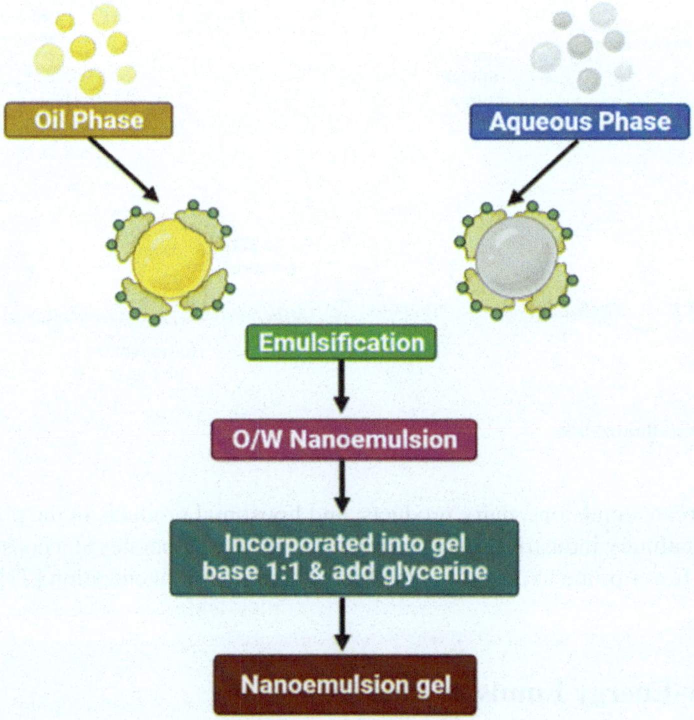

Fig. 1 Production of nanoemulsions gel

pressure changes that produce cavitation bubbles. The greater drops eventually split into smaller ones as the cavitation bubbles burst, creating shear forces in the process [19]. The ultrasonicator must operate for around 15 to 20 min until the droplet size becomes generally consistent, much like with the high-pressure homogenise [24]. The temperature in ultra sonication is maintained by water-jacket. It consists of probe made up of piso-electric quartz crystal for providing the energy. These probes expand and contract through changing of its voltage.

3. **Microfluidization**

A tool called a micro fluidizer is used in the patented mixing technique known as micro fluidization. The medicinal product is forced through the interaction chamber by means of high pressure, resulting in very small particles in the submicron range. To create a consistent nanoemulsion, the procedure is repeated numerous times to get the required particle size. A micro fluidizer employs shear forces and high pressure created by a pump to break down fibres. After being supplied into the intake, fibre suspension is pushed under intense pressure via a Y- or Z-shaped narrow channel. As a result, the suspension accelerates, producing a high shear rate that finally rips apart the fibres. According to a report, raising the pressure and the number of micro fluidizer cycles or reducing the chamber size enhances the degree of fibrillation. It is frequently

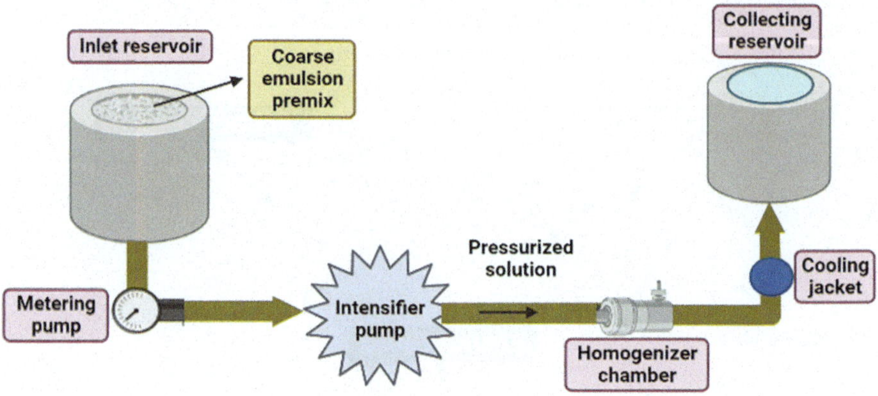

Fig. 2 Micro fluidization

used to create emulsions, dairy products, and liposomal products in the pharmaceutical and culinary industries. Because micro fluidization operates at a constant shear rate, it is fewer prone to clogging than high-pressure homogenization [77] (Fig. 2).

5 Low-Energy Emulsification Processes

1. **Phase Inversion Temperature**

The nano emulsion that are produced by this technique do not require any outside force as it involves production of dispersions when transition occur by altering its temperature. This method is also called persuasion method which is characterised into 4 phases: First phase includes the changing of only single variable for example temperature or optimal value. Second phase involves the change in more than one variable example composition and temperature. The Third phase involves the changing of low internal phase emulsion hence to convert it to external phase. Fourth phase is stabilized by aqueous crystal production for example nano droplets stabilization. When the surfactant's affinity for the water and oil phases is balanced in the PIT, an oil, water, and surfactant combination may be quickly chilled to create nanoemulsions [53]. The packing parameter (p), which is equal to the effective surfactant head group area divided by the effective surfactant tail group area, may be used to explain how a surfactant's behaviour changes with temperature [59]. O/W emulsions are more likely to develop at low temperatures because the hydrated surfactant head group occupies a higher cross-sectional area than the hydrophobic tail group (p 1) (Fig. 3).

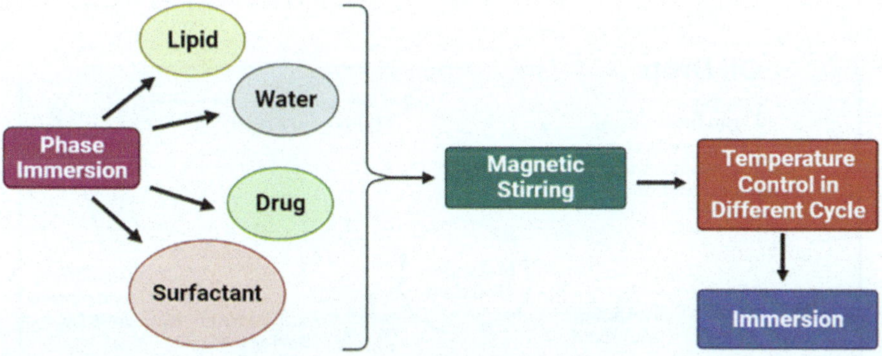

Fig. 3 Production of nanoemulsion by phase inversion

2. **Spontaneous Emulsification**

When immiscible liquids under non-equilibrium conditions come into contact, spontaneous emulsification may take place. Gradients in chemical potential between the phases cause this phenomena, which under some circumstances results in negative values of the emulsification's free energy. Droplet production can be accomplished in spontaneous emulsification without the need of external energy by touching two immiscible liquids that are not in equilibrium. As opposed to this, if immiscible liquids are in thermodynamic equilibrium, the energy to expand the interface, A (where A is the increase in interfacial area and T is the absolute temperature), is large and positive and cannot be offset by the small and positive entropy of dispersion, TS (where T is the absolute temperature and S is the increase in entropy) [66]. It contains 3 stages for the formation of nano emulsions: First, it includes organic solution formation that comprises of oil and surfactant in liquids miscible solvent and hydrophilic surfactant. The emulsion which is formed is converting organic phase into liquid phase under magnetic properties. In the third stage organic solvent is removed by heating. The nano emulsions size ranges between 50 and 100 nm [48] (Fig. 4).

6 Pros and Cons of High-Energy Methods Emulsification and Low-Energy Emulsification Processes

Microfluidization, high-pressure homogenization, or sonication are common examples of high-energy procedures that employ great mechanical energy to break apart macroscopic phases or droplets into smaller droplets [67]. This preparation process is unsuitable for many industrial applications due to the extremely high intensities needed to produce nanoemulsions with very minute droplets [70]. On the other hand, many low-energy techniques rely on the spontaneous generation of nanoemulsionsin particular system compositions or environmental circumstances as a result

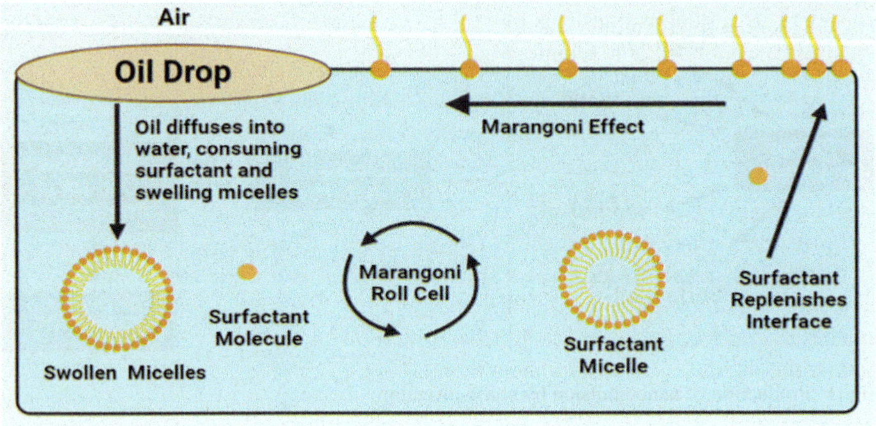

Fig. 4 Spontaneous emulsification

of modifications in the surfactant's optimal curvature [2, 4]. The composition can be maintained constant while the temperature is varied (the PIT technique, or phase inversion temperature), for example, to modify the curvature.

High-energy techniques are more reliable since there are less limitations on the surfactant that may be used. Additionally, they need less surfactant to create nanoemulsions. It has been demonstrated that high-energy approaches may create nanoemulsions with a dispersed phase volume percentage as high as 40% since the surfactant quantity is variable [26, 74]. But as was already said, high energy procedures need a lot more energy input than low energy ones. Furthermore, while high-energy techniques are inefficient from an energetic standpoint, they have a tendency to heat up the sample, which can lead to the breakdown of the material, particularly if biological components used to create the nanoemulsions are temperature-sensitive. However, although being energy-efficient, low-energy techniques are limited in the surfactant types that may be used and frequently need the use of two surfactants. Additionally, compared to high-energy procedures, a much larger amount of surfactant is needed. As a result, the scattered volume percentage is often lower than 25%. The final point is that low-energy nanoemulsions are less stable than high-energy techniques.

7 Properties of Nanoemulsions

1. **Stability**

Another way to define "stability" is the capacity of a nano emulsion to survive changes to its photochemical characteristics [47]. Nano emulsions are thermodynamically unstable due to the separated oil. When compared to the emulsified phases, the free energy of the aqueous phases is lower [48]. As a result, enough time is given, they always have a tendency to disintegrate, with the speed of change being influenced by the kinetic energy barriers with respect to height in the system. Nano emulsions may degrade via a variety of mechanisms, including phase inversion, separation on the basis of gravity, Ostwald ripening, coalescence, and flocculation [47]. It is crucial for a nano emulsion product to maintain physical and chemical stability when bring close to particular surrounding conditions during its production, storage, transit, and use (such as temperature, pH and mechanical forces). Nano emulsion might have to degrade in a different set of unique surrounding circumstances (such as when focusing the small intestine after ingestion). Nano emulsions' appearance, texture, and release characteristics can alter due to chemical and physical changes in their properties, which is crucial when developing them for specific purposes [47].

2. **Appearance or Transparency**

Because of the droplets' dimensions being close to the intensity of light (r), macro emulsions seem opaque and have a propensity to scatter light. Since the micro droplets in nano emulsions (10–100 nm in size) are lesser than the intensity of light which is visible (r), they are Crystal in nature or seem somewhat murky [20, 44, 48, 70]. A nano emulsion's overall appearance is determined by its optical characteristics, which depend on how light waves interact with it through its physical and characteristically properties [45]. A consumer's first sight of a nano emulsion-based output is typically its visible features, which has a significant bearing on whether they would use it. Nano emulsions' overall appearance have a effect by a variety of factors that has identical, cloudiness and colour [47]. When the beads are equally distributed to a complete nano emulsion, it seems homogeneous, but, when substantial bead accumulation, creaming, or greasing off takes place, it appears heterogeneous. Since the light that has been dispersed the oil droplets, ceremonial emulsions (200 nm–200 m frequently appear foggy or opaque. The droplet size in relation to the wavelength of the light (380–780 nm) determines whether nano emulsions are clear (d50 nm) or hazy (d50 nm–d200 nm) [45]. Generally speaking, the amount of light dispersion varies according on the nano emulsion's droplet size, quantity, and refractive index. The entity of any chromospheres that specifically engulf the visible light range of the electromagnetic spectrum (380–780 nm) determines the hue of nano emulsions [70].

3. **Rheology**

Rheology is a field of study that examines the flow and deformation of matter. It is a crucial factor in the production and usefulness of many substance. For several reasons, understanding the rheology of nano emulsions is crucial. First, the rheology of the resulting nano emulsion and the thickness of the distinct oil and aqueous phases determine how effectively droplets are disrupted inside a homogenizer [27]. Second, the rheology of the particular phases affects the shelf life of nano-emulsion-based products. For instance, the oiling rate of O/W nano emulsions typically declines as the aqueous phase's viscosity rises [41] Third, how a nano-emulsion moves affects the design and functioning of several crucial production processes, such as mixing inside a tank, flow via pipe, transit through a binding into product containers [38, 54]. Fourth, the rheology of nano-emulsion products affects a number of their sensory properties, including how thick, creamy, pourable, or flowable they appear to be. The rheological characteristics of nano-emulsions have changed in contrast to macroemulsions because of the microscopic size of the droplets [38, 48]. Droplet size and distributed volume are two factors that might affect rheology. The behaviour of a nano emulsion can be adjusted for a particular application. Nano-emulsions are far more elastic than macroemulsions. The food and cosmetic industries use a variety of gelling and depletion agents to change the viscosity of nano-emulsions to meet the needs of the finished product [76].

4. **Spread Ability**

The thermodynamic stability profile of the nanoemulsion was favourable. Under extreme circumstances, no appreciable changes in pH, appearance, zeta potential, or particle size were found. The improved formulation 5-fluorouracil chitosan-decorated nanoemulsion (5-FU-C-NE) demonstrated the greatest and most satisfying outcome at the extreme temperature (40 °C) and showed no significant changes in physicochemical parameters [50]. Special apparatus calculates spread ability of nano-emulsion.

$$S = ML/T$$

S = spread ability, T = time taken for the separation of the slide from each other

M = weight that is tied from upper part.

L = length of the slide.

5. **Drug Delivery**

There are different modes for the delivery of drug in the body like ocular, intravenous, oral route etc. compared to skin nano-emulsion also act as a barrier for drug administration through skin. Topical medication of nano-emulsion has great advantages in enhancing the solubility in oil and continuous phase. It also has great role in the agents used for ultrasound that contain per fluorocarbon [50].

8 Agriculture Nano-Emulsion

Numerous insect pests that attack different crops significantly reduce yield and production, wrecking the economies of many nations that rely heavily on agriculture. Compared to conventional synthetic insecticides that are harmful to both the environment and human health, green nanoemulsions are promising nanoproducts for the environmentally responsible and sustainable management of agricultural insects [68]. The presence of a substance with hydrophobicity groups has increased the electrostatic interactions at the interface of the droplets, significantly enhancing the properties of the nanoemulsion. It is also noteworthy that the properties of a nanoemulsion can vary depending on the polymer concentration used in it [49].

1. **Antimicrobial Agents**

Nano-emulsion of thymol with saponin, a water insoluble from the Quillaja tree, exhibits enhancement of plant growth and anti-microbial activity against microorganisms caused by spot disease *Xanthomonasaxonopodis*pv. Glycine in soyabean [39] thymol with saponin has increased effectiveness due to nano droplet size. In conjunction with targets in different directions, including easy penetration through bacterial membrane and increased cell lysis from the lipophilic moiety of saponin [72, 75] assessed the antibacterial action of anise oil against *L. monocytogenes* and *E. coli* in bulk and nano-emulsion form. There is a chance that volatile molecules with greater surface area and greater interfacial area will interact with the membrane of microorganisms [55]. According to diverse research, surfactant molecules can destabilise volatile chemicals at the interface, where they then adhere to microbial cells. Thymolnano-emulsion, according to [39] stimulates plant development by increasing its physical and chemical properties. To fully comprehend the function and technique of nano-emulsion in promoting plant growth, more research is required [72].

2. **Activity that is Larvicidal, Pesticidal, and Herbicidal**

Pesticide distribution uses the manufacturing of oil-in-water nano-emulsions. The world's most populous animal, arthropods, have the potential to control their larvae [1, 10, 14, 35, 52, 69]. This possibility exists or the bioactive substance can be contained in oil-based nano-emulsions. Nano-emulsions have been evolving as nano pesticides due to their bioactivity and the oil phase's capacity to become nano-sized. Numerous efforts have been undertaken to lessen the agricultural pest, which is a major source of crop loss. Numerous fatty acid methyl esters are used to control pests in lipophilic (glyphosate) nano-emulsions [7, 42]. A nano-emulsion of eucalyptus oil is specifically targeted at the freshly born larvae of *Pectinophoragossypiella* and *Eariasinsulana* by cotton bollworms, a destructive insect that infests cotton plants. Additionally, eucalyptus oil has a wide spectrum of insecticidal efficacy against *Sitophilusgranarius, Oryzaephilussurinamensis, Triboliumcastaneum, Callosobruchusmaculatus, Rhyzoperthadominica,* and *Callosobruchussurinamensis* [3]. Insecticidal action of jojoba seed oil nano-emulsion upon the wheat pest *S. oryzae* was observed by [58]. Because oil droplets are smaller now, they can more easily penetrate an insect's cuticle layer, and the extinction of grown insects is inversely correlated with oil concentration. Pests cause agricultural damage in storage buildings after harvest. Neem oil was used to create a nano-emulsion by [9] that is adequate upon *S. oryzae* (L.) and *T. castaneum* (Herbst) and can replace pesticides in warehouses with natural alternatives. By creating a nano-emulsion of *Manilkarasubsericea* extract, organic insecticides for sustainable development in agriculture are promoted [13, 16] loaded the water-insoluble-cypermethrin.

Insecticide action was found using an oil-in-water emulsion pesticide. The technology of nano-emulsions has also been successfully used to control weeds and herbs. In order to combat [7] observed water-soluble IPA nano-emulsion formulations. Thyme and marjoram essential oil macro- and nano-emulsions have also demonstrated herbicidal effect on Convolvulus arvensis and *Setariaviridis* seeds and seedlings [25] employed *Satureja*hortensis essential oil as herbicide for organic agriculture systems to manage weeds.

9 Future Possibilities

Over 60% of the people in India is employed in agriculture, which also put up to around 18% of the country's total Gross Domestic Product. Nanotechnology is an amazing fusion of different fields that has a wide range of implementation in many facets of mortal life, including agriculture. Due to its affordability and environmental friendliness, the formation of nanomaterials by plants and microorganisms has attracted a lot of attention. The ability of nanomaterials to function as effective antibacterial agents against a variety of plant diseases is very high. Additionally, nanoparticles have demonstrated their use as nano sensors and nano fertilizers in the encouragement of plant development. Although there are several applications for nanoparticles and nano-emulsions in agriculture, our understanding of how plant pathogen and nanoparticle interactions affect a plant's transcriptome, proteome, and metabolome is still in its infancy and needs to be clarified. Although. Although developing nations are very excited about using nanotechnology in agriculture, there are still certain obstacles to be cleared. In order to allow nanotechnology to spark a new up coming under which happy faces of the needy can be seen, various limits, including those related to the usage of particles and intellectual property rights, need to be resolved as quickly as feasible. As they influence many of the functional and sensational characteristics of nano-emulsions, their physicochemical characteristics are crucial. Nano-emulsions are often more balanced against gravitational disconnection and aggregation than ordinary emulsions since their size is similar to that of a droplet, which may be useful for boosting the durability. The implementation of the required works benefit greatly from increasing their endurance. Nanomaterials must be constructed to avoid early damage since the size of the droplet shrinks and the instability process intensifies. This might be accomplished by adding inhibitors and compounds for particular oil phases. The substances utilised throughout the synthesis process might range from fluids to solids, more or less, depending on the requirements. It is frequently feasible to generate highly viscous materials in nano emulsion which have relatively strong attracting or repeling nature between the oil droplets at low droplet concentrations than in ordinary emulsions. The development of commercial goods with attractive textural qualities, such as foods, cosmetics, hygieneor personal care items, may benefit from this phenomenon. Additionally, nano-emulsions' structure and composition can be changed. to make materials that have various optical qualities, from transparent to opaque. When the radius of drop is less than 20–30 nm, transparent nano-emulsions can be created, however turbid or opaque nano-emulsions cannot. Larger droplet sizes can result in the formation of nano emulsions. Therefore, in some industrial applications, nano emulsions might be superior to traditional emulsions.

10 Conclusion

The properties of green nano emulsions both chemical and physical are such that suit they greatly affect both sensory and functional capabilities with the added advantage of being both sustainable and biologically active component they have an added advantage over other conventional emulsions along with they have greater stability in comparison with the other emulsions when it comes to properties like gravitational separation and aggregation due to their smaller size. Plant mediated nanoparticles is a route of green synthesis that is cheap and environment friendly. Green Nanomaterials has wide range of promising effect in Cutting–edge Technologies like drug delivery etc. Its properties determine the stability and reproducibility and mechanism involve in its resolvement. Chemicals properties determine the toxicity of animal and plant system that helps to enhance the delivery of anti-microbial compounds for plant protection. For the treatment of plants, it is important to know the ratio and quantity of trace elements for the particular plant species. Along with this they have many applications in various different fields like agriculture, medicinal purposes along with this they also have great untapped potential in industries such as edible items and other consumer goods industries due to the properties like strong inter molecular adhesion and repulsion the main field that green nano emulsion show the most potential is in the field of agriculture the conventional methods of agriculture fertilizer delivery are both highly waste full and non-targeted they are just given in a very large quantity to the soil and the plants absorbs the nutrients according to their potential and they are left in the soil green nano emulsion in the long run have the ability to reduce the use of high quantity of fertilizers and have a more targeted approach the potential and the possibilities are endless as a field of science and application nanotechnology and green nano emulsion are in very early stage by the prospects are very intriguing and could play a role in large number of upcoming discoveries.

References

1. Anjali C, Sharma Y, Mukherjee A, Chandrasekaran N (2012) Neem oil (*Azadirachta indica*) nanoemulsion–a potent larvicidal agent against Culex quinquefasciatus. Pest Manag Sci 68(2):158–163. https://doi.org/10.1002/PS.2233
2. Anton N, Vandamme TF (2009) The universality of low-energy nano-emulsification. Int J Pharm 377(1–2):142–147. https://doi.org/10.1016/J.IJPHARM.2009.05.014
3. Aref SP, Valizadegan O, Farashiani ME (2015) Eucalyptus dundasii Maiden essential oil, chemical composition and insecticidal values against *Rhyzopertha dominica* (F.) and *Oryzaephilus surinamensis* (L.). J Plant Protect Res 55(1):35–41. https://doi.org/10.1515/JPPR-2015-0005

4. Aveyard R (1988) Emulsions and solubilization. In: Shinoda K, Friberg S (ed) Wiley-interscience, New York, 1986. pp. xiii + 174, price £43.25. J Chem Technol Biotechnol 42(3):243–244. https://doi.org/10.1002/JCTB.280420311
5. Baker S, Harini BP, Rakshith D, Satish S (2013) Marine microbes: Invisible nanofactories. J Pharm Res 6(3):383–388. https://doi.org/10.1016/J.JOPR.2013.03.001
6. Bachheti RK, Abate L, Deepti, Bachheti A, Madhusudhan A, Husen H (2021) Algae-, fungi-, and yeast-mediated biological synthesis of nanoparticles and their various biomedical applications. In: Kharisov B, Kharissova O (eds) Handbook of greener synthesis of nanomaterials and compounds, vol 1: fundamental principles and methods. Elsevier Inc. 50 Hampshire St., 5th Floor, Cambridge, MA, pp 701–734
7. Chaw Jiang L, Basri M, Omar D, Abdul Rahman MB, Salleh AB, Raja Abdul Rahman RNZ, Selamat A (2012) Green nano-emulsion intervention for water-soluble glyphosate isopropylamine (IPA) formulations in controlling *Eleusine indica* (*E. indica*). Pesticide Biochem Physiol 102(1):19–29. https://doi.org/10.1016/J.PESTBP.2011.10.004
8. Chen H, Khemtong C, Yang X, Chang X, Gao J (2011) Nanonization strategies for poorly water-soluble drugs. Drug Discovery Today 16(7–8):354–360. https://doi.org/10.1016/J.DRUDIS.2010.02.009
9. Choupanian M, Omar D, Basri M, Asib N (2017) Preparation and characterization of neem oil nanoemulsion formulations against *Sitophilus oryzae* and *Tribolium castaneum* adults. J Pestic Sci 42(4):158. https://doi.org/10.1584/JPESTICS.D17-032
10. Da Botas GS, Cruz RAS, De Almeida FB, Duarte JL, Araújo RS, Souto RNP, Ferreira R, Carvalho JCT, Santos MG, Rocha L, Pereira VLP, Fernandes CP (2017) *Baccharis reticularia* DC. and limonene nanoemulsions: promising larvicidal agents for *Aedes aegypti* (Diptera: Culicidae) control. Molecules 22(11):1990. https://doi.org/10.3390/MOLECULES22111990
11. Delmas T, Piraux H, Couffin AC, Texier I, Vinet F, Poulin P, Cates ME, Bibette J (2011) How to prepare and stabilize very small nanoemulsions. Langmuir 27(5):1683–1692. https://doi.org/10.1021/LA104221Q/SUPPL_FILE/LA104221Q_SI_001.PDF
12. Donsì F, Sessa M, Ferrari G (2011) Effect of emulsifier type and disruption chamber geometry on the fabrication of food nanoemulsions by high pressure homogenization. Ind Eng Chem Res 51(22):7606–7618. https://doi.org/10.1021/IE2017898
13. Du Z, Wang C, Tai X, Wang G, Liu X (2016) Optimization and characterization of biocompatible oil-in-water nanoemulsion for pesticide delivery. ACS Sustain Chem Eng 4(3):983–991. https://doi.org/10.1021/ACSSUSCHEMENG.5B01058/ASSET/IMAGES/MEDIUM/SC-2015-01058V_0010.GIF
14. Duarte JL, Amado JRR, Oliveira AEMFM, Cruz RAS, Ferreira AM, Souto RNP, Falcão DQ, Carvalho JCT, Fernandesa CP (2015) Evaluation of larvicidal activity of a nanoemulsion of *Rosmarinus officinalis* essential oil. Rev Bras 25(2):189–192. https://doi.org/10.1016/J.BJP.2015.02.010
15. Feng J, Esquena J, Rodriguez-Abreu C, Solans C (2020) Key features of nano-emulsion formation by the phase inversion temperature (PIT) method
16. Fernandes CP, de Almeida FB, Silveira AN, Gonzalez MS, Mello CB, Feder D, Apolinário R, Santos MG, Carvalho JCT, Tietbohl LAC, Rocha L, Falcão DQ (2014) Development of an insecticidal nanoemulsion with *Manilkara subsericea* (Sapotaceae) extract. J Nanobiotechnol 12(1):1–9. https://doi.org/10.1186/1477-3155-12-22/TABLES/2
17. Grumbach C, Krüger V, Czermak P (2022) A new control strategy for high-pressure homogenization to improve the safety of injectable lipid emulsions. Pharmaceutics 14(8). https://doi.org/10.3390/pharmaceutics14081603
18. Gupta A (2019) Nanoemulsions. In: Nanoparticles for biomedical applications: fundamental concepts, biological interactions and clinical applications. Elsevier Inc. https://doi.org/10.1016/B978-0-12-816662-8.00021-7
19. Gupta A, Eral HB, Hatton TA, Doyle PS (2016) Controlling and predicting droplet size of nanoemulsions: scaling relations with experimental validation. Soft Matter 12(5):1452–1458. https://doi.org/10.1039/C5SM02051D

20. Gupta A, Eral HB, Hatton TA, Doyle PS (2016) Nanoemulsions: formation, properties and applications. Soft Matter 12(11):2826–2841. https://doi.org/10.1039/C5SM02958A
21. Gupta A, Narsimhan V, Hatton TA, Doyle PS (2016) Kinetics of the change in droplet size during nanoemulsion formation. Langmuir: ACS J Surfaces Colloids 32(44):11551–11559. https://doi.org/10.1021/ACS.LANGMUIR.6B01862
22. Håkansson A, Innings F, Trägårdh C, Bergenståhl B (2013) A high-pressure homogenization emulsification model-improved emulsifier transport and hydrodynamic coupling. Chem Eng Sci 91:44–53. https://doi.org/10.1016/J.CES.2013.01.011
23. Håkansson A, Trägårdh C, Bergenståhl B (2009) Dynamic simulation of emulsion formation in a high pressure homogenizer. Chem Eng Sci 64(12):2915–2925. https://doi.org/10.1016/J.CES.2009.03.034
24. Handa M, Ujjwal RR, Vasdev N, Flora SJS, Shukla R (2021) Optimization of surfactant- and cosurfactant-aided pine oil nanoemulsions by isothermal low-energy methods for anticholinesterase activity. ACS Omega 6(1):559–568. https://doi.org/10.1021/ACSOMEGA.0C05033/ASSET/IMAGES/MEDIUM/AO0C05033_M001.GIF
25. Hazrati H, Saharkhiz MJ, Niakousari M, Moein M (2017) Natural herbicide activity of *Satureja hortensis* L. essential oil nanoemulsion on the seed germination and morphophysiological features of two important weed species. Ecotoxicol Environ Saf 142:423–430. https://doi.org/10.1016/J.ECOENV.2017.04.041
26. Helgeson ME, Moran SE, An HZ, Doyle PS (2012) Mesoporous organohydrogels from thermogelling photocrosslinkable nanoemulsions. Nat Mater 11(4):344–352. https://doi.org/10.1038/NMAT3248
27. Hidajat MJ, Jo W, Kim H, Noh J (2020) Effective droplet size reduction and excellent stability of limonene nanoemulsion formed by high-pressure homogenizer. Colloids Interfaces 4(1):5. https://doi.org/10.3390/COLLOIDS4010005
28. Husen A (2019) Natural product-based fabrication of zinc oxide nanoparticles and their application. In: Husen A, Iqbal M (eds) N*anomaterials* and plant potential. Springer International Publishing AG, Gewerbestrasse 11, 6330 Cham, pp 193–291. https://doi.org/10.1007/978-3-030-05569-1_7
29. Husen A (2019) Medicinal plant-product based fabrication nanoparticles (Au and Ag) and their anticancer effect. In: Kintzios SE, Barberaki M, Flampouri (eds) Plants that fight cancer, 2nd edn. Taylor & Francis/CRC Press pp 133–147
30. Husen A (2023) Secondary metabolites based green synthesis of nanomaterials and their applications. Springer Nature Singapore Pte Ltd., 152 Beach Road, #21–01/04 Gateway East, Singapore 189721, Singapore
31. Husen A (2023) Nanomaterials from agricultural and horticultural products. Springer Nature Singapore Pte Ltd., 152 Beach Road, #21–01/04 Gateway East, Singapore 189721, Singapore
32. Husen A, Siddiqi KS (2023) Advances in smart nanomaterials and their applications. Elsevier Inc., 50 Hampshire St., 5th Floor, Cambridge, MA 02139, USA
33. Iravani S (2011) Green synthesis of metal nanoparticles using plants. Green Chem 13(10):2638–2650. https://doi.org/10.1039/C1GC15386B
34. Jaiswal M, Dudhe R, Sharma PK (2015) Nanoemulsion: an advanced mode of drug delivery system. Biotech 5(2):123. https://doi.org/10.1007/S13205-014-0214-0
35. Jesus FLM, De Almeida FB, Duarte JL, Oliveira AEMFM, Cruz RAS, Souto RNP, Ferreira RMA, Kelmann RG, Carvalho JCT, Lira-Guedes AC, Guedes M, Solans C, Fernandes CP (2017) Preparation of a nanoemulsion with *Carapa guianensis* Aublet (Meliaceae) oil by a low-energy/solvent-free method and evaluation of its preliminary residual larvicidal activity. Evidence-Based Complementary Alternative Med. https://doi.org/10.1155/2017/6756793
36. Jintapattanakit A (2018) Preparation of nanoemulsions by phase inversion temperature (PIT) method. Pharmaceutical Sci Asia 45(1):1–12. https://doi.org/10.29090/PSA.2018.01.001
37. Kadziński M, Cinelli M, Ciomek K, Coles SR, Nadagouda MN, Varma RS, Kirwan K (2018) Co-constructive development of a green chemistry-based model for the assessment of nanoparticles synthesis. Eur J Oper Res 264(2):472–490. https://doi.org/10.1016/J.EJOR.2016.10.019

38. Kumar N, Verma A, Mandal A (2021) Formation, characteristics and oil industry applications of nanoemulsions: a review. J Petrol Sci Eng 206:109042. https://doi.org/10.1016/J.PETROL.2021.109042
39. Kumari S, Kumaraswamy RV, Choudhary RC, Sharma SS, Pal A, Raliya R, Biswas P, Saharan V (2018) Thymol nanoemulsion exhibits potential antibacterial activity against bacterial pustule disease and growth promotory effect on soybean. Sci Rep 8(1):1–12. https://doi.org/10.1038/s41598-018-24871-5
40. Larue C, Castillo-Michel H, Sobanska S, Cécillon L, Bureau S, Barthès V, Ouerdane L, Carrière M, Sarret G (2014) Foliar exposure of the crop Lactuca sativa to silver nanoparticles: evidence for internalization and changes in Ag speciation. J Hazard Mater 264:98–106. https://doi.org/10.1016/J.JHAZMAT.2013.10.053
41. Li M, Kang W, Li Z, Yang H, Jia R, He Y, Kang X, Zheng Z, Wang Y, Sarsenbekuly B, Gabdullin M (2021) Stability of oil-in-water (O/W) nanoemulsions and its oil washing performance for enhanced oil recovery. Phys Fluids 33(7):072002. https://doi.org/10.1063/5.0058759
42. Lim CJ, Basri M, Omar D, Abdul Rahman MB, Salleh AB, Raja Abdul Rahman RNZ (2013) Green nanoemulsion-laden glyphosate isopropylamine formulation in suppressing creeping foxglove (*A. gangetica*), slender button weed (*D. ocimifolia*) and buffalo grass (*P. conjugatum*). Pest Manage Sci 69(1):104–111. https://doi.org/10.1002/PS.3371
43. Makarov VV, Love AJ, Sinitsyna OV, Makarova SS, Yaminsk IV, Taliansky ME, Kalinina NO (2014) Green nanotechnologies: synthesis of metal nanoparticles using plants. Acta Naturae 6(1):35. https://doi.org/10.32607/20758251-2014-6-1-35-44
44. Mason TG, Wilking JN, Meleson K, Chang CB, Graves SM (2006) Nanoemulsions: formation, structure, and physical properties. J Phys: Condens Matter 18(41):R635. https://doi.org/10.1088/0953-8984/18/41/R01
45. McClements DJ (2002) Theoretical prediction of emulsion color. Adv Coll Interface Sci 97(1–3):63–89. https://doi.org/10.1016/S0001-8686(01)00047-1
46. McClements DJ (2012) Nanoemulsions versus microemulsions: terminology, differences, and similarities. Soft Matter 8(6):1719–1729. https://doi.org/10.1039/C2SM06903B
47. McClements DJ (2015) Food emulsions: principles, practices, and techniques, 3rd edn. Food Emulsions: Principles, Practices, and Techniques, Third Edition, pp 1–676. https://doi.org/10.1201/B18868/FOOD-EMULSIONS-DAVID-JULIAN-MCCLEMENTS
48. McClements DJ, Rao J (2011) Food-grade nanoemulsions: formulation, fabrication, properties, performance, biological fate, and potential toxicity. Crit Rev Food Sci Nutr 51(4):285–330. https://doi.org/10.1080/10408398.2011.559558
49. Mustafa IF, Hussein MZ (2020) Synthesis and technology of nanoemulsion-based pesticide formulation. Nanomaterials 10(8):1–26. https://doi.org/10.3390/NANO10081608
50. Nawaz A, Latif MS, Alnuwaiser MA, Ullah S, Iqbal M, Alfatama M, Lim V (2022) Synthesis and characterization of chitosan-decorated nanoemulsion gel of 5-fluorouracil for topical delivery. Gels 8(7):1–13. https://doi.org/10.3390/gels8070412
51. Nirmala MJ, Dhas SP, Saikrishna N, Raj US, Sai PS, Nagarajan R (2022) Green nanoemulsions: Components, formulation, techniques of characterization, and applications. Bio-Based Nanoemulsions Agri-Food Appl. https://doi.org/10.1016/B978-0-323-89846-1.00013-9
52. Oliveira AEMFM, Duarte JL, Cruz RAS, Souto RNP, Ferreira RMA, Peniche T, Conceição EC, Oliveira LAR, Faustino SMM, Florentino AC, Carvalho JCT, Fernandes CP (2017) Pterodon emarginatus oleoresin-based nanoemulsion as a promising tool for *Culex quinquefasciatus* (Diptera: Culicidae) control. J Nanobiotechnol 15(1):1–11. https://doi.org/10.1186/S12951-016-0234-5/TABLES/4
53. Shafiq S, Shakeel F, Talegaonkar S, Ahmad FJ, Khar RK, Ali M (2007) Development and bioavailability assessment of ramipril nanoemulsion formulation. Eur J Pharm Biopharm Official J Arbeitsgemeinschaft Fur Pharmazeutische Verfahrenstechnik 66(2). https://doi.org/10.1016/J.EJPB.2006.10.014
54. Saguy IS, Singh RP, Johnson T, Fryer PJ, Sastry SK (2013) Challenges facing food engineering. https://doi.org/10.1016/j.jfoodeng.2013.05.031

55. Salvia-Trujillo L, Rojas-Graü MA, Soliva-Fortuny R, Martín-Belloso O (2014) Impact of microfluidization or ultrasound processing on the antimicrobial activity against *Escherichia coli* of lemongrass oil-loaded nanoemulsions. Food Control Complete 37:292–297. https://doi.org/10.1016/J.FOODCONT.2013.09.015
56. Siddiqi KS, Husen A (2016) Fabrication of metal nanoparticles from fungi and metal salts: scope and application. Nanoscale Res Lett 11(98):1–15
57. Siddiqi KS, Husen A (2017) Recent advances in plant-mediated engineered gold nanoparticles and their application in biological system. J Trace Elem Med Biol 40:10–23
58. Sh A, Abdelrazeik AB (2015) Nanoemulsion of jojoba oil, preparation, characterization and insecticidal activity against *Sitophilus oryzae* (Coleoptera: Curculionidae) on wheat. Int J Agric Innov Res 4(1):2319–1473
59. Shafiq-un-Nabi S, Shakeel F, Talegaonkar S, Ali J, Baboota S, Ahuja A, Khar RK, Ali M (2007) Formulation development and optimization using nanoemulsion technique: a technical note. AAPS Pharm Sci Tech 8(2):E12. https://doi.org/10.1208/PT0802028
60. Singh JS (2015) Biodiversity: current perspective. Clim Change Environ Sustain 3(1):71. https://doi.org/10.5958/2320-642X.2015.00007.1
61. Singh JS (2015) Microbes play major roles in the ecosystem services. Clim Change Environ Sustain 3(2):163. https://doi.org/10.5958/2320-642X.2015.00018.6
62. Singh JS (2015) Microbes: the chief ecological engineers in reinstating equilibrium in degraded ecosystems. Agr Ecosyst Environ 203:80–82. https://doi.org/10.1016/J.AGEE.2015.01.026
63. Singh JS (2015) Book review. Appl Soil Ecol 92:45–46. https://doi.org/10.1016/J.APSOIL.2015.03.004
64. Singh JS, Abhilash PC, Gupta VK (2016) Agriculturally important microbes in sustainable food production. Trends Biotechnol 34(10):773–775. https://doi.org/10.1016/j.tibtech.2016.06.002
65. Singh R (2019) Microbial biotechnology: a promising implement for sustainable agriculture. https://doi.org/10.1016/B978-0-444-64191-5.00008-0
66. Solans C, Morales D, Homs M (2016) Spontaneous emulsification. Curr Opin Colloid Interface Sci 22:88–93. https://doi.org/10.1016/J.COCIS.2016.03.002
67. Solè I, Maestro A, González C, Solans C, Gutiérrez JM (2006) Optimization of nano-emulsion preparation by low-energy methods in an ionic surfactant system. Langmuir: ACS J Surf Colloids 22(20):8326–8332. https://doi.org/10.1021/LA0613676
68. Subramanya S, Sumanth K, Gupta PK, Chayapathy V, Keshamma E, Murugan K (2022) Formulation of green nanoemulsions for controlling agriculture insects. Bio-Based Nanoemulsions Agri-Food Appl. https://doi.org/10.1016/B978-0-323-89846-1.00002-4
69. Sugumar S, Clarke SK, Nirmala MJ, Tyagi BK, Mukherjee A, Chandrasekaran N (2014) Nanoemulsion of eucalyptus oil and its larvicidal activity against *Culex quinquefasciatus*. Bull Entomol Res 104(3):393–402. https://doi.org/10.1017/S0007485313000710
70. Tadros T, Izquierdo P, Esquena J, Solans C (2004) Formation and stability of nano-emulsions. Adv Coll Interface Sci 108–109:303–318. https://doi.org/10.1016/J.CIS.2003.10.023
71. Tiwari SB, Amiji MM (2006) Nanoemulsion formulations for tumor-targeted delivery. Nanotechnol Cancer Therapy. https://doi.org/10.1201/9781420006636-35/NANOEMULSION-FORMULATIONS-TUMOR-TARGETED-DELIVERY-SANDIP-TIWARI-MANSOOR-AMIJI
72. Topuz OK, Özvural EB, Zhao Q, Huang Q, Chikindas M, Gölükçü M (2016) Physical and antimicrobial properties of anise oil loaded nanoemulsions on the survival of foodborne pathogens. Food Chem 203:117–123. https://doi.org/10.1016/J.FOODCHEM.2016.02.051
73. Virkutyte J, Varma RS (2011) Green synthesis of metal nanoparticles: Biodegradable polymers and enzymes in stabilization and surface functionalization. Chem Sci 2(5):837–846. https://doi.org/10.1039/C0SC00338G
74. Wilking JN, Mason TG (2007) Irreversible shear-induced vitrification of droplets into elastic nanoemulsions by extreme rupturing. Phys Rev E Stat Nonlinear Soft Matter Phys 75(4):041407. https://doi.org/10.1103/PHYSREVE.75.041407/FIGURES/4/MEDIUM
75. Wojciechowski K, Orczyk M, Gutberlet T, Trapp M, Marcinkowski K, Kobiela T, Geue T (2014) Unusual penetration of phospholipid mono- and bilayers by Quillaja bark saponin biosurfactant.

Biochimica et Biophysica Acta (BBA) Biomembranes 1838(7):1931–1940. https://doi.org/10.1016/J.BBAMEM.2014.04.008
76. Wooster TJ, Golding M, Sanguansri P (2008) Impact of oil type on nanoemulsion formation and Ostwald ripening stability. Langmui: ACS J Surf Colloids 24(22):12758–12765. https://doi.org/10.1021/LA801685V
77. Yurdacan HM, Murat Sari M (2021) Functional green-based nanomaterials towards sustainable carbon capture and sequestration. Sustain Mater Transitional Altern Energy. https://doi.org/10.1016/B978-0-12-824379-4.00004-5

Nanoemulsions from Essential Oils: Preparation, Characterization, and Their Applications

Yilma Hunde, Archana Bachheti, Kundan Kumar Chaubey, Azamal Husen, and Rakesh Kumar Bachheti

Abstract Nanoemulsion technology has emerged as the most promising delivery channel for lipophilic components such as nutraceuticals, drugs, flavors, antioxidants, and antimicrobial agents. Nanoemulsions of essential oils commonly involve two main preparation methods. Preparation of nanoemulsions of essential oils is based on either non-spontaneous (high-energy) or spontaneous (low-energy) processes. Several studies reported that ultrasonic emulsification produced the smallest droplet sizes of nanoemulsions of essential oils. It is documented as an efficient and fast method of preparation of nanoemulsions. The potential functionality of nanoemulsions is because of the higher activity property of their surface area. Droplet sizes, zeta potentials, polydispersity index, viscosity, stability, and morphologies of particles are the characteristic properties of nanoemulsions. The characteristics and properties of nanoemulsions of essential oils depend not only on the components but also on the preparation methods. The nanoemulsions preparation method has several advantages over conventional emulsion techniques. Nanoemulsions are suitable formulations to

Y. Hunde · R. K. Bachheti
Centre of Excellence in Nanotechnology, College of Applied Science, Addis Ababa Science and Technology University, P.O. Box-16417, Addis Ababa, Ethiopia

Department of Industrial Chemistry, College of Applied Science, Addis Ababa Science and Technology University, P.O. Box-16417, Addis Ababa, Ethiopia

R. K. Bachheti
e-mail: rakesh.kumer@aastu.edu.et

A. Bachheti (✉)
Department of Environment Science, Graphic Era University, Dehradun, Uttarakhand 248002, India
e-mail: bachheti.archana@gmail.com

K. K. Chaubey
Division of Research and Innovation, School of Applied and Life Sciences, Uttaranchal University, Arcadia Grant, P.O. Chandanwari, Premnagar, Dehradun, Uttarakhand 248007, India

School of Basic and Applied Sciences, Sanskriti University, Mathura, Uttar Pradesh 281401, India

A. Husen
Wolaita Sodo University, P.O. Box: 138, Wolaita, Ethiopia

A. Husen et al. (eds.), *Current Trends in Green Nano-emulsions*, Smart Nanomaterials Technology, https://doi.org/10.1007/978-981-99-5398-1_2

serve as an additive in clear drinks and beverages due to their good optical transparent solution. The kinetic stability of colloidal systems is also another advantage of nanoemulsions of essential oils. This book chapter provides insight into the concise preparation, characterizations, and applications of nanoemulsions from essential oils.

Keywords Essential oils · Nanoemulsions · Sonications · Emulsification methods · Eco-friendly

1 Introduction

Environmental factors such as light, oxygen, and temperature are the main causes of the degradation of essential oils. Nano emulsification is an important technique to maintain the stability of essential oils during their applications [1]. It is usually difficult to handle essential oils and obtain their efficient activities through their direct incorporation into different industrial applications because of the high volatility, instability, and low solubility of essential oils in water [2–6]. Therefore, nanoemulsion formulation is the easiest method to handle the delivery system of essential oils in various fields. These techniques are also suitable to increase the chance of the availability of bioactive components of essential oils to target sites [5, 6].

Oils, water, emulsifiers, and energy are important inputs to successfully prepare nanoemulsions of essential oils [7]. Emulsifiers play a greater role in linking aqueous and oil phases together through their hydrophilic and lipophilic sites, respectively [8]. Nanoemulsions can be produced in various emulsification ways, such as oil-in-water (O/W) or water-in-oil (W/O), and multiple emulsions, such as water-in-oil–water (W/O/W) or oil-in-water-in-oil (O/W/O), Fig. 1, [9–11]. The fabrication of nanoemulsion of essential oils is possible using methods including high-energy processes: high pressure, micro fluidization, or sonication and low-energy processes: spontaneous emulsion, phase inversion, and emulsion inversion point [12]. Emulsion materials with droplet size less than 100 nm are classes of nanoemulsions.

Nanoemulsions are thermodynamically non-equilibrium systems [13]. However, they are kinetically stable heterogeneous colloidal dispersions with droplets of one

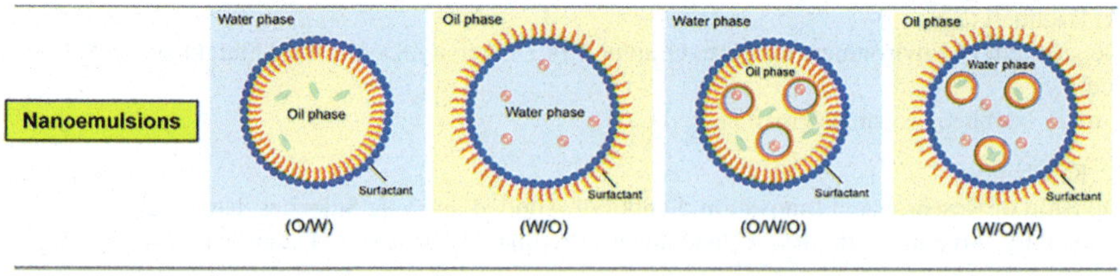

Fig. 1 Preparation of various nanoemulsions taken from [10]

phase in the immiscible continuous system bounded by the film of non-ionic surfactants with the size of particles below 100 nm [14]. Nanoemulsions with droplets size 10–100 nm are reported to act as carriers for lipophilic compounds such as nutraceuticals, drugs, flavors, antioxidants, and antimicrobial agents in food, pharmaceutical, and cosmetic industries [15, 16]. Researchers widely prefer nanoemulsions because the method increases the solubility of weak water miscible materials such as oils [17]. Nanoemulsions of essential oils of plants using non-ionic surfactants are known for their high stability and environmentally friendly properties [18]. Surfactants in the system stabilize nanoemulsion by reducing the surface tension at the interface of water and oil droplets [19].

Nowadays, nanoemulsion is attracting the strongest attention of scholars as an essential oils delivery system than micro- and conventional emulsions methods [20]. Previous studies also showed that nanoemulsions containing essential oils' chemical components possess potential applications in the industry to treat pathogenic microbial contamination of food [21]. Therefore, the nano-formulation of essential oils greatly extends the storage time and enhances the delivery chances of bioactive compounds of nanoemulsions [22]. Similarly, active chemical components of essential oils in the form of nano-pesticide emulsion in water effectively manage crop pests [23]. Encapsulation and film formulation are efficient techniques to increase the stability and efficacy of nanoemulsions of essential oils in various applications [24]. This book chapter summarizes the updated scientific reports on nanoemulsions of essential oils, their characterization, and some potential applications.

2 Preparation Methods of Nanoemulsions of Essential Oils

The method of nanoemulsions preparation determines their stability nature [25]. Nanoemulsions are prepared using high-energy input or surfactants, and the two systems are sometimes applied together [26]. The two basic nanoemulsions fabrication systems are high-energy and low-energy methods [27]. The need to prepare nanoemulsions of essential oils is to formulate O/W, W/O, O/W/O, or W/O/W emulsions throughout the process (Fig. 1) [21, 28].

2.1 High Energy Methods

High energy emulsification is the most commonly used technique for the synthesis of nanoemulsion of oils [29]. High energy approaches are favoured over conventional emulsions for forming nanoemulsions due to their kinetic stability, bioavailability, and optical transparency [30]. High-energy methods also use ultrasound generators, high-pressure homogenizers, and high-shear stirring, which are preferred because of their ease of formation with large-scale yield [31].

2.1.1 High Pressure Homogenization

A high pressure pumped through a resistive valve is applied to the oil–water-surfactant mixture to produce a nanoemulsion formulation [32]. Homogenization process under high-pressure agitation improves the solubility of essential oils in water for nanoemulsion formulation [33]. The challenging property of preparation of essential oil nanoemulsions easily using a high pressure homogenization system is the low polarity and high viscosity of components of essential oils [34]. High-pressure homogenization is mainly used for nanoemulsions in laboratories and industries due to its versatility and scale-up feasibility [30]. It is also reported for its popular flexibility and ease of tuning droplet size [35].

2.1.2 Ultrasonicators

Nanoemulsion of essential oils in water is prepared by aqueous to oil phase ratio of various components using high-speed ultrasonicator. Ultrasound-promoted high-energy input application for the mixing of two immiscible fluids using ultrasonicators to form emulsions is called the ultrasonication process [23, 36]. The ultrasonication method is a widely used approach to preparing nanoemulsions of essential oils [37]. As shown in Fig. 2, some ways of nanoemulsions formulation were reported either as the preparation of coarse emulsion and then treating the course emulsion with high energy pressure, such as the sonication process, or directly preparing the nanoemulsion from the crude materials using ultrasonication technique [38]. Most of the time, using the ultrasonication process to prepare nanoemulsions produces the desired results without any requirement for making coarse emulsions [39]. Higher creaming stability, smaller droplet sizes, and higher zeta potentials of nanoemulsions are prepared by using sonication process [40]. The nanoemulsions with smaller particle sizes always have better stability [41].

The ultrasonication method is employed to prepare smaller size nanomaterials [12]. The ultrasonication waves employ higher intensity of energy forces to convert larger emulsion droplets into nanosized ones [43]. Ultrasonication involves two processes: creating droplets in the acoustic field and the production of high disturbance in emulsion and micro-jets during asymmetric cavity collapse. An asymmetric cavity collapse results in the breakdown and disorder of droplets in the continuous phase [44]. One of the significant advantages of the ultrasonication system is that it usually uses less toxic solvents to prepare nanoemulsion of oils [45]. As other advantages, it affords nanoemulsions with the smallest size, the lowest polydispersity index, the highest stability, and the least amount of surfactants which can solve the drawbacks of low energy methods in nanoemulsion preparation [8]. Nanoemulsions of essential oils formed by ultrasonication methods have more stability for the extended shelf-life as compared with the nanoemulsions synthesized using other high-energy systems [46].

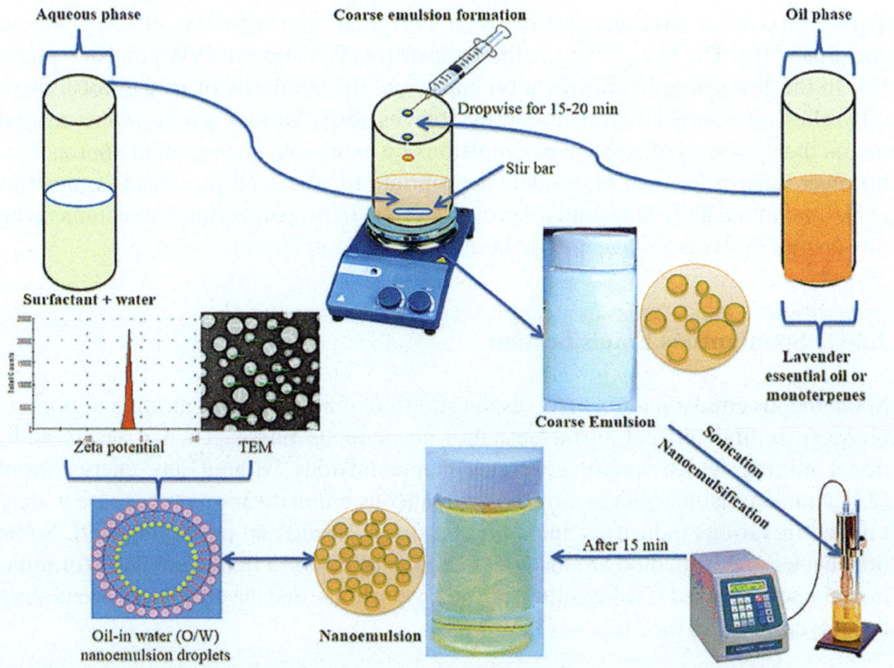

Fig. 2 Oil-in-water nanoemulsions using ultrasonication method taken from [42]

2.1.3 Micro-Fluidization

Micro-fluidization is an efficient process commonly used for the reduction of droplet sizes of emulsions [47]. Micro-fluidization process reported to show significant decrease of droplet size and viscosity of nanoemulsions of essential oils [16]. Some studies reported that active nanoemulsion delivery ways based on micro-fluidization approach for the prototype of essential oil with the prolonged stability condition by using saponin as biosurfactant [48]. Even though the desired size of droplets can be obtained in the micro-fluidization system, there are more conducive techniques to synthesizing nanoemulsions in large-scale production [49].

2.2 *Low-Energy Methods*

Low-energy methods of nanoemulsions preparation involve phase inversion temperature, phase transition, and self-emulsification methods [31]. Low-energy emulsification methods are mainly monitored by the physicochemical properties of surfactants in the preparation of nanomaterials. For example, in low-energy system, large amounts of surfactants are required during the whole process in the food industry.

This approach, in turn, changes the food's original odor, which is unacceptable to consumers [8]. The low-energy method transforms W/O into an O/W emulsion phase [5]. In the low-energy emulsification approach, the synthesis of nanoemulsions of oils relies on internal chemical energy sources [50]. Various studies were carried out on the synthesis of stable nanoemulsions by using low-energy input approaches; however, only a few use oil models appropriate for the food processing industries [51]. Of course, there is recently a growing effort in preparing nanoemulsions using low energy systems to save energy in industries (W. [52].

2.2.1 Spontaneous Emulsification

Spontaneous emulsification involves the addition of a small concentration of organic phase (essential oil and surfactant) into an aqueous phase. Then it needs additional mixing which finally generates nanoemulsions without any energy input [23]. Nanoemulsions obtained by the spontaneous emulsification process are widely utilized in various industries, including food processing and packaging [39]. Some drawbacks of this method are the low oil content dispersed in the emulsion formulation, the solvent used is water soluble in all proportions, and the difficulty of removing the solvent used in the process [32].

2.2.2 Phase Inversion Temperature (PIT)

In the phase inversion method, oil, water, and non-ionic surfactants are stirred slightly at room temperature to prepare nanoemulsions [53]. Three important steps as (1) formation of coarse emulsion by stirring the solution containing oil, water, and surfactant at room temperature, (2) heating the solution gradually around or above the PIT, and (3) forming O/W nanoemulsion from the previous solution by using rapid cooling or dilute into cold water by continuous stirring [7]. Phase inversion temperature approach maintains constant composition in the preparation of nanoemulsions [54]. Coalescence and Ostwald ripening rate are the two processes reported as driving force for the instability of nanoemulsions [55].

2.2.3 Phase Inversion Composition (PIC)

Phase inversion compositions mainly use essential oils and biodegradable surfactants in water to synthesize nanoemulsions over other low-energy systems [50]. Phase inversion composition approach maintains a constant temperature in nanoemulsions formation [54]. This method also demonstrated the formation of significant nanoparticles after the evaporation of solvent from nanoemulsions [56].

3 Characterization of Nanoemulsions Formation

The physicochemical stability of nanoemulsion is an essential parameter due to their droplets' stronger surface activity. The small droplet size of nanoemulsions is significant for their stability against sedimentation or creaming because of the Brownian motion, as the diffusion rate of droplets is higher than the sedimentation rate induced by gravity force [32, 57]. pH of the system is very important in the formulation of nanoemulsions of essential oils. This is because essential oils in water show different solubility results at various pH conditions [58]. The viscosity of nanoemulsions is characterized at room temperature through a rotational viscometer. Viscosity is a vital physicochemical agent for the evaluation of stability and efficient release of bioactive components of nanoemulsions [59]. The aggregation nature of droplets of nanoemulsions, i.e. flocculation is directly proportional to the viscosity of emulsions [51]. The nano-sized nature of emulsions gives unique rheological properties to the nanoemulsions [27]. The surface rheological properties influence the diffusion of active components of nanoemulsions [60]. Flow curves of nanoemulsions are also determined by a Modular Advanced Rheometer system [61].

The electrophoretic properties of prepared nanoemulsions of essential oils, including droplet size, zeta potentials, and polydispersity index, are characterized by the need for quality evaluation of nanomaterials applying the dynamic light scattering (DLS) techniques that utilize zeta nano-sizer instruments [42, 62]. The measurements of droplet sizes, zeta potentials, and polydispersity index of nanoemulsions of essential oils are usually determined for the determination of appropriate particles of nanoemulsions for the required area of application [63]. Particle size measurement indicates the stability phenomena of nanoemulsions against phase separation and creaming at different storage periods [64]. Polydispersity index determines the heterogeneity of sizes of perticles in the nanoemulsion [51]. Zeta potential of nanoemulsion is an important indicator for showing the electrostatic interaction between the surfaces of droplets. Zeta potential depicts the charge accumulation on the surface of the particle, which determines nanoemulsions chemical and biological properties [65]. The determined zeta potential results greater than $\pm$ 30 mV indicate the maximum stability of nanoemulsions of essential oils because of their stronger Van der Waals forces between particles, decreasing the agglomeration effects of the droplets of emulsions [66].

Electrical conductivity was analysed by a conductometer at specific frequency, used to characterize the stability and nature of the formulation of nanoemulsions of essential oils [64]. It is also reported to reveal the electrical charges present on the droplets in the dispersion medium [67]. The morphologies and structural network of nanoemulsions of essential oils are investigated by scanning electron microscopy (SEM) and tunnelling electron microscopy (TEM) [67, 68]. TEM is also reported to confirm the nanosizes of droplets of nanoemulsions measured by the dynamic light scattering (DLS) method [69]. In the sample preparation of the sample characterization procedure, the dispersions are ultrasonicated to avoid the aggregation of particles, and a drop of the most stable emulsion is applied on the grid and then dried in a vacuum-based oven and finally used in a machine for SEM and TEM characterization [70]. X-ray diffraction (XRD) is applied to characterize the packing structures of encapsulated and loaded essential oils in the nanoemulsions formulation [71].

4 Potential Applications of Nanoemulsions of Essential Oils

Various plants extract such as essential oils are the current focus area for several valuable and useful applications [72–74]. Nanoemulsions of essential oils have become an interesting formulation for various applications in different fields [75]. Industrial fields such as polymerization's reaction media, cosmetics, drug delivery system and personal care, agrochemicals, and consumer products are the major fields of application of nanoemulsion of essential oils [76]. Currently, nanoemulsions are also obtained strong research interest for their diversified applications in the cosmetics, pharmaceuticals, pesticides, paints, food, and environmental industries [23]. The weak flocculation, reduced coalescence, and no creaming or sedimentation properties of nanoemulsions on their storage due to their smaller droplet sizes make them for applications in the cosmetics industries [77]. Nanoemulsion is a promising system for incorporating important bioactive components of essential oils into food systems [78]. It is also reported that nanoemulsions are the important systems to incorporate the pleasant odor or flavor essential oils into food items [7].

Essential oils based nanoemulsions are reported to possess promising antimicrobial activities in clinical applications because of their non-phospholipid, stability, non-toxic, sustainability, and inexpensive properties [18]. Nanoemulsions formulation approach improves the delivery system of bioactive compounds of essential oils to the more efficient antibacterial activities in the food and beverage products [79]. During application, the efficiency of nanoemulsion of essential oils is greater than that of crude essential oils because of the improvement of solubility and bioavailability of active components [80]. For instance, nanoemulsion of terpene mixture and D-limonene showed potent antimicrobial activity against *Lactobacillus delbrueckii*, *Saccharomyces cerevisiae*, and *Escherichia coli* in fruit juice [81]. The citral nanoemulsions also exhibited significant inhibition activity against the growth of bacteria species such as *Listeria monocytogenes* and *Staphylococcus aureus* due to the potential activity of citral to disrupt and penetrate the lipid structure of the cell wall of bacteria and cause their death [58]. Optimization of the preparation of nanoemulsion of essential oils in the various areas of applications demonstrates highly improved activities [82]. The preparations of nanoemulsions of essential oils in the form of nanoencapsulation and loaded-film formulation are reported to optimize their delivery systems and elongate the shelf-life of bioactive components [24]. Nanoemulsions of oils can be delivered in oral, topical, and transdermal application forms [59] (Table 1).

Table 1 Preparation of nanoemulsions of essential oils, characterization, and applications

Main components of EOs (source)	Preparation methods	Characterization technique*	Applications	Key references
D-Limonene (Citrus fruits)	High-energy	ZS (27–99 nm) ZP (−20 to −30 mV)	Nanoencapsulation in food industry	[83]
Citral (citrus fruits)	High-energy	ZS (10–100 nm)	Antimicrobial activity	[58]
Eucalyptol, (+)-2-bornanone, α-pinene (*Salvia officinalis*)	High-energy	ZS (204.4 nm) PDI (0.12)	Antibacterial activity	[84]
Curcumin (rhizome of *Curcuma longa*)	High-energy	ZS (93.64 nm) ZP (−11.67 mV) PDI (0.263)	Wound healing and inflammation treatment	[85]
Eugenol (Cinnamon leaf)	High-energy	ZS (65 nm) PDI (0.138)	Antibacterial activity	[86]
EO (*Grammosciadium ptrocarpum*)	High-energy	ZS (41.46 nm) ZP (−15.10 mV) PDI (0.186)	Antimicrobial activity Antioxidant activity	[38]
Carbonyl compounds (aerial parts of Nettle)	High-energy	ZS (134.10 nm)	Antioxidant agents	[15]
EO (betel leaf)	High-energy	ZS (58 nm)	Antibacterial activity	[87]
EO (lemongrass, thyme, sage)	High-energy	ZS (50, 85, 42 nm, respectively) ZP (−44, −50, −79, respectively mV) PDI (0.57, 0.58, 0.69, respectively)	Antibacterial activity	[88]
p-cyamene and α-phellandrene (Dill)	High-energy	ZS (20 nm) PDI (2.0)	Larvicidal activity	[89]
EO (orange peel)	High-energy	ZS (12.68 nm) PDI (0.439)	Food and pharmaceutical additives	[4]
curcumin (*Curcuma longa* rhizome)	High-energy	ZS (49–105 nm) ZP (−20.91 mV) PDI (0.06)	Wound healing	[90]
Carvone and limonene (Caraway)	High-energy	ZS (113.2 nm) PDI (0.117)	Fruits preservation	[91]
EO (*Foeniculum vulgare*)	High-energy	ZS (44–105 nm) ZP (−14 to −34 mV) PDI (0.32–0.48)	Antidiabetic activity	[92]

(continued)

Table 1 (continued)

Main components of EOs (source)	Preparation methods	Characterization technique*	Applications	Key references
Phenolics (cumin seeds)	High-energy	PS (69–111 nm) ZP (−8 to −33 mV) PDI (0.23–0.57)	Antioxidant and hepatoprotective activities	[93]
Citral (lemongrass)	Low-energy	ZS (115 nm) ZP (−2.5 mV) PDI (0.20)	Phyto cosmetics and phytotherapics	[50]
Limonene, α-pinene, linalyl acetate (citrus fruit)	High-energy	ZS (62 nm) ZP (−14.9 mV) PDI (0.125)	Antioxidant activity	[67]
Estragole (Basil)	High-energy	ZS (41.60 nm) PDI (0.093)	Antibacterial activity	[94]
Lavender oil, camphene, and α-terpinyl acetate (*Lavandula spica*)	High-energy	ZS (104.55, 117.23, 105.23, respectively) ZP (−22.26, −27.10, −18.14 mV, respectively) PDI (0.125)	Antimicrobial activity Antioxidant activity	[42]
D-limonene (citrus fruits)	High-energy	ZS (269–428 nm) ZP (−37.71 to −51.79 mV) PDI (0.21 to 0.65)	Seafood preservation	[95]
Eugenol (clove leaves)	High-energy	ZS (130 nm) ZP (−12 mV) PDI (0.20)	Antibacterial activity	[96]
1,8-cineol and camphor (*Rosmarinus officinalis*)	Low-energy	ZS (89.87 nm) ZP (−12 mV) PDI (0.193)	Anti-inflammatory activity	[97]
EO (lemongrass)	High-energy	ZS (4.49 nm) ZP (−20.9 to −62.2 mV)	Food preservation	[98]
Chlorophenols (aerial part of nettle)	Low-energy	ZS (102 nm) PDI (0.192)	Food packaging	[99]

(continued)

Table 1 (continued)

Main components of EOs (source)	Preparation methods	Characterization technique*	Applications	Key references
α-pinene, ger-macrene D, β-caryophyllene, α-copaene, sabinene, and D-limonene (resin of *Araucaria heterophylla*)	High-energy	ZS (106 nm) ZP (+4.91 mV)	Anti-inflammatory, and antipyretic activities	[66]
1,8-cineole (eucalyptus), α-citral and β-citral (lemongrass)	High-energy	ZS (68.15 nm, 90.56 nm, respectively) ZP (−9.74 mV, −8.77 mV, respectively) PDI (0.19, 0.22, respectively)	Antifungal activity	[33]

*PDI = polydispersity index, ZP = zeta potential, ZS = Zeta size of droplets

5 Some Limitations and Opportunities of Nanoemulsions of Essential Oils

One of the drawbacks of nanoemulsion formulation is the lack of easily maintaining long-term stability for commercial purposes due to their production of colloidal dispersions with smaller droplet sizes [13, 48]. The advanced instruments and process methods cost a lot for reductions of droplets size of nanoemulsion [75]. The unacceptable taste of final products limits the applications of nanoemulsions of essential oils in the food industry because of the excess amount of surfactants used, especially for low-energy formulation methods [10]. These limitations allow researchers from different disciplines to design new nano-engineered delivery materials with cost-effective and high-stability properties [5, 23]. Preparations of nanoemulsions of essential oils are based on eco-friendly approaches, which also attracts big opportunities to replace synthetic emulsifiers such as surfactants with natural alternatives [36]. Some drawbacks of nanoemulsion preparations may bring an opportunity to develop improved formulation methods and new products with unique functionality [36]. The improved efficient and controllable delivery system development in preparation for the nanoemulsion of essential oils can be extended to other formulation processes. An additional opportunity for producing nanoemulsion of essential oils is their vast market demand due to their preference in various food industries [100].

6 Conclusions

Even though there are several methods of preparation of nanoemulsions of essential oils, high-energy approaches are the most commonly used systems. Nanoemulsions of essential oils are promising methods to be utilized in diverse areas of applications. The use of nanoemulsions of essential oils in various areas such as food, paint, pharmaceutical, pesticides, and cosmetics industries highly attracted the interest of researchers from different disciplines. The demand for substituting limited and synthetic or semi-synthetic resources with sustainable and natural additives in the consumed products makes nanoemulsions of essential oils more important in various fields of applications. Creaming, flocculation, coalescence, and Ostwald ripening are mechanisms to be controlled during nanoemulsions preparation for managing the instability of nanoemulsions. Due to Brownian motion, nanoemulsions of essential oils with smaller particle sizes are highly stable relative to the creaming of droplets. Due to the smaller size of droplets, less chance of agglomeration and higher stability of particles are observed in the nanoemulsions of essential oils prepared by ultrasonication methods. Nanoemulsion is applied as a potential method to encapsulate bioactive compounds of essential oils from environmental stresses such as temperature, oxygen, and light effects. Optimization of nanoemulsions of essential oils before their applications is highly required to maximize their activities. Therefore, nanoemulsions of essential oils have potential uses in wide applications. However, the instability problem of most nanoemulsions of essential oils in storage for longer periods of time requires strong attention from researchers.

References

1. Ali B, Jamil A, Majeed U, Khan QF, Iqbal KJ, Shoemaker CF, Zhong F (2015) Essential oil encapsulations: uses, procedures, and trends. RSC Adv. https://doi.org/10.1039/C5RA06556A
2. Gonfa YH, Gelagle AA, Hailegnaw B, Kabeto SA, Workeneh GA, Tessema FB, Tadesse MG, Wabaidur SM, Dahlous KA, Abou Fayssal S, Kumar P, Adelodun B, Bachheti A, Bachheti RK (2023) Optimization, characterization, and biological applications of silver nanoparticles synthesized using essential oil of aerial part of *Laggera tomentosa*. Sustainability 15(1):797
3. Gonfa YH, Tessema FB, Gelagle AA, Getnet SD, Tadesse MG, Bachheti A, Bachheti RK (2022) Chemical compositions of essential oil from aerial parts of *Cyclospermum leptophyllum* and its application as antibacterial activity against some food spoilage bacteria. J Chem. https://doi.org/10.1155/2022/5426050
4. Hashtjin AM, Abbasi S (2015) Nano-emulsification of orange peel essential oil using sonication and native gums. Food Hydrocolloids 44:40–48. https://doi.org/10.1016/j.foodhyd.2014.08.017
5. Pavoni L, Perinelli DR, Bonacucina G, Cespi M, Palmieri GF (2020) An overview of micro- and nanoemulsions as vehicles for essential oils: formulation, preparation and stability. Nanomaterials 10(1):1–24. https://doi.org/10.3390/nano10010135
6. Xue J (2015) Essential oil nanoemulsions prepared with natural emulsifiers for improved food safety

7. Komaiko JS, Mcclements DJ (2016) Formation of food-grade nanoemulsions using low-energy preparation methods: a review of available methods. Compr Rev Food Sci Food Saf 15:331–352. https://doi.org/10.1111/1541-4337.12189
8. Long Y, Huang W, Wang Q, Yang G (2020) Green synthesis of garlic oil nanoemulsion using ultrasonication technique and its mechanism of antifungal action against *Penicillium italicum*. Ultrason Sonochem 64:104970. https://doi.org/10.1016/j.ultsonch.2020.104970
9. Guilherme I, Sen K, Rinaldi C (2020) Effect of sodium alginate and different types of oil on the physical properties of ultrasound-assisted nanoemulsions. Chem Eng Process 153:1–8. https://doi.org/10.1016/j.cep.2020.107942
10. Koshani R, Mahdi S (2019) Ultrasound-assisted preparation of different nanocarriers loaded with food bioactive ingredients. Adv Coll Interface Sci 270:123–146. https://doi.org/10.1016/j.cis.2019.06.005
11. Taha A, Ahmed E, Ismaiel A, Ashokkumar M, Xu X, Pan S, Hu H (2020) Ultrasonic emulsification: an overview on the preparation of different emulsifiers-stabilized emulsions. Trends Food Sci Technol. https://doi.org/10.1016/j.tifs.2020.09.024
12. Dasgupta N, Ranjan S (2018) An introduction to food grade nanoemulsions. Springer Nature Singapore Pte Ltd
13. Gutiérrez JM, González C, Maestro A, Solè I, Pey CM, Nolla J (2008) Nano-emulsions: new applications and optimization of their preparation. Curr Opin Colloid Interface Sci 13:245–251. https://doi.org/10.1016/j.cocis.2008.01.005
14. Kumar H, Kumar V (2018) Ultrasonication assisted formation and stability of water-in-oil nanoemulsions: optimization and ternary diagram analysis. Ultrason Sonochem. https://doi.org/10.1016/j.ultsonch.2018.07.022
15. Gharibzahedi SMT, Rostami H, Yousefi S (2015) Formulation design and physicochemical stability characterization of nanoemulsions of Nettle (*Urtica dioica*) essential oil using a model-based methodology. J Food Process Preserv 39(6):2947–2958. https://doi.org/10.1111/jfpp.12546
16. Salvia-Trujillo L, Rojas-Graü A, Soliva-Fortuny R, Martín-Belloso O (2015) Physicochemical characterization and antimicrobial activity of food-grade emulsions and nanoemulsions incorporating essential oils. Food Hydrocolloids 43:547–556. https://doi.org/10.1016/j.foodhyd.2014.07.012
17. Gupta M, Gupta D (2022) Essential oils: as potential larvicides. J Drug Delivery Ther 12(3):193–201
18. Pathania R, Khan H, Kaushik R, Khan MA (2018) Essential oil nanoemulsions and their antimicrobial and food applications. Curr Res Nutr Food Sci 6(3):626–643. https://doi.org/10.12944/CRNFSJ.6.3.05
19. Zhang S, Zhang M, Fang Z, Liu Y (2017) Preparation and characterization of blended cloves/cinnamon essential oil nanoemulsions. Food Sci Technol 75:316–322. https://doi.org/10.1016/j.lwt.2016.08.046
20. Rao J, Chen B, McClements DJ (2019) Improving the efficacy of essential oils as antimicrobials in foods: mechanisms of action. Annu Rev Food Sci Technol 10:365–387. https://doi.org/10.1146/annurev-food-032818-121727
21. Singh N, Jaipur F (2017) An overview of prospective application of nanoemulsions in food stuffs and food packaging. ASIO J Microbiol Food Sci Biotechnol Innov 1:1–7
22. Bilia AR, Guccione C, Isacchi B, Righeschi C, Firenzuoli F, Bergonzi MC (2014) Essential oils loaded in nanosystems: a developing strategy for a successful therapeutic approach. Evid-Based Complement Altern Med 2014:1–15
23. Safaya M, Rotliwala YC (2019) Nanoemulsions: a review on low energy formulation methods, characterization, applications, and optimization technique. Mater Today: Proc. https://doi.org/10.1016/j.matpr.2019.11.267
24. Deepika, Chaudhari AK, Singh A, Das S, Dubey NK (2021) Nanoencapsulated Petroselinum crispum essential oil: characterization and practical efficacy against fungal and aflatoxin contamination of stored chia seeds. Food Biosci 42:101117. https://doi.org/10.1016/j.fbio.2021.101117

25. Sugumar S, Singh S (2016) Nanoemulsion of orange oil with non ionic surfactant produced emulsion using ultrasonication technique: evaluating against food spoilage yeast. Appl Nanosci. https://doi.org/10.1007/s13204-015-0412-z
26. Vinaya KN, John AM, Mangsatabam M (2021) Ultrasound-assisted synthesis and characterization of sesame oil based nanoemulsion. IOP Conf Ser: Mater Sci Eng 1114:1–9. https://doi.org/10.1088/1757-899X/1114/1/012085
27. Jadhav AJ, Holkar CR, Karekar SE, Pinjari DV, Pandit AB (2014) Ultrasound assisted manufacturing of paraffin wax nanoemulsions: process optimization. In: Ultrasonics sonochemistry. Elsevier B.V. https://doi.org/10.1016/j.ultsonch.2014.10.024
28. Raviadaran R, Han M, Manickam S, Chandran D (2018) Ultrasound-assisted water-in-palm oil nano-emulsion: influence of polyglycerol polyricinoleate and NaCl on its stability. Ultrason Sonochem. https://doi.org/10.1016/j.ultsonch.2018.12.012
29. Mei Z, Xu J, Sun D (2011) O/W nano-emulsions with tunable PIT induced by inorganic salts. Colloids Surf, A 375(1–3):102–108. https://doi.org/10.1016/j.colsurfa.2010.11.069
30. Espitia PJP, Fuenmayor CA, Otoni CG (2019) Nanoemulsions: synthesis, characterization, and application in bio-based active food packaging. Compr Rev Food Sci Food Saf 18(1):264–285. https://doi.org/10.1111/1541-4337.12405
31. Nuchuchua O, Sakulku U, Uawongyart N, Puttipipatkhachorn S, Soottitantawat A, Ruktanonchai U (2009) In vitro characterization and mosquito (*Aedes aegypti*) repellent activity of essential-oils-loaded nanoemulsions. AAPS Pharm Sci Tech 10(4):1234–1242. https://doi.org/10.1208/s12249-009-9323-1
32. Gadhave A (2014) Nanoemulsion-formation, stability and applications nanoemulsions: formation, stability and applications. Int J Res Sci Adv Technol 2(3):38–43
33. Gündel S, da S, de Godoi SN, Santos RCV, da Silva JT, Leite LB, de M, Amaral AC, Ourique AF (2020) In vivo antifungal activity of nanoemulsions containing eucalyptus or lemongrass essential oils in murine model of vulvovaginal candidiasis. J Drug Delivery Sci Technol 57:101762. https://doi.org/10.1016/j.jddst.2020.101762
34. Jin W, Xu W, Liang H, Li Y, Liu S, Li B (2016) Nanoemulsions for food: properties, production, characterization, and applications. In: Emulsions. Elsevier Inc. https://doi.org/10.1016/b978-0-12-804306-6.00001-5
35. Besseling R, Arribas-Bueno R, van Tuijn R, Gerich A (2021). Realtime droplet size monitoring of nano-emulsions during white paper realtime droplet size monitoring of nano-emulsions during high pressure. Inprocess-LSP 1–11. https://doi.org/10.13140/RG.2.2.30640.28166
36. Silva EK, Rosa MTMG, Meireles MAA (2015) Ultrasound-assisted formation of emulsions stabilized by biopolymers. Curr Opin Food Sci 5:50–59. https://doi.org/10.1016/j.cofs.2015.08.007
37. Liu T, Gao Z, Zhong W, Fu F, Li G, Guo J, Shan Y (2022) Preparation, characterization, and antioxidant activity of nanoemulsions incorporating lemon essential oil. Antioxidants 11(4):1–13. https://doi.org/10.3390/antiox11040650
38. Ghadetaj A, Almasi H, Mehryar L (2018) Development and characterization of whey protein isolate active films containing nanoemulsions of *Grammosciadium ptrocarpum* Bioss. essential oil. Food Packaging Shelf Life 16:31–40. https://doi.org/10.1016/j.fpsl.2018.01.012
39. Nejadmansouri M, Mohammad S, Hosseini H, Niakosari M, Yousefi GH, Golmakani MT (2016) Physicochemical properties and storage stability of ultrasound-mediated WPI-stabilized fish oil nanoemulsions. Food Hydrocolloids. https://doi.org/10.1016/j.foodhyd.2016.07.011
40. Shamsara O, Mahdi S, Muhidinov ZK (2017) Fabrication, characterization and stability of oil in water nano-emulsions produced by apricot gum-pectin complexes. Int J Biol Macromol 103:1285–1293. https://doi.org/10.1016/j.ijbiomac.2017.05.164
41. Xu T, Gao CC, Feng X, Wu D, Meng L, Cheng W, Zhang Y, Tang X (2020) Characterization of chitosan based polyelectrolyte films incorporated with OSA-modified gum arabic-stabilized cinnamon essential oil emulsions. Int J Biol Macromol 150:362–370. https://doi.org/10.1016/j.ijbiomac.2020.02.108

42. Badr MM, Badawy MEI, Taktak NEM (2021) Characterization, antimicrobial activity, and antioxidant activity of the nanoemulsions of *Lavandula spica* essential oil and its main monoterpenes. J Drug Delivery Sci Technol 65:102732. https://doi.org/10.1016/j.jddst.2021.102732
43. Jesser E, Yeguerman C, Gili V, Santillan G, Murray AP, Domini C, Werdin-González JO (2020) Optimization and characterization of essential oil nanoemulsions using ultrasound for new ecofriendly insecticides. ACS Sustain Chem Eng 8(21):7981–7992. https://doi.org/10.1021/acssuschemeng.0c02224
44. Carpenter J, Saharan VK (2017) Ultrasonic assisted formation and stability of mustard oil in water nanoemulsion: effect of process parameters and their optimization. Ultrason Sonochem 35:422–430. https://doi.org/10.1016/j.ultsonch.2016.10.021
45. Nair A, Mallya R, Suvarna V, Khan TA, Omri A (2022) Nanoparticles attractive carriers of antimicrobial essential oils. Antibiotics 11:1–45. https://doi.org/10.3390/antibiotics11010108
46. Ramisetty KA, Shyamsunder R (2011) Effect of ultrasonication on stability of oil in water emulsions. Internal J Drug Deliv 3:133–142
47. Sharif HR, Abbas S, Majeed H, Safdar W, Shamoon M, Khan MA, Shoaib M, Raza H, Haider J (2017) Formulation, characterization and antimicrobial properties of black cumin essential oil nanoemulsions stabilized by OSA starch. J Food Sci Technol 54(10):3358–3365. https://doi.org/10.1007/s13197-017-2800-8
48. Sedaghat Doost A, Devlieghere F, Dirckx A, Van der Meeren P (2018) Fabrication of Origanum compactum essential oil nanoemulsions stabilized using Quillaja Saponin biosurfactant. J Food Process Preserv 42(7):1–12. https://doi.org/10.1111/jfpp.13668
49. Kundu P, Arora K, Gu Y, Kumar V, Mishra IM (2019) Formation and stability of water-in-oil nano-emulsions with mixed surfactant using in-situ combined condensation-dispersion method. Can J Chem Eng 97:2039–2049. https://doi.org/10.1002/cjce.23481
50. Pereira SF, Barroso A, Mour RHV (2021) A low energy approach for the preparation of nano-emulsions with a high citral-content essential oil. Molecules 26:1–11
51. Yuliani S, Syakir TRMM (2018) Changes in characteristics of nanoemulsion of cinnamon oil and their relationships with instability mechanisms during storage. J Food Process Preserv. https://doi.org/10.1111/jfpp.13745
52. Zhang W, Qin Y, Chang S, Zhu H, Zhang Q (2020) Influence of oil types on the formation and stability of nano-emulsions by D phase emulsification. J Dispersion Sci Technol. https://doi.org/10.1080/01932691.2020.1737538
53. Ohtani SG, Rasert WP (2014) Nano-emulsions; emulsification using low energy methods. Japan J Food Eng 15(3):119–130
54. Sole I, Maestro A, Pey CM, Gonzalez C, Solans C, Gutiérrez JM (2006) Nano-emulsions preparation by low energy methods in an ionic surfactant system. Colloids Surf A Physicochem Eng Aspects 288:138–143. https://doi.org/10.1016/j.colsurfa.2006.02.013
55. Izquierdo P, Esquena J, Tadros TF, Dederen C, Garcia MJ, Azemar N, Solans C (2002) Formation and stability of nano-emulsions prepared using the phase inversion temperature method. Langmuir 18:26–30
56. Calderó G, García-celma MJ, Solans C (2011) Formation of polymeric nano-emulsions by a low-energy method and their use for nanoparticle preparation. J Colloid Interface Sci 353:406–411. https://doi.org/10.1016/j.jcis.2010.09.073
57. Schalbart P, Kawaji M (2013) Comparison of paraffin nanoemulsions prepared by low-energy emulsification method for latent heat storage. Int J Therm Sci 67:113–119. https://doi.org/10.1016/j.ijthermalsci.2012.12.007
58. Lu W, Huang D, Wang CCR, Yeh C, Tsai J, Huang Y, Li P (2017) Preparation, characterization, and antimicrobial activity of nanoemulsions incorporating citral essential oil. J Food Drug Anal. https://doi.org/10.1016/j.jfda.2016.12.018
59. Saani SM, Abdolalizadeh J, Heris SZ (2019) Ultrasonic/sonochemical synthesis and evaluation of nanostructured oil in water emulsions for topical delivery of protein drugs. Ultrason Sonochem 55:86–95. https://doi.org/10.1016/j.ultsonch.2019.03.018

60. Brusewitz C, Schendler A, Funke A, Wagner T, Lipp R (2007) Novel poloxamer-based nanoemulsions to enhance the intestinal absorption of active compounds. Int J Pharmaceutics 329:173–181. https://doi.org/10.1016/j.ijpharm.2006.08.022
61. Martin-Piñero MJ, García MC, Santos J, Alfaro-Rodriguez MC, Muñoz J (2020) Characterization of novel nanoemulsions, with improved properties, based on rosemary essential oil and biopolymers. J Sci Food Agric 100(10):3886–3894. https://doi.org/10.1002/jsfa.10430
62. Maghamian N, Goli M, Najarian A (2021) Ultrasound-assisted preparation of double nanoemulsions loaded with glycyrrhizic acid in the internal aqueous phase and skim milk as the external aqueous phase. Food Sci Technol 141:110850. https://doi.org/10.1016/j.lwt.2021.110850
63. Amjadi S, Almasi H, Ghadertaj A, Mehryar L (2021) Whey protein isolate-based films incorporated with nanoemulsions of orange peel (*Citrus sinensis*) essential oil: preparation and characterization. J Food Process Preserv 45(2):1–12. https://doi.org/10.1111/jfpp.15196
64. Moradi S, Barati A (2019) Essential oils nanoemulsions: Preparation, characterization and study of antibacterial activity against *Escherichia coli*. Int J Nanosci Nanotechnol 15(3):199–210
65. Firoozi M, Rezapour-Jahani S, Shahvegharasl Z, Anarjan N (2020) Ginger essential oil nanoemulsions: preparation and physicochemical characterization and antibacterial activities evaluation. J Food Process Eng 43(8). https://doi.org/10.1111/jfpe.13434
66. Elshamy AI, Ammar NM, Hassan HA, Al-Rowaily SL, Ragab TI, El Gendy AENG, Abd-ElGawad AM (2020) Essential oil and its nanoemulsion of *Araucaria heterophylla* resin: chemical characterization, anti-inflammatory, and antipyretic activities. Ind Crops Prod 148:112272. https://doi.org/10.1016/j.indcrop.2020.112272
67. Kaur H, Pancham P, Kaur R, Agarwal S, Singh M (2020) Synthesis and characterization of *Citrus limonum* essential oil based nanoemulsion and its enhanced antioxidant activity with stability for transdermal application. J Biomater Nanobiotechnol 11(04):215–236. https://doi.org/10.4236/jbnb.2020.114014
68. Almasi H, Azizi S, Amjadi S (2020) Development and characterization of pectin films activated by nanoemulsion and Pickering emulsion stabilized marjoram (*Origanum majorana* L.) essential oil. Food Hydrocolloids 99:105338. https://doi.org/10.1016/j.foodhyd.2019.105338
69. Guerra-Rosas MI, Morales-Castro J, Ochoa-Martínez LA, Salvia-Trujillo L, Martín-Belloso O (2016) Long-term stability of food-grade nanoemulsions from high methoxyl pectin containing essential oils. Food Hydrocolloids 52:438–446. https://doi.org/10.1016/j.foodhyd.2015.07.017
70. Hassanshahian M, Saadatfar A, Masoumipour F (2020) Formulation and characterization of nanoemulsion from *Alhagi maurorum* essential oil and study of its antimicrobial, antibiofilm, and plasmid curing activity against antibiotic-resistant pathogenic bacteria. J Environ Health Sci Eng 18(2):1015–1027. https://doi.org/10.1007/s40201-020-00523-7
71. Das S, Singh VK, Dwivedy AK, Chaudhari AK, Upadhyay N, Singh A, Deepika, Dubey NK (2020) Fabrication, characterization and practical efficacy of *Myristica fragrans* essential oil nanoemulsion delivery system against postharvest biodeterioration. Ecotoxicol Environ Saf 189:110000. https://doi.org/10.1016/j.ecoenv.2019.110000
72. Abate L, Bachheti A, Bachheti RK, Husen A (2021) Antibacterial properties of medicinal plants: recent trends, progress, and challenges. In: Traditional herbal therapy for the human immune system. CRC Press, pp 13–54
73. Bachheti RK, Bachheti A, Satyan RS (2018) Chemical composition of the essential oil from *Schinus molle* L. (Peruvian pepper). Der Pharma Chemica 10(10):139–147
74. Gonfa YH, Tessema FB, Bachheti A, Tadesse MG, Eid EM, Abou Fayssal S, Adelodun B, Choi KS, Širić I, Kumar P, Bachheti RK (2022) Essential oil composition of aerial part of *Pluchea ovalis* (Pers.) DC., silver nanoparticles synthesis, and larvicidal activities against fall armyworm. Sustainability 14(23):15785
75. Sharma N, Bansal M, Visht S, Sharma PK, Kulkarni GT (2014) Nanoemulsion: A new concept of delivery system anoemulsion: a new concept of delivery system. Chronicles Young Sci 1(2):1–6

76. Jung H, Goo W, Nam S (2009) Stability of oil-in-water nano-emulsions prepared using the phase inversion composition method. J Ind Eng Chem 15:331–335. https://doi.org/10.1016/j.jiec.2009.01.001
77. Tadros T, Izquierdo P, Esquena J, Solans C (2004) Formation and stability of nano-emulsions. Adv Coll Interface Sci 109:303–318. https://doi.org/10.1016/j.cis.2003.10.023
78. Walker R, Deckera EA, M DJ (2014). Function opportunities and obstacles in the food industry. Food Function 1–14. https://doi.org/10.1039/c4fo00723a
79. Lou Z, Chen J, Yu F, Wang H, Kou X, Ma C, Zhu S (2017) The antioxidant, antibacterial, antibiofilm activity of essential oil from *Citrus medica* L. var. sarcodactylis and its nanoemulsion. Food Sci Technol 80:371–377. https://doi.org/10.1016/j.lwt.2017.02.037
80. Melanie M, Miranti M, Kasmara H, Malini DM, Husodo T, Panatarani C, Joni IM, Hermawan W (2022) Nanotechnology-based bioactive antifeedant for plant protection. Nanomaterials 12(4):1–32. https://doi.org/10.3390/nano12040630
81. Ghani S, Barzegar H, Noshad M, Hojjati M (2018) The preparation, characterization and in vitro application evaluation of soluble soybean polysaccharide films incorporated with cinnamon essential oil nanoemulsions. Int J Biol Macromol 112:197–202. https://doi.org/10.1016/j.ijbiomac.2018.01.145
82. Ebrahimi P, Ebrahim-Magham B, Pourmorad F, Honary S (2013) Ferulic acid lecithin-based nano-emulsions prepared by using spontaneous emulsification process. Iran J Chem Chem Eng 32(3):17–25
83. Li P, Chiang B (2012) Process optimization and stability of D-limonene-in-water nanoemulsions prepared by ultrasonic emulsification using response surface methodology. Ultrason Sonochem 19(1):192–197. https://doi.org/10.1016/j.ultsonch.2011.05.017
84. Yazgan H (2020) Investigation of antimicrobial properties of sage essential oil and its nanoemulsion as antimicrobial agent. LWT 130:109669. https://doi.org/10.1016/j.lwt.2020.109669
85. Ahmad N, Ahmad R, Al-qudaihi A, Alaseel E (2019) Preparation of a novel curcumin nanoemulsion by ultrasonication and its comparative effects in wound healing and the treatment of inflammation. RSC Adv 9:20192–20206. https://doi.org/10.1039/c9ra03102b
86. Ghosh V, Saranya S, Mukherjee A, Chandrasekaran N (2013) Cinnamon oil nanoemulsion formulation by ultrasonic emulsification: investigation of its bactericidal activity. J Nanosci Nanotechnol 13(1):114–122. https://doi.org/10.1166/jnn.2013.6701
87. Roy A, Guha P (2018) Formulation and characterization of betel leaf (*Piper betle* L.) essential oil based nanoemulsion and its in vitro antibacterial efficacy against selected food pathogens. J Food Process Preserv 42(6):1–7. https://doi.org/10.1111/jfpp.13617
88. Acevedo-Fani A, Salvia-Trujillo L, Rojas-Graü MA, Martín-Belloso O (2015) Edible films from essential-oil-loaded nanoemulsions: physicochemical characterization and antimicrobial properties. Food Hydrocolloids 47:168–177. https://doi.org/10.1016/j.foodhyd.2015.01.032
89. Osanloo M, Sereshti H, Sedaghat MM, Amani A (2017) Nanoemulsion of Dill essential oil as a green and potent larvicide against *Anopheles stephensi.* Environ Sci Pollut Res 1–8
90. Algahtani MS, Ahmad MZ, Nourein IH, Albarqi HA, Alyami HS, Alyami MH, Alqahtani AA, Alasiri A, Algahtani TS, Mohammed AA, Ahmad J (2021) Preparation and characterization of curcumin nanoemulgel utilizing ultrasonication technique for wound healing: in vitro, ex vivo, and in vivo evaluation. Gels 7:1–17
91. El-Moghazy TFA (2021) Comparison of the efficacy of micro and nano emulsions of caraway essential oil on quality of red chili fruits (*Capsicum annuum* L.) during cold storage. Int J Modern Agric 10(2):3692–3708
92. Mostafa DM, El-Alim SHA, Asfour MH, Al-Okbi SY, Mohamed DA, Awad G (2015) Transdermal nanoemulsions of *Foeniculum vulgare* Mill. essential oil: preparation, characterization and evaluation of antidiabetic potential. J Drug Deliv Sci Technol 29:99–106. https://doi.org/10.1016/j.jddst.2015.06.021
93. Mostafa DM, Kassem AA, Asfour MH, Al Okbi SY, Mohamed DA, Hamed TES (2015) Transdermal cumin essential oil nanoemulsions with potent antioxidant and hepatoprotective activities: in-vitro and in-vivo evaluation. J Mol Liq 212:6–15. https://doi.org/10.1016/j.molliq.2015.08.047

94. Ghosh V, Mukherjee A, Chandrasekaran N (2013) Ultrasonic emulsification of food-grade nanoemulsion formulation and evaluation of its bactericidal activity. Ultrason Sonochem 20(1):338–344. https://doi.org/10.1016/j.ultsonch.2012.08.010
95. Wu C, Wang L, Hu Y, Chen S, Liu D, Ye X (2016) Edible coating from citrus essential oil-loaded nanoemulsions: physicochemical characterization and preservation performance. RSC Adv 6(25):20892–20900. https://doi.org/10.1039/c6ra00757k
96. De Meneses AC, Sayer C, Puton BMS, Cansian RL, Araújo PHH, Oliveira DD (2019) Production of clove oil nanoemulsion with rapid and enhanced antimicrobial activity against gram-positive and gram-negative bacteria. J Food Process Eng. https://doi.org/10.1111/jfpe.13209
97. Borges RS, Keita H, Ortiz BLS, dos Santos Sampaio TI, Ferreira IM, Lima ES, de Jesus Amazonas da Silva M, Fernandes CP, de Faria Mota Oliveira AEM, da Conceição EC, Rodrigues ABL, Filho ACMP, Castro AN, Carvalho JCT (2018) Anti-inflammatory activity of nanoemulsions of essential oil from *Rosmarinus officinalis* L.: in vitro and in zebrafish studies. Inflammopharmacology 26(4):1057–1080. https://doi.org/10.1007/s10787-017-0438-9
98. Salvia-Trujillo L, Rojas-Graü A, Soliva-Fortuny R, Martín-Belloso O (2013) Physicochemical characterization of lemongrass essential oil-alginate nanoemulsions: effect of ultrasound processing parameters. Food Bioprocess Technol 6(9):2439–2446. https://doi.org/10.1007/s11947-012-0881-y
99. Kalateh-Seifari F, Yousefi S, Ahari H, Hosseini SH (2021) Corn starch-chitosan nanocomposite film containing nettle essential oil nanoemulsions and starch nanocrystals: optimization and characterization. Polymers 13(13):1–26. https://doi.org/10.3390/polym13132113
100. Dasgupta N, Ranjan S, Gandhi M (2019) Nanoemulsions in food: market demand. Environ Chem Lett. https://doi.org/10.1007/s10311-019-00856-2

Citrus Fruit Nanoemulsions and Their Applications

Esraa A. Elhawary, Ashaimaa Y. Moussa, Omayma A. Eldahshan, and Abdel Nasser B. Singab

Abstract Essential oils biological activities are more than any time before needed and sought after. From their antimicrobial, antioxidant to antitumor and neuroprotective activities, citrus essential oils are endued with wide health and nutritional benefits, yet their limiting low solubility were a hurdle against their widespread use. Nano emulsions represent an attractive controlled strategy to deliver the lipophilic natural volatile components with ultra-efficacy due to their minute size and stable kinetic properties. Various applications demand the use of *Citrus* nano emulsions formulations; for instance, sustainable food industry requires safe and powerful antimicrobial agents for long term storage as well as flavoring and coloring agents. In medicine, the low bioavailability of Citrus oil was always a concern and restricted its wide use as a nutraceutical agent despite its proven health profits. This chapter focuses on the different methods applied in the preparation of citrus nano emulsions, their types and properties with a special insight over their role in food, therapy and medicine as well as cosmetics.

Keywords Essential oil · Nanoemulsions · Uses · Application

1 Introduction

Nanoemulsions are type of emulsions in which water and oil phases are being stabilized by surfactants. Surfactants act by decreasing surface tension together with co-surfactant. Nano emulsions are mostly used in food, drug and toiletries industries, they are preferred due to their high stability and low opacity rate as target delivery systems. Their unique physicochemical properties allow for their wide utilization in medical, agricultural, food, cosmeceutical, pharmaceutical and diagnostic applications [4, 18, 19]. Nanoemulsions help in using essential oils in food products by

E. A. Elhawary · A. Y. Moussa · O. A. Eldahshan · A. N. B. Singab (✉)
Department of Pharmacognosy, Faculty of Pharmacy, Ain Shams University, Abbasia, Cairo 11566, Egypt
e-mail: dean@pharma.asu.edu.eg

A. Husen et al. (eds.), *Current Trends in Green Nano-emulsions*, Smart Nanomaterials Technology, https://doi.org/10.1007/978-981-99-5398-1_3

allowing more dispersibility of such essential oils in places of microbial growth leading to enhanced antimicrobial activity [13, 14, 30].

Essential oils had gained much attention in the area of food preservation due to their potent antimicrobial activities against wide array of microbes that affect human health. Essential oils usually comprise part of many plants' secondary metabolites, they are volatile compounds with pleasant aroma and thus have been used as food flavoring agents to impart pleasant odors. Food-borne microbes are one of the hot topics worldwide affecting food quality and leading to the rapid deterioration of many crops and industrial products. Food preservation usually takes place through many techniques including; drying out, warm treatment, adding anti-microbial agents, pH and environmental changes such techniques are applied in many food processing factories. An important part of food preservation is played by the antimicrobial agents added to foods either from natural or artificial sources. One study addressed the effect of adding garlic essential oil (*Allium sativum*) and the essential oil of wild garlic (*Allium vineale*) as food additive. Part of the artificial antimicrobials added to foods viz. esters and organic acids has been legally prohibited from use due to their various health hazards [25, 30, 37].

Genus *Citrus* with its members is part of family Rutaceae, which includes about 140 genera and 1300 species. *Citrus* comprises a lot of important species including; *Citrus sinensis* (Orange), *Citrus reticulate* (Mandarin), *Citrus aurantifolia* (Limes), *Citrus limon* (Lemon), *Citrus paradisi* (Grapefruit), *Citrus bergamia* (Bergamot), *Citrus junos* (Yuzu) and *Citrus japonica* (Kumquat) (Fig. 1). Many *Citrus* species are native to Asia especially the Himalayan hills in India, Northern Myanmar, Southern China and Southeast Asia. *Citrus* represents important genera of fruits worldwide, used as condiment, in the pastry industry, and bloody orange in many sweet and savory recipes from Europe. *Citrus* fruits have a high nutritional value and are rich with many secondary metabolites viz. polyphenolics, phenolic acids, flavonoids and their derivatives, limonoids, essential oils and vitamins (vitamin C and carotenoids) [8].

The nutritive and health-related effects of *Citrus* are wide and are mainly as antioxidants, anti-inflammatory agents and cytotoxic. However, their low solubility, stability, and bioavailability challenge their potential industrial formulations. This book chapter represents the use of citrus fruit formulations like nanonization and encapsulation in nano-scale carriers to facilitate their commercial use and applications [11].

Recently the idea of food safety had gained much attention from both customers and food industry members due to increased public awareness regarding health. Natural components are of great importance for food and medical industrial applications owing to their little or no side effects, cost-effectiveness and eco-friendly nature compared to synthetic counterparts. Plant-based natural antifungal agents have become ideal alternatives to commercial synthetic chemical preservatives to enhance food quality and safety. Citrus essential oils had gained much interest in this field due to their wide spectrum of activity against various bacteria, fungi, viruses and even insects [29, 40]. Thus, the citrus essential oils were widely used as food

Fig. 1 Diagram showing the most common *Citrus* species

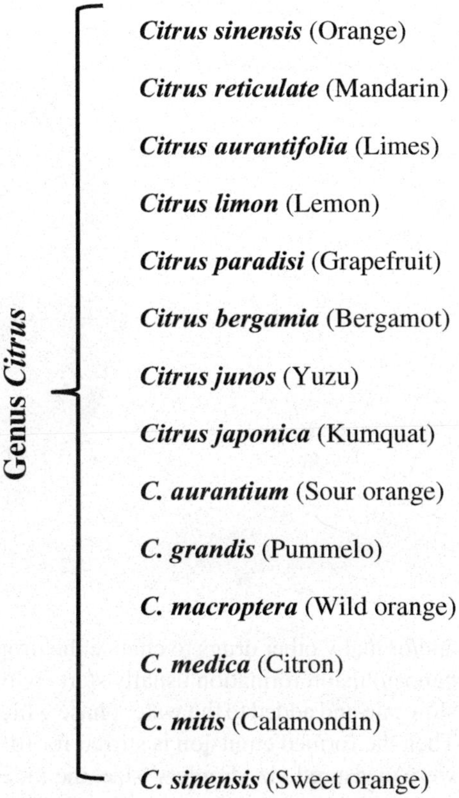

preservatives, in food formulations, packaging, and preservation to enhance both food quality and safety [8].

2 Methods of Nano Emulsion Preparation (Fig. 2)

1. **Spontaneous emulsification method**

In this method the emulsion is formed by mixing two liquids at ambient temperature. The first liquid is represented by water while the second is composed of oil, surfactant, and a water-miscible liquid. The two liquids here are stable all alone while this thermodynamic stability reach non-equilibrium upon mixing them together. The fast transfer of the water soluble components from the lipophilic to the hydrophilic phase leads to significant elevation of the interfacial area which results in metastable emulsion state. As a result of that, solvent diffusion gets really fast and the turbulence increases as well giving rise to Nano droplets. The nanoemulsion formulation may contain admixed constituents like surfactants, monomers, macromolecules, polymers

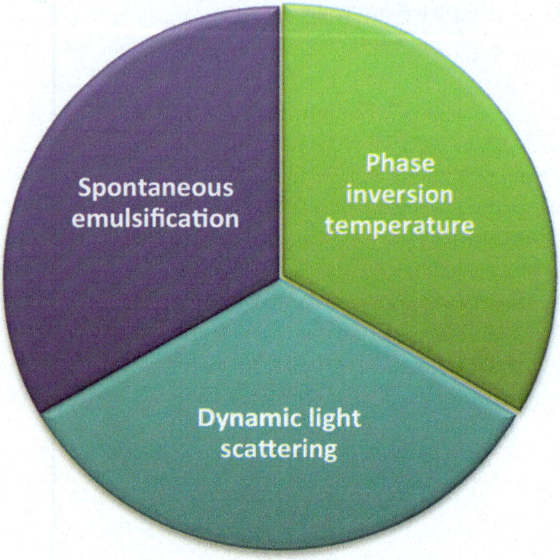

Fig. 2 Diagram showing different methods of nanoemulsion formulation

and/or many other drugs to change the droplets size [3]. The process of spontaneous nanoemulsion formation usually starts with mixing both the oil and the surfactant at slow rate and added to the water phase which led to rapid formation of nanoemulsion. Then the formed emulsion is stirred for fifteen minutes then the emulsion properties were measured viz. droplets' size, the index of polydispersity and the concentration of each droplet [3].

2. **The phase inversion method**

Herein the emulsion constituents are mixed from the beginning. Water and surfactant are mixed together from the start then the oil phase is added gradually after that till milky emulsion is formed. The temperature of this emulsion system was closely checked and being under control in a heater and with stirring. The emulsion heat gets slight rise by 10 °C and after that the formed mixture gets abruptly diluted with H_2O with 3× the water volume used thus nano-emulsions are produced [3].

3. **Dynamic light scattering**

The Hydrodynamic diameters were determined through dynamic light scattering with the help of a Malvern NanoZS instrument [3].

3 Types of Nanoparticle Delivery Systems (Fig. 3)

1. **Polymeric nanoparticles**

Polymeric nanoparticles are usually referred to as nanocarriers composed of biocompatible and biodegradable polymers. They differ in their final form, according to their method of preparation, being either nanocapsules or nanospheres. Nanospheres are in the form of matrix with uniform drug distribution in the matrix system while in the other system; the biomolecule is found in the most inner part and is coated with the polymeric material. Different kinds of biomolecules can be loaded viz. low molecular weight drugs, amino acids, proteins, plasmid and DNA. Different factors control the drug delivery potential of this nanoparticles like size of the particles, surface to charge scattering, and the solubility in water or oil phases of the nanoparticles [7].

2. **Micelles**

Micelles represent a type of nanocarriers where the lipophilic biomolecule is being encapsulated and covered with a water loving polymer to facilitate its bioavailability [7].

3. **Chitosan and dextran sulfate**

Chitosan and dextran sulfate can help in the encapsulation of many drugs to improve their therapeutic characters. The naringenin-containing nanoparticles formed of chitosan and dextran sulfate were spherically shaped and showed cytotoxicity when tested for breast cancer using MTT assay after 24 h incubation thus this delivery system was proved suitable for the encapsulation of lipid soluble molecules [7].

4. **Lipid-based nanoparticles**

Fig. 3 Diagram showing different types of nanoparticle delivery systems

Lipid nanoparticles are formed of lipids in either solid state only or in both solid and liquid states. Free fatty acids, fatty alcohols, triglycerides, steroids, and waxes are the types of lipids usually used for these nanoparticles in addition to phospholipids, glycolipids, and sphingolipids. The main advantages of such lipid particles relies on their biodegradability, biocompatibility, safety, solubility enhancement for the carried drug and controlled drug release [7].

5. **Nanosuspensions**

This system is composed of the biomolecule/drug plus a stabilizer forming nanosuspension with particles size less than 1 μm. These nanosuspensions may help in the solubilization of drugs with low bioavailability and poor pharmacokinetics. Furthermore, such system can play a role in directed treatment and enhancing drug solubility for IV administration of rutin instead of oral route [7].

6. **Co-delivery systems**

In many diseases and drug delivery experiments the need for a combination therapy represents one of the popular therapeutic options. Naringenin-hesperetin nano dispersion was prepared with PVP by solvent evaporation method. Drug recrystallization in the nanocarrier may weaken its physical stability in the delivery system while the interaction between the drug molecules and the used polymer may help to enhance the system stability. Upon using PVP polymer it helps in increasing the biomolecule stability via hydrogen bonding interaction [7].

4 Benefits of Preparing Nano Emulsions from Plant Extracts

Essential oils prepared from plants are usually utilized as potent natural antimicrobial agents especially for food-borne pathogens such as bacteria and fungi [15]. Their lipophilic nature hinders their proper distribution and bioavailability thus formulating them inside a nanoemulsion may help overcoming this issue [30].

5 Applications of Citrus Nano Emulsions

The idea of covering fruits and vegetables with edible coatings had gained much interest as a natural, safe and effective way of postharvest preservation of such crops.

1. **Applications in agriculture and food industry (Table** 1)

An edible chitosan coat was used for the nanoencapsulation of *Citrus* essential oil (92% d-limonene) in order to increase its safety and efficiency. The nano emulsion

Table 1 Applications of *Citrus*-based nano emulsions in food, medicine and cosmetics industries

Food industry	Medical applications	Cosmetics industry
An edible chitosan coat from nanoencapsulation of *Citrus* essential oil [38]	5-Demethyltangeretin (5DT) is a unique citrus flavonoid nanoemulsion had improved bioavailability and activity against intestinal cancer cells [43]	Hesperidin from *Citrus sinensis* in nanoemulsions then incorporated into a cream formulation for a tropical climate use to moisturize skin [35]
Pectin-based edible coating containing orange peel essential oil ↑ quality parameters of orange slices stored at 4 °C for 17 days [31]	Different rutin nanoemulsions ↑ bioavailability of rutin ↑ its solubility ↑ its absorption Cytotoxicity (IC_{50} 11.8 μM) against prostate carcinoma cells [1]	
Lemon essential oil encapsulated using chitosan → prolonged release of the essential oil in vitro [16]	A nanoemulsion from the essential oil of *Citrus medica* L. var. *sarcodactylis* showed potent antioxidant and antimicrobial activities 2× the pure essential oil in all assays 60% radical scavenging activity (26% for the pure oil) 80% inhibition rate against *S. aureus* (40% for the pure oil) MIC 0.48 mg/mL [24]	
The essential oil nano emulsion of sweet orange (*Citrus sinensis*) ↑ shelf-life of the same species juice (up to 3 months stability and bactericidal activity) [6]	*Citrus* essential oils (lemon and bergamot) nano emulsions Potent activity against (*Aspergillus niger*) ↓ mycelial radial growth ↓ spore germination MIC (0.35–5.50 μg/g) [32]	
β-carotene isolated from the peel of orange (*Citrus reticulate*) was formulated as nano emulsion and used as a natural food colorant in food industry [5]	*Citrus medica* L. var. *sarcodactylis* loaded into nano emulsion showed: Strong antibacterial activities (*E. coli, B. subtilis* and *S. aureus*) MIC (0.31–0.63 μL/mL). MBC (0.31–1.25 μL/mL) Weak antifungal activity (*A. niger* and *P. citrinum*) (3.6–27.8%) [21]	

(continued)

Table 1 (continued)

Food industry	Medical applications	Cosmetics industry
A nano emulsion loaded with D-limonene as nano emulsion was utilized as a natural fungicide and insecticide for fungi and pest associated with crops against *Pyricularia oryzae* (rice blast), *Rhizoctonia solani* (rice sheath blight), *Colletortrichum gloeosporiodes* (pepper anthracnose) and *Phomopsis amygdali* (peach shoot blight) [12]	Lemon essential oil nano emulsion showed higher activity on food-borne bacteria except *K. pneumoniae* compared to lemon essential oil (100% concentration) [41]	
Nano emulsions prepared from orange, grapefruit, mandarin, and lemon essential oils were tested as potential antioxidant and antimicrobial agents in the rainbow trout fillets preservation when stored at 4 ± 2_°C in refrigerator Nano emulsions helped in: ↓ fishy odor ↓ biochemical parameters ↓ growth of bacteria [10]	Nano emulsion of *Citrus sinensis* together with sodium alginate as edile coating for tomatoes preservation over 15 days storage period ↑ Tomatoes firmness (33%) 3× less weight loss compared to the uncoated ones [9]	
Nano emulsion from edible citrus peel essential oils showed antioxidant activity for fish fillets fatty acids stored for 16 days [36]	*Citrus aurantium* L. bloom essential oil nano emulsion showed cytotoxic and apoptotic activity on human lung (A549 cells) (IC_{50} 152 μg/mL) [26]	
Nanoemulsion (eugenol essential oil + limonin) showed antifungal activity against *P. italicum* (MIC = 160 μg/mL and inhibition rate 59.21%) The limonin-loaded eugenol emulsion protected the citrus fruits from mold infection for longer periods (infection rate of 29.2% after 5 days) [20]	Grapefruit peel essential oil nanoemulsion exhibited antibacterial activity against *Staphylococcus aureus*, *Enterococcus faecalis*, *Klebsiella pneumonia* and *Salmonella paratyphi* A as food-borne bacteria [28]	

(continued)

Table 1 (continued)

Food industry	Medical applications	Cosmetics industry
Various limonoid-type derivatives (halogenated/seven-membered lactam derivatives of obacunone and halogenated/oxime esters/oxime ethers/seven-membered lactam derivatives of limonin) showed potent green acaricidal activity and acted as sustainable alternatives for traditional resistant pesticides against *Tetranychus cinnabarinus* [17]	Finger citron essential oil encapsulated in a nanoemulsion formulation ↑ Solubility ↓ Unstability Prolonged digestion in the colon ↓ Gas production in colon ↑ Degradation of starch ↑Good gut microbiota (*Lactobacillus*, *Lactococcus* and the other probiotic bacteria) [23]	
Grapefruit (*Citrus paradisi*) peel phenolic nano-emulsion in mustard oil helped in stabilizing the active compound inside mustard oil and thus extended oxidative stability of mustard oil [27]		
Whey protein isolate -based films containing *Citrus sinensis* peel essential oil loaded nanoemulsions showed antioxidant and antimicrobial and acted as a suitable food packaging system for the preservation of food products [2]		
Mandarin (*Citrus reticulata* L.) essential oil loaded into chitosan nanoparticles showed antibacterial properties (*S. aureus* and *E. coli*) thus used in pork preservation [34]		
Nanoemulsion from (cinnamaldehyde, carvacrol and eugenol essential oils) helped 'Newhall' navel orange (*Citrus sinensis*) to resist against *P. digitatumin* infection [39]		
Lemon essential oil nanoemulsion presented enhanced bioavailability and exhibited good antioxidant activity [22]		

was produced by interaction between the chitosan functional groups and the essential oil through hydrogen bonding and electrostatic interactions [38].

Pectin-based edible coating microemulsion and nano emulsion, containing orange peel essential oil, were prepared using two essential oil concentrations (0.5 and 1.0%) and used as coatings to enhance the quality characteristics of orange cuttings stored at 4 °C for 17 days. The differences in pH and acidity in the treated samples were significantly higher than the controls and pectin-coated slices. Higher antibacterial and antifungal effects were observed in the nano emulsion-coated samples. The nano emulsion formulations containing the orange essential oil significantly extended the shelf-life for the orange slices compared to the pectin-coated and control groups [31].

Lemon essential oil was encapsulated using chitosan in order to decrease its light and heat sensitivity in addition to oxidation risk. The encapsulated essential oil nanoparticles showed an encapsulation efficiency of (85%). In vitro release studies demonstrated that the resulted nanocapsules had a rough surface without the spherical shape and with higher essential oil stability [16].

In order to preserve sweet orange (*Citrus sinensis*) juice from spoilage, the essential oil was obtained through cold press then was nano encapsulated in chitosan nano emulsion to enhance its stability and organoleptic properties. The sweet orange essential oil was analyzed through GC/MS, and 19 compounds were identified with 95% predominance of *d*-limonene and the encapsulated nano emulsion showed up to 3 months stability and bactericidal activity against *E. coli* compared to the pure essential oil [6].

The peel of orange (*Citrus reticulate*) usually gets removed as waste was used to extract β-carotene. The resulting β-carotene pure compound was utilized to form a stable nano emulsion using ultrasonic homogenizer with the purpose of using it as a natural food colorant in food industry. The droplet size and surface charge of the produced nano emulsion were 143.7 nm and −38.2 mV, respectively. The addition of the produced nano emulsion to the citrus juice significantly enhanced the juice color leading to an exemption of any other synthetic color. β-carotene formed as nano emulsion led to improvement in its bio accessibility and retinol activity equivalent (RAE) in the fruit juice. Analysis of the orange peel extract was done using UV–Vis spectrophotometer and HPLC. The β-carotene and the total carotenoid content extracted from *Citrus reticulate* peel were 23.57 and 27.61 μg/g, respectively [5].

A nano emulsion loaded with *d*-limonene (major essential oil component of lemon and other citrus fruits) was prepared as a natural fungicide and insecticide for fungi and pest associated with crops where it showed significant fungicidal and insecticidal activities against *Pyricularia oryzae, Rhizoctonia solani, Colletortrichum gloeosporiodes* and *Phomopsis amygdali.* Here the nano emulsion helped in decreasing the lipophilicity of *d*-limonene which usually limits its activity thus giving more opportunity of combating fungi and insects attacking crops [12].

Trout fillets contains usually a number of significantly valuable unsaturated fatty acids that are used as an indicator of its quality thus it is of great concern to preserve them well in order to protect such fatty acids from oxidation. Nano emulsions prepared from orange, grapefruit, mandarin, and lemon essential oils were tested as potential antioxidant and antimicrobial agents in the rainbow trout fillets preservation when preserved at 4 ± 2 °C. The results showed that the fillets shelf life was recorded to be 10 days (control group), 12 days when using tween 80, 14 days for orange/ lemon treated fillets and 16 days mandarin/grapefruit treated fillets. Nano emulsions helped in removing the fishy odor and the overall physical properties and acceptability. The nano emulsions also reduced the biochemical characters and decreased bacteria growth. The mandarin and grapefruit nano emulsion treated groups showed the lowest bacterial counts [10].

In this current study, nano emulsion prepared from citrus peel volatile oils had antioxidant effect on the fatty acid content of rainbow trout fillets under storage. The fillets were preserved with the prepared nano emulsion and stored for sixteen days. Samples of fats from the fish fillets showed that fatty acids were converted to the esters of fatty acids, and they were analyzed by GC. The following fatty acids (docosahexaenoic acid, stearic acid, vaccenic acid, oleic acid, eicosenoic acid, linoleic acid, linolenic acid, eicosapentaenoic acid, palmitoleic acid and palmitic acid) represented the main fats in the fish fillets. The treated groups showed higher representation of both polyunsaturated and monounsaturated fatty acids while the saturated fatty acids were lower in all groups compared to control. Thus adding citrus nano emulsions to the fish fillets positively affected their goodness and helped in preserving the contents of fatty acids from oxidation and deterioration upon storage [36].

The proharvest quality of citrus fruits is usually affected by *Penicillium italicum* which is a disease-causing fungus causing serious economic loss. A nano emulsion was prepared from the eugenol essential oil together with limonin acting as added antimicrobial component. The nano emulsion was performed through a high-pressure microfluidizer and its antifungal effect against *P. italicum* was tested. The antifungal activity measured by both the MIC values and the inhibition rate. The nano emulsion showed an MIC value and inhibition rates of 160 μg/mL and 59.21%, respectively. The nanoemulsion helped in protecting the citrus fruits by 29.2% after 5 days of storage [20].

Different limonoid-type derivatives viz. the seven-membered lactam, oxime esters, oxime ethers and halogenated derivatives of obacunone and limonin) were produced using citrus waste. The idea behind the synthesis of such derivatives is to develop stronger green insecticidal agents that can act as a replacement for the pre-existing agents. A biocompatible nano emulsion was utilized to develop the limonin-containing nanoemulsion, which showed significant insecticidal activity against *Tetranychus cinnabarinus*, well known for being resistant to many used insecticides [17].

The peel of *Citrus paradisi* (grapefruit) is filled with polyphenolics which have many health benefits but unfortunately these polyphenolics have poor stability against oxidation, light and temperature thus they have limited role in food industry. In this study, the feasibility of nano-emulsion prepared from the phenolics of the grapefruit peel in mustard oil was tested (the nanoemulsion was prepared through ultrasonication). Response surface methodology was used to test the nanoemulsion properties where; the sonication time (9.5 min) and droplet size (29.73 ± 1.62 nm) gave the most stable nano-emulsion [27].

Citrus sinensis peel essential oil was loaded into whey protein (WPI)-derived nanofilm polymers in order to develop both emulsions and nano emulsions (2.5 and 5% of WPI). The formed emulsions were evaluated via Fourier transform IR spectroscopy, Electron Microscopy, X-ray and tensile analyses. The formed nanoemulsion showed pronounced antioxidant and antimicrobial effects outweighing those of the emulsion alone without the essential oil. The developed nanoemulsion can be used as food packaging films for preserving food products against oxidation and microbes [2].

Mandarin (*Citrus reticulata* L.) essential oil is usually reported for being potent antimicrobial agent. The essential oil of mandarin was loaded into chitosan nanoparticles, and the characteristics of the nanoparticles, its antibacterial properties and role in pork preservation was evaluated. The prepared nanoparticles showed potent encapsulation efficiency (67–82%). The mechanism of action relied on the damage of microbial cell membranes caused by the essential oil-chitosan nano emulsions. The affected microbes were *S. aureus* and *E. coli*, which led to the change of bacterial cell morphology [34].

The use of nano emulsion technology was mainly to allow for the formation of natural fruit resistance against pathogens that act on citrus fruits after harvest (postharvest infection). A combination of different essential oils, prepared as nano emulsions, (cinnamaldehyde, carvacrol and eugenol nanoemulsion) were used to help 'Newhall' navel orange (*Citrus sinensis*) to resist against *P. digitatumin* infection. The prepared nano emulsions induced protection via decreasing the deposition of H_2O_2 and MDA in this citrus fruit, elevating PAL, C_4H, $_4CL$ and CAD activity and thus leading to elevated percentage of polyphenolic components. Different sets of genes were either up regulated or down regulated in the fruits treated with the nano emulsion. Genes encoding shikimate *O*-hydroxycinnamoyltransferase (HCT), caffeoyl-CoA *O*-methyltransferase (CCoAOMT), caffeic acid 3-*O*-methyltransferase/acetylserotonin *O*-methyltransferase (COMT), cinnamyl-alcohol dehydrogenase (CAD) and peroxidase (POD) were examples of such genes that were affected by the nano emulsion. A number of newly identified ^{2}ry metabolites were traced from the nano emulsion-treated fruits versus the untreated ones. The identified compounds were 175 compounds belonging to the phenylpropanoid pathway and are related to infection resistance. The identified compounds included coniferaldehyde, caffeic acid, coniferyl alcohol, *L*-phenylalanine and caffeoyl quinic acid. On the other side, many ^{1}ry metabolites gets accumulated in the treated fruits as well viz. amino acids, sugars, organic acids, lipids and alcohols [39].

Lemon essential oil was formulated into nano emulsion in order to enhance its bioavailability. The physicochemical characterization and antioxidant activities were evaluated. The lemon essential oil based nano emulsion showed moderate antioxidant activity thus providing a good scientific theory for the industrial applications [22].

2. **Applications in medicine (Table** 1)

5-Demethyltangeretin (5DT) represents a unique citrus flavonoid with potent anticancer activity, but unfortunately it has low solubility in water and poor bioavailability which affects its use as a drug. Emulsion-based delivery systems were adopted to enhance the bioavailability of 5DT and its uptake by intestinal cancer cells. The amount of 5DT absorbed by intestinal cancer cells was quantified by HPLC using an electrochemistry detector. A cytotoxicity assay showed that 5DT nanoemulsion showed potent cancer cells growth inhibition compared to the same drug with larger droplets [33, 42, 43].

In order to increase the bioavailability of rutin and to enhance its solubility and absorption, different nano emulsions containing rutin were prepared through aqueous titration method. The prepared nano emulsions were tested for surface morphology, droplet size, in vitro release profile and zeta potential thus the formulations were adjusted. Their cytotoxic activities were evaluated through MTT assay and antioxidant potential using human prostate cancer cell line. The results showed that the rutin nano emulsions showed (IC_{50} 11.8 μM) against the cancer cells. The prepared nano emulsion showed potent cytotoxic activity against prostate cancer cell line [1].

A nano emulsion was prepared from the essential oil of *Citrus medica* L. var. *sarcodactylis* through spontaneous emulsification method. This nano emulsion was rich with d-limonene (50% of the essential oil) then it was tested for its antioxidant and antimicrobial activities using different radical scavenging assays and the pour plate method where the prepared nano emulsion showed superior activities (up to twofold) compared to the pure essential oil in all assays with nearly 60% radical scavenging activity of the nano emulsion formulation compared to 26% for the pure oil and the inhibition rate of pure essential oil against *S. aureus* was 40% while the nano emulsion was 80% while at 0.48 mg/mL, the inhibition rate of the nano emulsion was 100% with MIC 0.48 mg/mL [24].

Nanoemulsion-based delivery systems for different *Citrus* essential oils (lemon and bergamot) against the spore-forming fungus (*Aspergillus niger*) was evaluated. The inhibition of mycelial radial growth and spore germination were used as indicators of antifungal activity of the nanoemulsions, which were prepared at 3 wt% essential oil, using Tween 80 or whey protein isolate (1 wt%) as emulsifiers, and sunflower oil (1 wt%) as ripening inhibitor. The nanoemulsions were physically stable over seven days of accelerated aging at 35 °C. The MIC of free cinnamon leaf and of both citrus essential oils were 0.35 and 5.50 μg/g, respectively. The encapsulation in nanoemulsions generally decreased the antifungal activity, likely because of the nanoemulsion acting as a hydrophobic sink for the main constituents of citrus essential oils. The emulsifier played a fundamental role in the resulting antifungal activity, with WPI-based nanoemulsions being more effective in inhibiting the mycelial growth and the spore germination of *A. niger* than T80-based ones. The antifungal action was correlated to the morphological alterations observed in *A. niger*, such as the loss of cytoplasm in fungal hyphae and hyphal tip [32].

In this study, *Citrus medica* L. var. *sarcodactylis* also known as finger citron was loaded into nanoemulsion using a mixture of cremophor EL/1, 2-propanediol (1:1 m/m), cremophor EL/glycerol (1:1 m/m), tween 80/ethanol (3:2 m/m), tween 80/PEG-400 (3:2 m/m) and tween 80/1, 2-propanediol (3:1 m/m) together as surfactants in order to ensure the nanoemulsion stability. The developed nanoemulsion showed strong antibacterial activities against *E. coli*, *B. subtilis* and *S. aureus* and weak antifungal activity against *A. niger* and *P. citrinum* through MIC (0.31–0.63 μL/mL), MBC (0.31–1.25 μL/mL) and mycelial growth assays (3.6–27.8%) compared to the pure essential oil [21].

Food-borne microbes usually cause huge economic and health effects leading to financial losses affecting human health through food poisoning. The effect of the lemon essential oil (10 and 100% concentrations) versus its nanoemulsion on food-borne pathogens (*Staphylococcus aureus*, *Klebsiella pneumoniae*, *Enterococcus faecalis* and *Salmonella paratyphi* A) was evaluated. The lemon essential oil was rich with limonene (52%), *p*-cymene (14%) and *β*-pinene (13%). The oil nanoemulsion showed higher activity on food-borne bacteria except *K. pneumoniae* compared to lemon essential oil (100% concentration) while the 10% concentration presented the highest inhibition effect on *S. paratyphi* A thus the nano emulsification significantly enhanced the bactericidal activity of lemon essential oil against food-borne pathogens [41]. This research on the nano emulsion of *Citrus sinensis* targeted such food-borne bacteria viz. *Salmonella typhi* and *Listeria monocytogenes*. The essential oil of this sweet orange and sodium alginate were prepared together as an edible coating for tomatoes preservation over 15 days storage period. The tomatoes firmness increased up to 33% and showed threefold less weight loss compared to the uncoated ones [9].

Citrus aurantium L. bloom essential oil nano emulsion was tested for its potential cytotoxic and apoptotic activity on human lung (A549 cells) followed by in vivo activity testing on mice health parameters. The essential oil sample showed the presence of linalyl acetate, limonene and *α*-terpineol as the main components. The prepared nano emulsion presented cytotoxic activity against A549 cells with IC_{50} 152 μg/mL. The mechanism of action depends on overexpression of Cas-3 and thus leading to apoptotic cells death. The mice were given the nano emulsion at a dosage of 10 and 20 mg/kg body weight for 30 days and they showed no histopathological alteration in the liver and kidney while exhibited enhancement in the jejunum morpho-structural architecture and hepatic antioxidant redox potential. The prepared nano emulsion showed promising alternative to prevent lung cancer progression and it provides a new way to enhance the therapeutic value of the plant phytochemicals [26].

This current research compared the antimicrobial activity of the grapefruit peel essential oil and its nano emulsion. The essential oil was tested using GC/MS where d-limonene represented 80% of the oil then its antibacterial activity against *Staphylococcus aureus*, *Enterococcus faecalis*, *Klebsiella pneumonia* and *Salmonella paratyphi* A as food-borne bacteria was evaluated using disc diffusion method and compared to its nano emulsion activity. Both the pure essential oil and its nano emulsion showed bacteriostatic activity however the activity of the nano emulsion was more powerful [28].

Finger citron essential oil is highly utilized in pharmaceutical and food industries. This citrus essential oil was encapsulated in a nano emulsion formulation to overcome its insolubility and instability and prolong its digestion in the colon. The effects of the developed nano emulsion on human gut microbiota were evaluated herein through in vitro fermentation. The use of low dosage of this nano emulsion significantly increased the production of acetate, CO_2 and H_2 as end products in basal medium containing starch as sole carbon source, while the high concentration of the nano emulsion significantly inhibited the gas production. Moreover, the nano emulsion enhanced the degradation of starch and modified gut microbiota community after 24 h anaerobic incubation thus nourished the growth of *Lactobacillus*, *Lactococcus* and the other probiotic bacteria, while suppressing the pathogenic *Bilophila* bacteria [23].

3. **Applications in cosmetics (Table** 1)

Hesperidin is a type of flavonoid isolated from orange (*Citrus sinensis*). Hesperidin is a well-known nutraceutical based on its antioxidant, chelating, and anti-ageing properties. In this study, hesperidin was evaluated as a therapeutic agent against dark eye circles. Besides its best formulation method was examined. The formed nano emulsions were formulated into a cream suitable for tropical climate. The resulting cream formulation was evaluated in vitro on artificial skin from cultured cells. The results showed that one of the nanonized hesperidin formulations was the most skin-friendly and might be used in cosmetics [35].

6 Conclusion and Future Prospects

In conclusion, many techniques, including spontaneous emulsification, phase inversion temperature, and dynamic light scattering, have been used to successfully prepare citrus nanoemulsions. There are various advantages to making nanoemulsions from plant extracts, particularly citrus fruit extracts. Improved bioavailability, increased solubility, and tailored delivery of active substances to certain locations inside the body are a few of these. Citrus nano emulsions have been used in a variety of fields, including food, medicine, cosmetics, and agriculture. Citrus nano emulsions have been demonstrated to enhance crop growth, productivity, and quality in agriculture. They have been utilised as natural preservatives, flavour enhancers, and colouring agents in the food sector. Citrus nano emulsions have showed promise in the treatment of a number of ailments and as drug delivery methods. They have been incorporated in cosmetics as all-natural moisturisers, antioxidants, and anti-aging agents. The prospects for citrus nano emulsions in the future are bright. Their potential uses and advantages will become more obvious with more research. Technology developments will also make it possible to prepare and transport substances in more effective and affordable ways, expanding the use of citrus nano emulsions for a variety of purposes. Overall, the application of citrus nano emulsions provides a promising new direction for innovation across numerous industries.

References

1. Ahmad M, Akhtar J, Hussain A, Arshad M, Mishra A (2017) Development of a new rutin nanoemulsion and its application on prostate carcinoma PC3 cell line. EXCLI J 16:810
2. Amjadi S, Almasi H, Ghadertaj A, Mehryar L (2021) Whey protein isolate-based films incorporated with nanoemulsions of orange peel (Citrus sinensis) essential oil: preparation and characterization. J Food Process Preserv 45:e15196
3. Anton N, Vandamme TF (2009) The universality of low-energy nano-emulsification. Int J Pharm 377:142–147
4. Bachheti RK, Abate L, Hunde Y, Tadesse MG, Bachheti A, Pandey DP, Sharama A, Husen A (2023) Essential oils from medicinal plants and their role in nanoparticles synthesis, characterization, and applications. In: Secondary metabolites from medicinal plants. CRC Press, pp 177–198
5. Barman K, Chowdhury D, Baruah PK (2020) Development of β-carotene loaded nanoemulsion using the industrial waste of orange (Citrus reticulate) peel to improve in vitro bioaccessibility of carotenoids and use as natural food colorant. J Food Process Preserv 44:e14429
6. Bento R, Pagán E, Berdejo D, de Carvalho RJ, García-Embid S, Maggi F, Magnani M, de Souza EL, García-Gonzalo D, Pagán R (2020) Chitosan nanoemulsions of cold-pressed orange essential oil to preserve fruit juices. Int J Food Microbiol 331:108786
7. Bhia M, Motallebi M, Abadi B, Zarepour A, Pereira-Silva M, Saremnejad F, Santos AC, Zarrabi A, Melero A, Jafari SM (2021) Naringenin nano-delivery systems and their therapeutic applications. Pharmaceutics 13:291
8. Bora H, Kamle M, Mahato DK, Tiwari P, Kumar P (2020) Citrus essential oils (CEOs) and their applications in food: an overview. Plants 9:357
9. Das S, Vishakha K, Banerjee S, Mondal S, Ganguli A (2020) Sodium alginate-based edible coating containing nanoemulsion of Citrus sinensis essential oil eradicates planktonic and sessile cells of food-borne pathogens and increased quality attributes of tomatoes. Int J Biol Macromol 162:1770–1779
10. Durmus M (2020) The effects of nanoemulsions based on citrus essential oils (orange, mandarin, grapefruit, and lemon) on the shelf life of rainbow trout (Oncorhynchus mykiss) fillets at 4±2 C. J Food Saf 40:e12718
11. Elmeligy S, Hathout RM, Khalifa SA, El-Seedi HR, Farag MA (2021) Pharmaceutical manipulation of citrus flavonoids towards improvement of its bioavailability and stability. A mini review and a meta-analysis study. Food Biosci 44:101428
12. Feng J, Wang R, Chen Z, Zhang S, Yuan S, Cao H, Jafari SM, Yang W (2020) Formulation optimization of D-limonene-loaded nanoemulsions as a natural and efficient biopesticide. Colloids Surf A 596:124746
13. Gonfa YH, Gelagle AA, Hailegnaw B, Kabeto SA, Workeneh GA, Tessema FB, Tadesse MG, Wabaidur SM, Dahlous KA, Abou Fayssal S, Kumar P, Adelodun B, Bachheti A, Bachheti RK (2023) Optimization, characterization, and biological applications of silver nanoparticles synthesized using essential oil of aerial part of *Laggera tomentosa*. Sustainability 15(1):797
14. Gonfa YH, Tessema FB, Bachheti A, Tadesse MG, Eid EM, Abou Fayssal S, Adelodun B, Choi KS, Širić I, Kumar P, Bachheti RK (2022) Essential oil composition of aerial part of Pluchea ovalis (Pers.) DC., silver nanoparticles synthesis, and larvicidal activities against Fall Armyworm. Sustainability 14(23):15785
15. Gonfa YH, Tessema FB, Gelagle AA, Getnet SD, Tadesse MG, Bachheti A, Bachheti RK (2022) Chemical compositions of essential oil from aerial parts of Cyclospermum leptophyllum and its application as antibacterial activity against some food spoilage bacteria. J Chem. https://doi.org/10.1155/2022/5426050
16. Hasani S, Ojagh SM, Ghorbani M (2018) Nanoencapsulation of lemon essential oil in Chitosan-Hicap system. Part 1: study on its physical and structural characteristics. Int J Biol Macromol 115:143–151

17. Huang X, Lv M, Ma Q, Zhang Y, Xu H (2021) High value-added application of natural products in crop protection: semisynthesis and acaricidal activity of limonoid-type derivatives and investigation of their biocompatible O/W nanoemulsions as agronanopesticide candidates. J Agric Food Chem 69:14488–14500
18. Husen A (2023) Secondary metabolites based green synthesis of nanomaterials and their applications. Springer Nature Singapore Pte Ltd., 152 Beach Road, #21–01/04 Gateway East, Singapore 189721, Singapore
19. Husen A, Siddiqi KS (2023) Advances in smart nanomaterials and their applications. Elsevier Inc., 50 Hampshire St., 5th Floor, Cambridge, MA 02139, USA
20. Li Y, Zhao R, Li Y, Zhou Z (2021) Limonin enhances the antifungal activity of eugenol nanoemulsion against penicillium italicum in vitro and in vivo tests. Microorganisms 9:969
21. Li Z-H, Cai M, Liu Y-S, Sun P-L (2018) Development of finger citron (Citrus medica L. var. sarcodactylis) essential oil loaded nanoemulsion and its antimicrobial activity. Food Control 94:317–323
22. Liu T, Gao Z, Zhong W, Fu F, Li G, Guo J, Shan Y (2022) Preparation, characterization, and antioxidant activity of nanoemulsions incorporating lemon essential oil. Antioxidants 11:650
23. Liu W, Li Z, Yang K, Sun P, Cai M (2021) Effect of nanoemulsion loading finger citron (Citrus medica L. var. Sarcodactylis) essential oil on human gut microbiota. J Funct Foods 77:104336
24. Lou Z, Chen J, Yu F, Wang H, Kou X, Ma C, Zhu S (2017) The antioxidant, antibacterial, antibiofilm activity of essential oil from Citrus medica L. var. sarcodactylis and its nanoemulsion. LWT 80:371–377
25. Mossa A-TH, Mohafrash SM, Ziedan E-SH, Abdelsalam IS, Sahab AF (2021) Development of eco-friendly nanoemulsions of some natural oils and evaluating of its efficiency against postharvest fruit rot fungi of cucumber. Ind Crops Prod 159:113049
26. Navaei Shoorvarzi S, Shahraki F, Shafaei N, Karimi E, Oskoueian E (2020) Citrus aurantium L. bloom essential oil nanoemulsion: synthesis, characterization, cytotoxicity, and its potential health impacts on mice. J Food Biochem 44:e13181
27. Nishad J, Dutta A, Saha S, Rudra SG, Varghese E, Sharma R, Tomar M, Kumar M, Kaur C (2021) Ultrasound-assisted development of stable grapefruit peel polyphenolic nano-emulsion: optimization and application in improving oxidative stability of mustard oil. Food Chem 334:127561
28. Özogul Y, Özogul F, Kulawik P (2021) The antimicrobial effect of grapefruit peel essential oil and its nanoemulsion on fish spoilage bacteria and food-borne pathogens. LWT 136:110362
29. Pascual-Villalobos M, Díaz I, Martín F, Cantó-Tejero M, Villora G, Guirao P (2022) Plant bioactive volatile products and their efficiency in aphid control. Ind Crops Prod 183:114975
30. Pathania R, Kaushik R, Khan MA (2018) Essential oil nanoemulsions and their antimicrobial and food applications. Curr Res Nutr Food Sci J 6:626–643
31. Radi M, Akhavan-Darabi S, Akhavan HR, Amiri S (2018) The use of orange peel essential oil microemulsion and nanoemulsion in pectin-based coating to extend the shelf life of fresh-cut orange. J Food Process Preserv 42:e13441
32. Ribes S, Fuentes A, Talens P, Barat JM, Ferrari G, Donsì F (2017) Influence of emulsifier type on the antifungal activity of cinnamon leaf, lemon and bergamot oil nanoemulsions against Aspergillus niger. Food Control 73:784–795
33. Severino P, Andreani T, Chaud MV, Benites CI, Pinho SC, Souto EB (2015) Essential oils as active ingredients of lipid nanocarriers for chemotherapeutic use. Curr Pharmaceut Biotechnol 16:365–370
34. Song X, Wang L, Liu T, Liu Y, Wu X, Liu L (2021) Mandarin (Citrus reticulata L.) essential oil incorporated into chitosan nanoparticles: characterization, anti-biofilm properties and application in pork preservation. Int J Biol Macromol 185:620–628
35. Stanisic D, Liu LH, Dos Santos RV, Costa AF, Durán N, Tasic L (2020) New sustainable process for hesperidin isolation and anti-ageing effects of hesperidin nanocrystals. Molecules 25:4534
36. Uçar Y (2020) Antioxidant effect of nanoemulsions based on citrus peel essential oils: prevention of lipid oxidation in trout. Eur J Lipid Sci Technol 122:1900405

37. Wang S, Liang X, Zhao W, Mi X, Zhang C, Zhang W, Cheng Y, Wang L, Jiang Y (2022) Preparation of nanoemulsion of grapefruit seed extract and evaluation of its antibacterial activity. J Food Process Preserv 46:e16197
38. Wu C, Wang L, Hu Y, Chen S, Liu D, Ye X (2016) Edible coating from citrus essential oil-loaded nanoemulsions: physicochemical characterization and preservation performance. RSC Adv 6:20892–20900
39. Yang R, Miao J, Chen X, Chen C, Simal-Gandara J, Chen J, Wan C (2022) Essential oils nano-emulsion confers resistance against Penicillium digitatum in 'Newhall' navel orange by promoting phenylpropanoid metabolism. Ind Crops Prod 187:115297
40. Yang R, Miao J, Shen Y, Cai N, Wan C, Zou L, Chen C, Chen J (2021) Antifungal effect of cinnamaldehyde, eugenol and carvacrol nanoemulsion against Penicillium digitatum and application in postharvest preservation of citrus fruit. LWT 141:110924
41. Yazgan H, Ozogul Y, Kuley E (2019) Antimicrobial influence of nanoemulsified lemon essential oil and pure lemon essential oil on food-borne pathogens and fish spoilage bacteria. Int J Food Microbiol 306:108266
42. Zheng B, Zhang X, Peng S, McClements DJ (2019) Impact of curcumin delivery system format on bioaccessibility: nanocrystals, nanoemulsion droplets, and natural oil bodies. Food Funct 10:4339–4349
43. Zheng J, Li Y, Song M, Fang X, Cao Y, McClements DJ, Xiao H (2014) Improving intracellular uptake of 5-demethyltangeretin by food grade nanoemulsions. Food Res Int 62:98–103

Nanoemulsions Synthesis from Seed Oil, Characterization, and Their Applications

Yenework Nigussie, Mesfin Getachew Tadesse, Archana Bachheti, Kundan Kumar Chaubey, and Rakesh Kumar Bachheti

Abstract Nanoemulsions are kinetically stable colloidal particle systems with an average droplet size range from 100 to 500 nm. Nanoemulsions are prepared by using different techniques using discontinuous phases from various seed oil with different types of emulsifiers and continuous phases. Of these different formulation techniques, the ultrasonication technique was the most frequent technique. The synthesized nanoemulsions can be characterized by using various parameters like droplet size, poly dispersive index, percentage transmittance, microscopic structure, refractive index, viscosity, zeta potential, and pH value. Nanoemulsions have excessive potential applications in pharmaceutics, foods, cosmetics, and biological activities such as antibacterial, and insecticidal due to their beautiful properties, such as small sizes, high surface area per unit volume, enhanced dispersion of active lipophilic components, and improved absorption. Thus it was observed that nanoemulsion proved itself as a promising alternative for improving the bioavailability of the drug. This chapter aims to give an overview of the nanoemulsions synthesis from seed oil, characterization, and their applications.

Keywords Characterization · Formulation · Nanoemulsion

Y. Nigussie · M. G. Tadesse · R. K. Bachheti (✉)
Division of Research and Innovation, School of Applied and Life Sciences, Uttaranchal University, Arcadia Grant, P.O. Chandanwari, Premnagar, Dehradun 248007, Uttarakhand, India
e-mail: rkbachheti@gmail.com

A. Bachheti
Department of Environment Science, Graphic Era University, Dehradun 248002, Uttarakhand, India

K. K. Chaubey
Department of Industrial Chemistry, College of Applied Sciences, P.O. Box-16417, Addis Ababa, Ethiopia

Department of Allied Sciences, Graphic Era Hill University (G.E.H.U), Society Area, Clement Town, Dehradun 248002, India

School of Basic and Applied Sciences, Sanskriti University, Mathura, Uttar Pradesh 281401, India

A. Husen et al. (eds.), *Current Trends in Green Nano-emulsions*, Smart Nanomaterials Technology, https://doi.org/10.1007/978-981-99-5398-1_4

1 Introduction

Nanoemulsions are correspondingly recognized as mini emulsions, submicron emulsions, ultrafine emulsions and colloidal particulate systems. The typical droplet size range of nanoemulsions are from 1 to 1000 nm [1]. Regularly, the typical droplet diameter is between 100 and 500 nm, and the timely typical droplet size range of nanoemulsions is between 50 and 1000 nm. Commercial nanoemulsions typical droplet size range is from 0.2 to 100 μm. From this nanoemulsion particles can occur as oil-in-water (O/W) and water-in oil (W/O) forms, from the main particle is either oil or water, correspondingly [2]. In recent time nanoemulsion are categorize into three these are O/W, W/O which differ by core particles dispersed in the aqueous phase and bi-continuous (microdomains of water and oil are interspersed within the system). These three categories can be achieved for transformation by varying the components of the nanoemulsions. Both oil-in-water and water-in-oil nanoemulsions are other types of particles of nanoemulsions that present at the same time in one system in multiple emulsions. Nanoemulsions are considered kinetically stable particle sizes with 100 nm of one liquid phase dispersed in another immiscible phase [3, 4]. The dispersed phase of nanoemulsion can be stabilized by decreasing the surface tension between the two immiscible liquids by using an appropriate surfactant and co-surfactant as an interfacial film to form a particular phase. Emulsifying agents with varied features (non-ionic or ionic) had been used for nanoemulsions. Among them which widely used were sorbitan esters, polysorbates from nonionic surfactants, polymers (Spans and Tweens) and ionic surfactants (potassium laurate, sodium lauryl sulphate, sodium dioctyl sulfosuccinate and tragacanthins), cationic surfactants (quaternary ammonium halide) and zwitterions surfactants [5, 6]. Nanoemulsions have a good transparent appearance, high surface area, excellent stability, and tunable rheology are the characteristic properties of nanoemulsion, based on the above excellent properties of nanoemulsions, they are real alternatives for cosmetic, food, and pharmaceutical industries, and drug delivery applications [7]. When nanoemulsions are not stable droplets are floating on the surface, cohesion between droplets, and lastly creaming and separation. To avoid colloidal suspension the droplet's diameter needs to be greater than 0.1 μm.

Nanoemulsions are used for the capability to integrate together with lipophilic and hydrophilic drugs solubilization; improved bioavailability for many drugs; improved the degree of absorption; minimized inconsistency in absorption; defense from oxidation and hydrolysis in oil-in-water nanoemulsions; as non-toxic and non-irritant vehicles for skin and mucous membrane delivery and release control through permeation of drug through the liquid film [8]. Generally, nanoemulsion is preferable to microemulsion due to; is prepared using lower concentration surfactant, containing from 3 to 10% can be enough [9]. The aim of this chapter is to provide updated information on the formulation, and characterization of nanoemulsion by using different techniques and applications of nanoemulsion in different industries and fields.

2 Synthesis and Characterization of Nanoemulsions

Synthesis of nanoemulsions has extreme potential applications in pharmaceutics, foods, cosmetics, and biological activities such as antibacterial, and insecticidal due to their attractive properties, such as small sizes, and high surface area per unit volume. Therefore, nanoemulsions can be formulated via two wide methods these are low energy techniques and high-energy techniques [3, 4]. High-energy methods of nanoemulsion synthesis use a lot of energy (~108 W/kg) to break down huge droplets to 100 nm in size [3, 4]. High energy techniques suggest a robust method to formulate nanoemulsions with dispersed phase volume fraction as high as 40% with brute force technique [10]. Nevertheless, a lot of shear makes them incompetent and exposed to heat effects. Examples of high-energy techniques of nanoemulsion synthesis are high-pressure homogenizers, microfluidizer, and ultrasonication. on the other hand, low energy techniques of nanoemulsion synthesis feat the low interfacial tension property of a system to decrease particle size with energy input that can be attained by a magnetic stirrer (~103 W/kg), and nanoemulsions can deliver an easy and scalable route to without the use of excessive shear. High speed homogenizer, spontaneous emulsification, phase inversion temperature and phase inversion composition are examples of low energy emulsification. Oil (for solubilization of the lipophilic molecules), water, surfactant, and co-surfactant (improve the act of surfactant) are the key components for the formulation of nanoemulsions. According to Solans et al. [11] study, various methods have been realized to formulate nanoemulsion droplet diameter from 100 to 600 nm, which is suitable range for the majority of applications. Commonly, water-in-oil nanoemulsions are synthesis by using low hydrophilic–lipophilic balance (HLB) mainly from 3 to 8 HLB range of surfactants whereas, oil-in-water nanoemulsions are formulated by using high hydrophilic–lipophilic balance particularly between 8 and 18 HLB range of surfactants.

According to Moksha et al. [12] scholarship nanoemulsions were prepared by ultrasonication technique by using 80 mg (5% v/v) of coconut oil and span 80 (0.5% v/v) was placed in a beaker and stirred continuously to mix the solution. The subsequent solution was added into the aqueous phase which 5% v/v of ethanol as co-solvent and 5 present tween 80 as an emulsifier with constant stirring at 3000 r/m for 3 min. After formulation of the nanoemulsion, various physicochemical parameters of the nanoemulsion were optimized such as particle size, poly-dispersive index, percentage transmittance, morphology, zeta potential, solubility examination, and viscosity.

Moksha et al. [12] study shows that the poly-dispersive index and typical particle size of formulated nanoemulsion were measured by means of photon correlation spectroscopy method using particle size analyzer at room temperature by using tween 80 and span 80 as surfactants and the result was found to be 79.0 ± 5.7, and 0.220 ± 0.05 nm respectively. Absorption of formulated nanoemulsion was improved may be attributed in terms of a large specific area due to the small particle size of the droplet. The other property is zeta potential which is an indication of electrostatically stabilized nanoemulsion. When the zeta potential value is higher the suspension is to be stable as the charged particles repel each other and this potency overcomes the natural tendency to aggregate. From this zeta potential value of this study was found -15.54 ± 0.21 mV, due to the negative charge value of zeta potential artemether nanoemulsions indicating the presence of the anionic groups of the fatty acids and glycols in the surfactant and co-surfactant. Thus, there are negligible probabilities of aggregation of nanoemulsion in the biological environment during its shelf life.

A digital refractometer was used to characterize the refractive index value of the nanoemulsion sample by putting a few spots of formulation nanoemulsion on the transparency of refractometer at 25 °C [13]. The optimized nanoemulsion of refractive index value was 1.320 ± 0.05, from this there is no significant difference ($P \pm 0.05$) value of nanoemulsion, which was a symptom of the isotropic nature of nanoemulsion. The optimized percentage transmittance value of artemether nanoemulsion was found to be $98.20 \pm 0.7\%$ that nearer to 100% and shows that the formulation nanoemulsion was pure and clear. TEM images of particles in formulated nanoemulsion were detected as uniformly distributed, and spherical in shape with droplet diameter less than 100 nm [12].

Grounded on Kumar et al. [1] reported that nanoemulsion were synthesized by ultrasound-assisted method from sesame oil using Tween 20 surfactant with variable concentrations of oil and surfactant such as 1:1, 1:1.5, 1:2, 1:3, 1:4 and 1:5, and 30 mL of deionized water was added drop-wise in the internal phase combination to get nanoemulsion. Then the mixture of emulsion was stirred by a magnetic stirrer for 10 min at 540 r/m to get constant droplet distribution. The nanoemulsions were sonicated by ultrasonicator with 20 kHz frequency for 10, 20, 30 min respectively for individual composition. The typical particle size and PSD of nanoemulsion were characterized by using the dynamic light scattering instrument. Dynamic light scattering technique was used to measure the suspension of particles size distribution from 3 nm to 5 μm and the profile was detected as a single and narrow peak as it is mono-dispersed [14]. DLS data is detected as the plot of concentration and average droplet size [15]. For determining the average particle size accurately by DLS technique the samples should be translucent and diluted. The examination was done at 25.1 °C with 173° scattering angle. From this study, the best average droplet size range was 189.5 nm achieved at the composition of 1:3 oil to surfactant concentration. Extremely different droplet size is the result of high polydisperse emulsion. Higher stability is obtained on the lower poly-dispersive index. For sample 1:3, the poly-dispersive index was found to be the lowest at 0.514, followed by a poly-dispersive index of 0.555 for sample 1:2 oil to surfactant ratio. The value of zeta potential for 1:4, 1:3, and 1:2 showed high values, such as −23.4, −19.5 and −18.6 respectively and showed their stability. Stability test can be performed by centrifugation of the composition at 3000 rpm for 10 min and can select better stability compositions over the others (Fig. 1) [1].

Wang et al. [16] scholarship described that nanoemulsions were formulated by using dynamic high-pressure micro fluidization methods at diverse circumstances at 0–140 Mpa by using 0.03–0.18 mg/mL emulsifier with *Eucommia ulmoides* seed oil as aqueous phase solution. Tween 80/ethanol (2:1 v/v) and the from 2 to 12% volume fraction of *Eucommia ulmoides* seed oil were mixed at various proportions and stirred

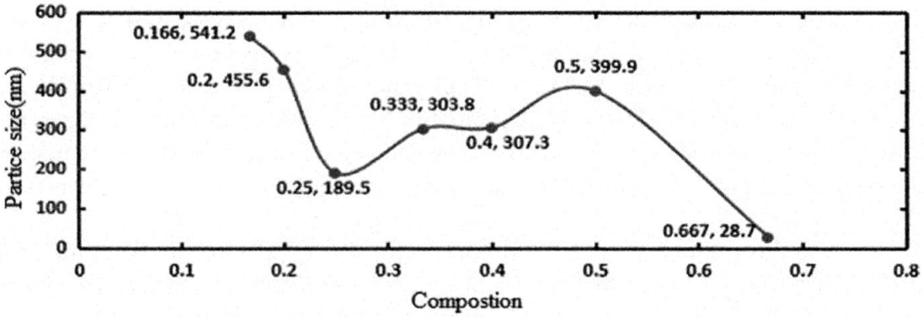

Fig. 1 Graph of average droplet size with variable composition (v/v) [1]

at 700 r/m magnetically for 10 min. Then the definite volume of distilled water was added dropwise, and the combination was, treated by ultraturrax at 10,000 rpm for 2 min. Consequently, the formulated crude emulsion was passed to prepare *Eucommia ulmoides* seed oil nanoemulsion. Rheology is the branch of physics that lessons how materials deform or flow in response to applied forces or stresses. The rheological properties of nanoemulsions are affected by the number density of the droplets, surfactants and shape, and interactions between the constituent droplets (Fig. 2) [17].

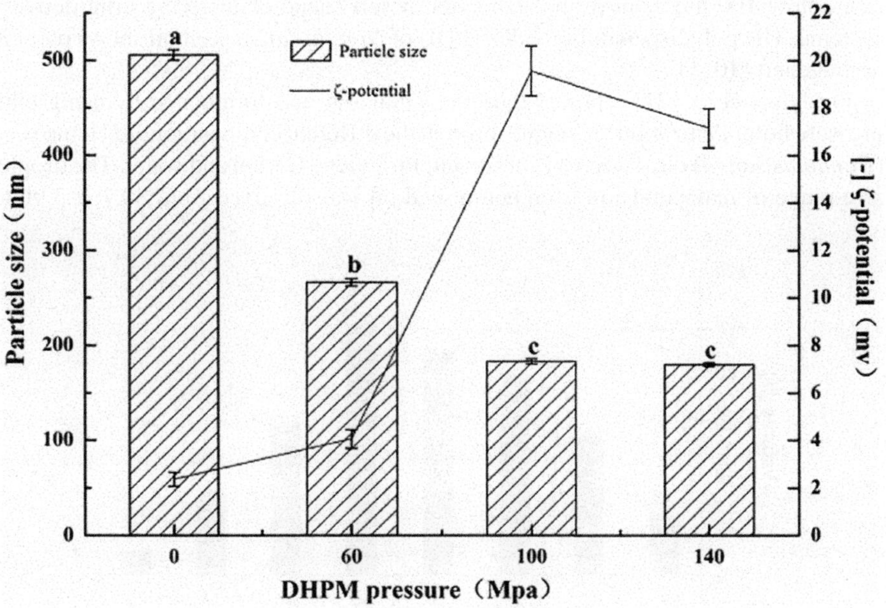

Fig. 2 Effect of droplet size and zeta-potential of nanoemulsion on dynamic high-pressure microfluidization method [16]

According to Vijayalakshmi et al. [18] nanoemulsion was synthesized using basil oil (*Ocimum basilicum*), Tween 80 and water through an ultrasonication technique that emulsified for 15 min. Stable basil oil nanoemulsion of average particle size from different concentrations was 29.3 nm droplet diameter. Based on this study the best particle charge value of the synthesized nanoemulsion was 3.70 ± 0.41 mV on the composition of 1:3 ratio and the native pH value of the nanoemulsion was 5.53 (Figs. 3 and 4).

According to Jeonghee et al. [19] report typical particle size can be minimized from 237 to 101 nm when the surfactant-to-oil combination was rise from 0.75 to 2. The poly-dispersive index decreased from 0.41 to 0.14 when the surfactant to oil proportion was a rise from 0.75 to 1.5, and then rise to 0.19 when the surfactant to oil was increased further to 2. Average droplet size is also based on the volume of surfactant, when the surfactant to oil ratio is equal to 0.75 the nanoemulsions had bimodal distributions.

Grounded on Priani et al. [20] nanoemulsion can also formulated by using ultrasonic bath. This nanoemulsion could be synthesized by the combination of *Nigella sativa* seed oil, co-surfactant (polyethylene glycol (PEG400)) and surfactant with diverse combination of oil and surfactant for self- nanoemulsifying drug delivery system. After mixing the component for few minutes, the mixture was homogenized at 40 °C for 5 min. After homogenization, the mixture was sonicated in bath sonicator for 15 min. The particle size of self- nano emulsifying drug delivery system containing *Nigella sativa* seed oil was 65.4 nm. The systems with an average droplet size below 100 nm achieve the standards of self- nano emulsifying drug delivery systems. The polydispersibility index (PDI) of *Nigella sativa* seed oil shows particle homogeneity (0–1).

Qushawy et al. [21] reported that nanoemulsion was formulated by using high pressure homogenization technique through the different compositions of Hemp seed (*Cannabis sativa*) oil, Tween 80 surfactant, Propylene Glycol and water. The droplet size range of nanoemulsion from hemp seed oil was observed from 80.7 ± 3.96 to

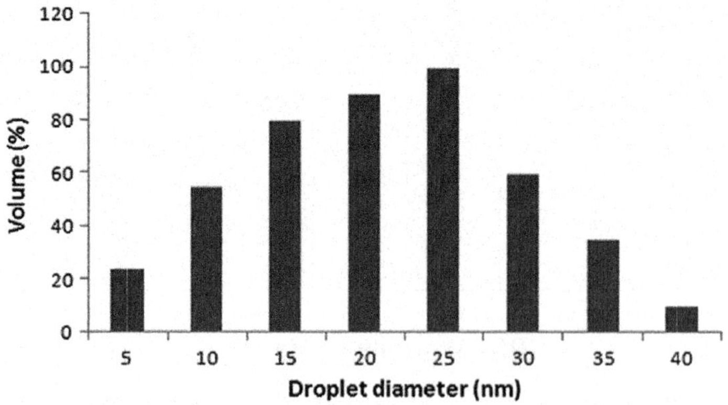

Fig. 3 Particle size of *Ocimum basilicum* oil nanoemulsion by volume [18]

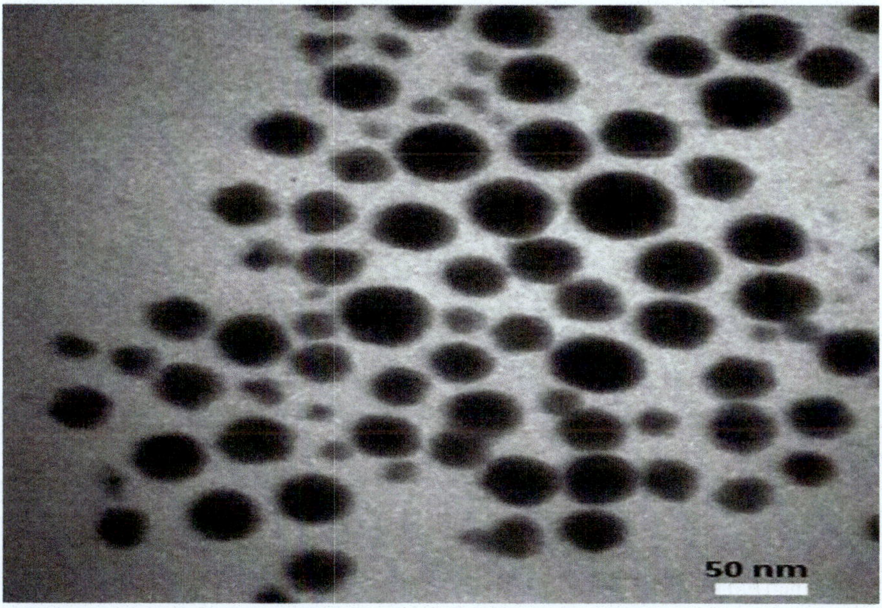

Fig. 4 TEM of *Ocimum basilicum* oil nanoemulsion with 1:3 (v/v) proportion of surfactant and oil [18]

140.9 ± 5.62 nm, polydispersity index of hemp seed oil nanoemulsions between 0.249 ± 0.06 and 0.493 ± 0.02 this result demonstrating a narrow size distribution and particle size of nanoemulsion was homogenous, zeta potential of all hemp seed oil nanoemulsions between −10.32 ± 1.35 and −21.94 ± 1.13 mV, this investigation was established that concentration of Tween 80 increased the value of zeta potential or the surface charge also increased. Minor particle size and greater surface area of oil droplets can be obtained by means of increasing Tween 80 concentration. pH value of hemp seed oil nanoemulsions was determined with a pH meter, and the value was between 5.76 ± 0.32 and 6.57 ± 0.41 pH, just like zeta potential value the pH value of nanoemulsion was enhanced with increasing concentration of Tween 80, from this opinion better stability of nanoemulsion could be obtained from better concentrations of surfactant. The viscosity of the prepared nanoemulsion formulation ranged from hemp seed oil between 3.16 ± 0.28 and 8.35 ± 0.34 cP. Lower viscosity was connected viscosity which may be attributed to higher surfactant. The electrical conductivity of the formulated nanoemulsion was between 170.93 ± 5.79 and 212.35 ± 8.46 mS/cm. TEM image of hemp seed oil nanoemulsions the best composition formulation indicates good distribution with spherical appearance and, small particle size in the nano range (Fig. 5) [21].

Based on McClements, (2010) report nanoemulsion could be synthesized by emulsion inversion point. Emulsion inversion point was formulated through water-in-oil to oil-in-water or vice versa. As its name indicates, by addition of oil or water into the water continuous phase or oil continuous phase respectively, during this time

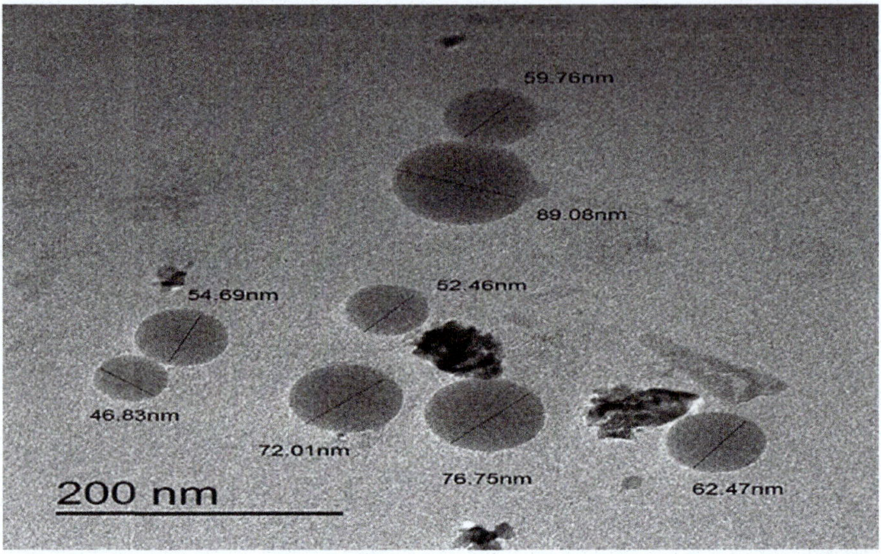

Fig. 5 The TEM image of greatest combination nanoemulsion preparation from hemp seed oil [21]

transition between distributed phases and continuous might be attained. When the addition of dispersed liquid phase reaches a critical point, the inversion mechanism occurs automatically. The emulsion inversion point could affect by surfactant-to-phase relation and degree of the distributed phase addition to the continuous phase.

McClements, 2010 reported that nanoemulsion can be emulsified by spontaneous emulsification. Spontaneous emulsification (SE) was a more suitable technique and formulated oil-in-water emulsion systems with surfactant and oil were mixed together with the help of a stirrer and then oil was dropped into the water phase. Physical and chemical properties of surfactant, type of fluids used, their solubility, degree of stirring, and surface to oil ratio (SOR) are features that disturb particle size and emulsions stability. The mechanism of SE technique system was examined on dispersed and continuous phases are established together as a distinctive phase. Whereas mixing was continued, surfactant used could pass over the water phase so that it can act as a bridge between water and oil with the help of turbulent strength. Additionally, the dispersed phase start to deport itself from the water phase. This motion can generate an enormous interfacial area between two phases. Then appropriately dispersed droplets are organized within a continuous phase spontaneously.

Niharika et al. [22] reported that nanoemulsion can also be formulated by low energy approach particularly spontaneous emulsification method by using pea protein and Tween 80 for vitamin D delivery application. The preparation procedure was 1 mg/mL of vitamin D in canola oil dissolved with Tween 80 primarily as organic phase. Then the solution was gradually titrated into distilled water with constant stirring 800 r/m on magnetic stirrer at ambient temperature [22]. Variable SOR equal to

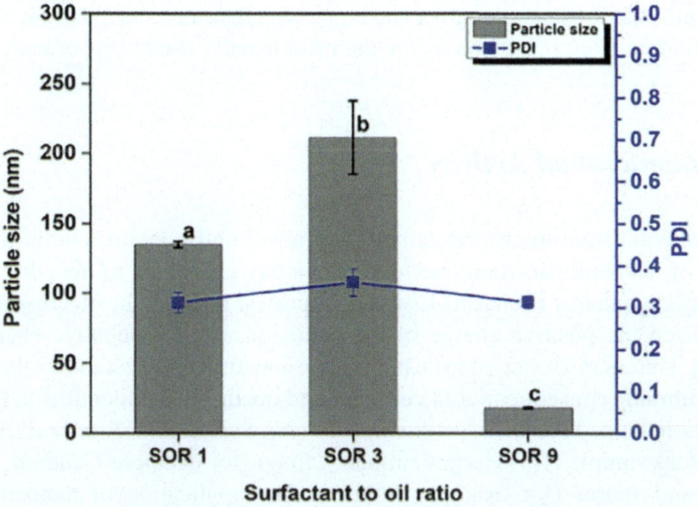

Fig. 6 SOR on the particle size and poly-dispersive index [22]

1.0, 3.0, 9.0 were adjusted and permitted to stir for 60 min followed by the addition of pea protein into the pre-formulated emulsions synthesized by 1:1 (v/v) SOR composition under constant mixing for 60 min to formulate the pea protein-Tween 80 complex nanoemulsion. To optimize concentration for emulsion production diverse SOR ratio equal to 1 with 3% w/v Tween 80 consequence was detected. After formulation of nanoemulsion average particle size and zeta-potential of the nanoemulsions were characterized by using dynamic light scattering at ambient temperature. Nanoemulsions were thinned by 1 mg/mL using double distilled water before analysis to avoid multiple scattering. From this as shown in Fig. 6, numerous surfactant molecules were presented to cover oil droplets at a surfactant to oil ratio of 8, and resulted in a minor nanoemulsion droplets size of 18.1 $\pm$ 0.7 nm. A surfactant-to-oil ratio of 1 for the formulation of emulsion also resulted in a relatively stable droplet size of 134.8 $\pm$ 2.2 nm.

3 Application of Nanoemulsion from Seed Oil

3.1 In Biotechnology

Nanoemulsions are applicable in biotechnology which have hydrophobic in nature, including dehydrogenases, oxidases, lipases, and esterases frequently purpose in the cells in micro-environments. From these numerous enzymes activate regularly stabilized by polar hydrophilic lipids and other natural amphiboles in biological systems. Nanoemulsions have been used as enzymatic catalysis for a variety of reactions, such as the synthesis of esters, peptides and sugar acetals transesterification, and steroid transformation, from these lipases are the most usually used class of enzymes [23].

3.2 Antimicrobial Activity

Nanoemulsion as antimicrobial activity is a novel and hopeful revolution because particles of nanoemulsions are thermodynamically motivated to fuse through lipid-containing organisms. From this fusion is improved through the electrostatic attraction between the positive charge of the emulsion and the negative charge of the pathogen. The energy released from both active constituent is destabilize the pathogen lipid membrane, consequential in cell lysis and death The nanoemulsion has a wide range action defined bacteria (for example *E. coli*, *Salmonella*, *S. aureus*), enveloped viruses (for example HIV, Herpes simplex), fungi (for example *Candida*, Dermatophytes), and spores (for example anthrax). The application of nanoemulsion to examine as an antimicrobial agent was encouraged through the recognized difficulty of growth of antimicrobial-resistant strains practiced with the use of present agents due to the prevalent, and occasionally unsuitable use of antibiotics, disinfectants and antiseptics. Additional research and progress of new antimicrobial agents targeting exact pathogens at the same time as being safe for the host. Since the mechanism of action of nanoemulsion seems to be nonspecific disruption of bacterial cell membranes, nanoemulsion would not consequence in the development of resistant strains. Nanoemulsion as an antimicrobial agent is a promising and new revolution. Karthikeyan et al. [24] showed that soybeans oil with cetylpyridinium chloride nanoemulsion was effective in antimicrobial activity of nanoemulsion which was greater than that of its component on cariogenic planktonic and biofilm organisms and showed 83% inhibition on *S. mutans* and *L. casei* micro-organisms. Formulated nanoemulsion from basil oil (*Ocimum basilicum*) using Tween80 surfactant and water through ultrasonic emulsification technique was assessed for antibacterial activity against *Escherichia coli* through kinetics of killing experimentation [18].

3.3 Drug Delivery Activity

Nanoemulsions have been used in furthermost topical, visual, intravenous, and oral drug delivery. Artemether nanoemulsion was better drug loading capacity of the nanoemulsion that exhibited a drug content of $98.42 \pm 0.87\%$. Nanoemulsion was used for lipophilic nature of drug to solvate water, zeta potential and rheology to synthesize aqueous solutions that can be simply distributed to patients. It is also used as a vehicle barrier against the management of drugs via the skin. Generally, formulations nanoemulsions are regularly applicable to manage drugs to improve the solubility of nonpolar compounds, pharmacokinetic profile and minimize adversative effects practiced through patients. Emulsions can be managed by patients through

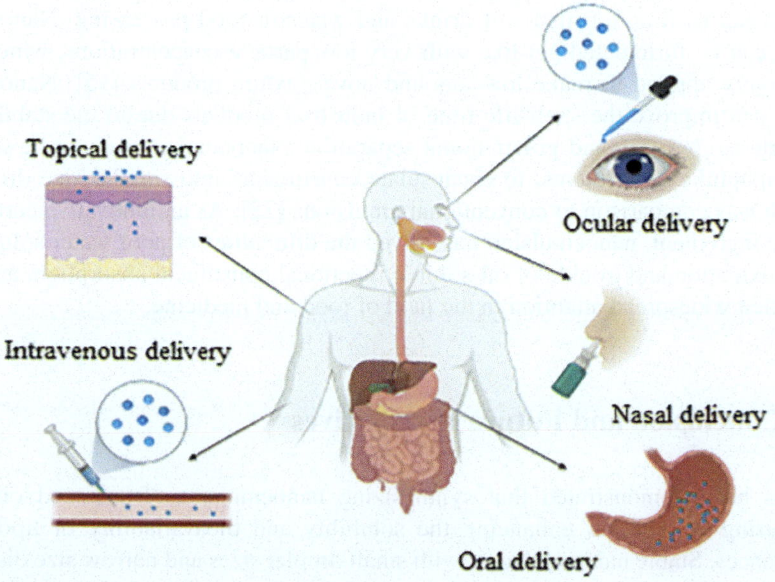

Fig. 7 Nanoemulsion management for drug delivery [25]

numerous directions that would be briefly familiarized as showed on the figure below. Generally, submicron size of nanoemulsion can be purely targeted to the tumor area (Fig. 7).

3.4 In Cosmetics

Nanoemulsion is applicable as means of transportation for measured delivery on the skin. For example, nanoemulsion formulation as kemira nanogel-based carrier system is applicable for cosmetic purposes which improve skin construction and diffusion of API. Nanoemulsion also delivers good skin feel [26]. Topical management itself has several benefits and by joining it with nanoemulsion, this formulation may impart a better way of drug delivery system. It can avoid the hepatic first-pass metabolism of the drug and related toxicity effects [27].

3.5 In Food Industry

Nanoemulsions have a broad range of applications due to their flexibility in several fields including beverage and food industries. Due to their small size, thermodynamic stability, transparency and optically transparent properties, nanoemulsion have been used as fortified soft drinks and water in food processing. Nanoemulsions can be formulated gel-like with very low particle concentrations, which can be simply applied to make low fats and novel texture products [15]. Nanoemulsions can improve the shelf-life time of industrial products due to the stability of particle aggregation and gravitational separation. Nanoemulsions could be solubilize lipophilic constituents, to encapsulate constituents into smaller size droplets, which was comparison to conventional emulsions [28]. As an innovative carrier of active ingredient, nanoemulsion has solved the difficulties of poor water solubility, easy oxidation and trouble of oil-soluble functional constituent absorption, and has attracted widespread attention in the field of food and medicine.

4 Conclusion and Future Perspectives

It has been demonstrated that synthesizing nanoemulsions from seed oil is a promising method for enhancing the solubility and bioavailability of lipophilic substances. Stable nanoemulsions with small droplet sizes and narrow size distributions have been created using a variety of techniques, such as high-pressure homogenization and ultrasound-assisted emulsification. Furthermore, these nanoemulsions have been characterized to learn more about their characteristics and stability utilizing methods like dynamic light scattering, transmission electron microscopy, and rheological analysis. The uses of seed oil-based nanoemulsions in a variety of industries, including food, cosmetics, and medicines, have also been investigated. Nanoemulsions have been utilized in the food sector to enhance the sensory qualities and shelf life of food goods. They have been utilized in cosmetics as carriers for active compounds and to enhance the skin absorption of these substances. Nanoemulsions have demonstrated potential in the pharmaceutical industry as drug-delivery devices for improving the solubility and bioavailability of medicines that are not readily soluble.

To fully utilize the potential of seed oil-based nanoemulsions, numerous obstacles still need to be overcome. To address the environmental issues related to the use of synthetic surfactants, it is important to investigate the use of natural and biodegradable surfactants and co-surfactants in addition to optimizing the formulation and processing parameters. Furthermore, the safety and toxicity of these nanoemulsions need to be thoroughly evaluated to ensure their potential use in various applications. In general, the creation of nanoemulsions from seed oil is a promising area of study with potential uses in many different industries. Unquestionably, further study in this field will result in the creation of cutting-edge, revolutionary nanoemulsion-based goods with enhanced functions.

References

1. Kumar NV, John AM, Mangsatabam M, Philip MA (2021) Ultrasound-assisted synthesis and characterization of sesame oil based nanoemulsion. In: IOP conference series: materials science and engineering, vol 1114, p 012085
2. Shah P, Bhalodia D, Shelat P (2010) Nanoemulsion: a pharmaceutical review. Syst Rev Pharm 1
3. Gupta A, Eral HB, Hatton TA, Doyle PS (2016) Controlling and predicting droplet size of nanoemulsions: scaling relations with experimental validation. Soft Matter 12:1452–1458
4. Gupta A, Eral HB, Hatton TA, Doyle PS (2016) Nanoemulsions: formation, properties and applications. Soft Matter 12:2826–2841
5. Kralova I, Sjöblom J (2009) Surfactants used in food industry: a review. J Dispersion Sci Technol 30(9):1363–1383
6. Haba E, Bouhdid S, Torrego-Solana N, Marqués AM, Espuny MJ, García-Celma MJ, Manresa A (2014) Rhamnolipids as emulsifying agents for essential oil formulations: antimicrobial effect against Candida albicans and methicillin-resistant Staphylococcus aureus. Inter J Pharm 476(1–2):134–141
7. Ravi TP, Padma T (2011) Nanoemulsions for drug delivery through different routes. Res Biotechnol 2:1–13
8. Gurpreet K, Singh SK (2018) Review of nanoemulsion formulation and characterization techniques. Indian J Pharmaceut Sci 80:781–7890
9. Sonneville-Aubrun O, Simonnet JT, L'alloret F (2004) Nanoemulsions: a new vehicle for skincare products. Adv Coll Interface Sci 108:145–149
10. Badruddoza AZM, Godfrin PD, Myerson AS, Trout BL, Doyle PS (2016) Core shell composite hydrogels for controlled nanocrystal formation and release of hydrophobic active pharmaceutical ingredients. Adv Healthc Mater 5:1960–1968
11. Solans C, Izquierdo P, Nolla J, Azemar N, Garcia-Celma MJ (2005) Nano-emulsions. Curr Opin Colloid Interface Sci 10:102–110
12. Moksha L, Bhardwaj A, Mehta S, Mehta A (2014) Development and characterization of nanoemulsion as carrier for the enhancement of bioavailability of artemether. Artif Cells Nanomed Biotechnol Early Online 1–11
13. Harika KB, Debnath S (2015) Formulation and evaluation of nanoemulsion of amphotericin. Int J Novel Trends Pharmaceut Sci 5:114–122
14. Yun Zhang PW (2003) Studying the interaction of gel carbohydrate antigen and protein by quartz-crystal microbalance. J Am Chem Soc 125:9292–9293
15. Mcclements DJ (2007) Critical review of technique and methodologies for characterization of emulsion stability. Crit Rev Food Sci Nutr 47:611–649
16. Wang S, Wang X, Liu M, Zhang L, Ge Z, Zhao G, Zong W (2020) Preparation and characterization of *Eucommia ulmoides* seed oil O/W nanoemulsion by dynamic high pressure microfluidization. LWT Food Sci Technol
17. Acharya DP, Hartley PG (2012) Progress in microemulsion characterization. Curriculum Opin Colloid Interface Sci 17:274–280
18. Vijayalakshmi G, Mukherjee A, Chandrasekaran N (2013) Ultrasonic emulsification of food-grade nanoemulsion formulation and evaluation of its bactericidal activity. Ultrason Sonochem 20:338–344
19. Jeonghee S, Decker EA, McClements DJ (2017) Original article Utilisation of spontaneous emulsification to fabricate lutein loaded nanoemulsion-based delivery systems: factors influencing particle size and colour. Int J Food Sci Technol 52:1408–1416
20. Priani SE, Maulidina SS, Darusman F, Purwanti L, Mulyanti D (2020) Development of self -nanoemulsifying drug delivery system for black seed oil (*Nigella sativa* L.). In: International Journal of Physics: conference series, vol 1469, no 1. IOP Publishing, p 012022
21. Qushawy M, Mortagi Y, Alshaman R, Mokhtar HI, Hisham FA, Alattar A, Liang D, Enan ET, Eltrawy AH, Alamrani ZH, Alshmrani A, Sawsan AZ (2022) Formulation and characterization

of O/W nanoemulsions of Hemp seed oil for protection from *Steatohepatitis*: analysis of Hepatic free fatty acids and oxidation markers. Pharmaceuticals 15:864
22. Niharika W, Zhang S, Wismer W, Chen L (2022) A low energy approach to develop nanoemulsion by combining pea protein and Tween 80 and its application for vitamin D delivery. Food Hydrocolloids Health 2:100078
23. Sukanya G, Mantry S, Anjum S (2013) Review on nanoemulsions. Int J Innovative Pharma Sci Res 192–205
24. Karthikeyan R, Amaechi BT, Ralph RH, Lee VA (2012) Antimicrobial activity of nanoemulsion on cariogenic planktonic and biofilm organisms. Arch Oral Biol Science Direct 15–22
25. Wilson RJ, Yang Li, Guangze Y, Chun-Xia Z (2022) Review on nanoemulsions for drug delivery. Particuology 64:85–97
26. Guglielmini G (2008) Nanostructured novel carrier for topical application. Clin Dermatol 26:341–346
27. Ali KA, Roy S (2019) Emulsification technology for cosmetics preparation. Nanotechnology, 401–418. CRC Press
28. Silva HD, Cerqueira MÂ, Vicente AA (2012) Nanoemulsions for food applications: development and characterization. Food Bioprocess Technol 5:854–867
29. McClements DJ, Rao J (2010) Food-grade nanoemulsions: formulation, fabrication, properties, performance, biological fate, and potential toxicity. Food Sci Nutr 51:285–330

Food and Agriculture

Role of Green Nanoemulsion in Controlling Food Spoilage

Pranchal Rajput, Chetan Shrivastava, Atreyi Pramanik, Aashna Sinha, Anis Kumar Pal, Kundan Kumar Chaubey, Sujata Jayaraman, Bidhi Kundu, Akanksha Shakya, Rakesh Kumar Bachheti, and Archana Bachheti

Abstract The prevalence of outbreaks of food borne diseases is the main problem on a global scale. There is rising interest in using natural resources rather than synthetic ones because of health concerns. Green nanotechnology is regarded as a cutting-edge strategy that uses nano products to reduce threats to human health and the environment. Also, this technique uses eco-friendly, biocompatible, and nontoxic materials. Micro and nanoemulsions are among the numerous types of nanosystems that hold promise as eco-friendly techniques to increase the bioavailability of bioactive chemicals with low stability and poor solubility. Technological barriers to the creation of functional foods include the limited solubility and stability of bioactive chemicals and their sensitivity to heat processing during food preparation and oxidative destruction during storage. Nanoemulsions have been used in the food sector for the controlled release of antibacterial, flavouring, and nutraceutical ingredients. By incorporating bioactive ingredients with antimicrobial properties into nanoemulsions, it may be possible to stop the spread of disease and food-spoilage bacteria. The aim of this

---

Authours Pranchal Rajput, Chetan Shrivastava are equally contributed.

P. Rajput · C. Shrivastava · A. Pramanik · A. Sinha · A. K. Pal · K. K. Chaubey (✉) · S. Jayaraman
Division of Research and Innovation, Uttaranchal University, P.O. Chandanwari, Prem Nagar, Arcadia Grant, Dehradun 248007, Uttarakhand, India
e-mail: kundan2006chaubey@gmail.com

K. K. Chaubey
School of Basic and Applied Sciences, Sanskriti University, Mathura 281401, Uttar Pradesh, India

B. Kundu · A. Shakya
Department of Biotechnology, Institute of Applied Sciences and Humanities, GLA University, Mathra 281406, UP, India

R. K. Bachheti
Department of Industrial Chemistry, Centre of Excellence in Nanotechnology, Addis Ababa Sciences and Technology University, P.O. Box-16417, Addis Ababa, Ethiopia

A. Bachheti
Department of Environment Science, Graphic Era University, Dehradun 248002, Uttarakhand, India

A. Husen et al. (eds.), *Current Trends in Green Nano-emulsions*, Smart Nanomaterials Technology, https://doi.org/10.1007/978-981-99-5398-1_5

chapter is to give an overview of the role of green nanoemulsions in controlling food spoilage and their application in food preservation.

Keywords Nanoemulsion · Food preservation · Food safety · Pesticide · Agriculture

1 Introduction

One of the leading public health concerns today is the safety of the world's food supply. Most food items, including bread, dairy, meat, fruits, and vegetables, are currently highly polluted owing to the presence of several toxic species of bacteria and fungus and the toxins they produce, which can cause severe disease and even death in humans. As a result, people are seeking food items that are safe, high quality, and healthful. Moreover, in tropical and subtropical locations, food contamination by different bacteria and their toxic metabolites can change the nutritional value of the food in addition to degrading it [35]. Because of this, significant focus has been placed on enhancing the quality, safety, and security of food systems against microbial deterioration and the toxins that go along with it.

According to many studies, certain bacterial infections [23] and other fungal pathogens cause food deterioration by converting food into toxic by-products known as mycotoxins, which are bad for human health [40]. Several synthetic preservatives have been widely used to suppress bacterial and fungal development. Nevertheless, the indiscriminate use of these preservatives may have various adverse effects, including developing microbial resistance, toxicity to non-target species, and loss of environmental sustainability. Thus, modern customers seek "safer alternatives" with a green image and perhaps nontoxic effects on people and animals to assure microbiological food safety [16].

Specifically in the cosmetics, pharmaceutical, agricultural, and food industries, nanotechnology is expanding quickly across numerous industries [24, 26]. Most of these pursuits centre on the creation of fatty acids, tastes, colors, and pharmaceuticals, all lipophilic compounds [5]. As emulsions have been made from various materials and additives for a long time, producing markets and profitability, using nanotechnology/nanoparticles in emulsion manufacturing is essential [1]. In Nanoemulsions (NEs), soybean oil (elongated chain triglycerides oil) is used as the distributed phase for food applications [38]. With a large increase in these compounds' bioavailability, these emulsions are employed to create preservatives, antioxidants, and medication compositions. NEs, which range in size from 20 to 500 nm, are sometimes referred to as mini emulsions, sub-micron emulsions, or ultrafine emulsions (Table 1). The aim of this chapter is to give a general overview of the usage of green nanoemulsions as a possible means of preventing food spoilage along with importance of food preservation, and the causes of food spoilage.

Table 1 Differences between macro-, nano-, and micro emulsions

Features	Macroemulsions	Nanoemulsions	Microemulsions	References
	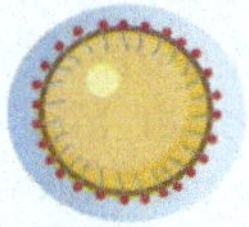	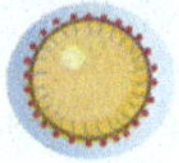	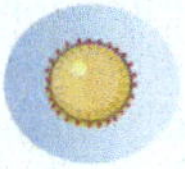	
Droplet size	>1 μm	10–200 Nm	4–200 Nm	[28, 50]
Shape	Spherical	Spherical	Spherical, lamellar	
Appearance	Formulation dependent	Transparent/ translucent to milky	Transparent	
Surfactant load	Fairly low	Medium (<1 To >10%)	Fairly high (10–30%)	
Stability	Kinetically stable	Kinetically stable	Thermodynamically stable	
Method of preparation	Conventional homogenizations	High/low energy emulsification	High/low energy emulsification	

2 Food Spoilage

2.1 *Mechanism*

The process by which food loses its edibility is called food deterioration. Food safety and food deterioration are related. It is possible to spot the early signs of food deterioration using the flavour, smell, texture, colour, or meal itself. Numerous chemical, physical and microbiological mechanisms can cause food to deteriorate. These pathways are not necessarily mutually exclusive because the decline brought on by one mechanism can encourage another. Air, temperature, pH, nutrition, and various chemicals are the primary factors that cause food to deteriorate [21].

2.2 *Physical Spoilage*

Physical food deterioration is described as food that has undergone physical alteration or instability. Physical spoilage occurs when components or ingredients physically separate from one another, moisture is lost or gained, or moisture migrates between different components [41]. Moisture content, temperature, glass transient temperature, crystal development, and crystallisation are the main elements influencing physical deterioration [54].

3 Microbial Food Contamination

According to several studies, bacterial pathogens like *Klebsiella spp.*, *Campylobacter spp.*, *Escherichia spp.*, *Bacillus spp.*, *Salmonella spp.*, and *Listeria spp.* cause food commodities, particularly meat, fruits, vegetables, and other high-fat-containing goods, to rot or go bad the fastest [34, 37]. Recently, the most prevalent illnesses in humans associated with chicken and pork products in developing and developed countries are campylobacteriosis caused by Campylobacter coli, and Campylobacter jejuni [27]. Another most significant bacteriological pathogen after *Campylobacter spp.* causes vomiting, diarrhea, fever, stomach pains, headaches, and blood in stools, *Salmonella spp.* has been isolated from various foods. According to the World Health Organization, 750,000 Americans were infected with *Salmonella typhi* in 2016, which led to more than 52,000 losses [23]. Another significant pathogen that contaminates food and causes listeriosis is *Listeria monocytogenes* [58]. In addition to bacterial contamination, several fungal species, including *Penicillium*, *Cladosporium*, *Alternaria*, *Fusarium*, *Mucur*, *Rizopus*, and *Aspergillus* have also been implicated in the rotting of fresh produce, meats, stored grains, and other necessary food items [3, 17, 31]. In addition to fungal pathogens, food spoilage degrades food products by creating poisonous secondary metabolites, known as mycotoxins that harm human health [40]. Additionally, *Aspergillus*, *Penicillium*, and *Fusarium* frequently contaminate pre-harvest and post-harvest stored food goods, secreting many mycotoxins that cause large monetary losses [12]. Aflatoxin, deoxynivalenol, ochratoxins, zearelenone, and fumonisins are the main mycotoxins that pose health risks to mammals when consumed through contaminated food among the 400 that have been recorded [4]. Food product spoilage is rarely looked at until it has caused an outbreak. Therefore, there is a critical need for in-depth study of natural goods, including essential oils (Eos), to safeguard public health through the long-term preservation of various food commodities [39].

3.1 Principal Microbes Involved in Food Rotting

Usually, microorganisms infect food goods' surfaces through soil, water, and air. Moreover, biopolymers found in food, such as protein, polysaccharides, and lipids, are broken down and absorbed by bacteria as necessary nutrients for growth and reproduction, leading to food spoiling and safety risks for humans. However, the primary bacteria responsible for food deterioration are strongly tied to food kinds. For instance, *Salmonella spp.*, *Listeria*, and *Escherichia coli* [51].

3.2 Food Oxidative Deterioration

The oxidation of nutrients during food production, storage, and transportation is another significant factor in food deterioration. Fresh juice, freshly cut fruit, and various vegetables, for instance, can change color quickly when exposed to air, which can be linked to the intricate browning reaction. Meanwhile, oxidation is the primary factor contributing to the degeneration of polyphenols, protein, amino acids, and other nutrients [51].

4 Control Food Spoilage by Bioactive Food Fresh-Keeping Films

Bioactive food fresh-keeping films (FFKFs) are a kind of film-like materials made from naturally occurring carbohydrate polymers used as the matrix material and created by physical or chemical cross-linking techniques. According to the film-forming matrix, the bioactive FFKFs are primarily separated into protein-, polysaccharide-, and multi matrix-based films [51].

The oxidation of nutrients and microbial contamination are the two leading causes of food spoilage. Bioactive (FFKFs) or functional edible coatings (FECs), which refer to a class of nature-based polymer-based preservation materials with benefits of green, nontoxic, edible, and degradable characteristics, can effectively maintain the freshness of food items, have attracted much attention. Food items' shelf life can be increased by limiting microbiological contamination and oxidative degradation using bioactive FFKFs and FECs [51].

5 Nanoemulsions

The physicochemical properties of edible coatings on various food items may be improved with the use of NEs, such as the pectin-based edible coating on freshly cut orange slices, which improves the quality and sensory qualities of the product [44].

5.1 Formulation of NEs

Three main ingredients—oil phase, aqueous phase, and stabilizers—are utilized to create food-grade NEs. Typically, triglycerides such triacylglycerols (TAGs), diacylglycerols (DAGs), monoacylglycerols (MAGs), essential oils, mineral oils, fatty acids, and other nutraceuticals are used to create the oil phase in n-emulsion formulation [25, 36]. Triacylglycerol oils, including flaxseed, sunflower, safflower, olive, maize, soybean, and fish oil, are suggested for use in the food sector due to their low cost and enhanced functional and nutritional qualities. For creating a n-emulsion, polar substances such as alcohols, carbohydrates, proteins, minerals, acids, and bases are combined with water to generate the aqueous phase. The amount and nature of these components determine the polarity, refractive index, pH, density, and strength of the aqueous phase, all of which impact the stability and physicochemical characteristics of the NEs [56].

Only two phases—oil and aqueous—are used, and their combination results in the creation of a transient emulsion that is quickly decomposable. Hence, stabilizers such emulsifying agents are utilised to preserve the stability of n-emulsions and prevent their collapse. Emulsifying agents may be broadly categorised into two classes: hydrophilic colloids like acacia, bentonite, and vegum, and surfactants like spans and tweens. These substances produce a stable NE by reducing surface tension, preventing coalescence, and being nontoxic [32].

5.1.1 Oil Phase

The properties of n-emulsions are significantly influenced by the physical and chemical properties of the oil phase, including viscosity, water solubility, density, polarity, refractive index, and chemical stability [47]. A range of nonpolar molecules, including waxes, mineral oils, free fatty acids (FFA), MAGs, DAGs, TAGs, and numerous lipophilic nutraceuticals, can be employed to create the oil phase needed to create food-grade n-emulsions. The most popular TAG oils used in n-emulsions are those derived from soybean, safflower, maize, flaxseed, sunflower, olive, algae, or fish principally because they are inexpensive and have good nutritional value.

5.1.2 Water Phase

Water can be combined with a range of polar molecules, carbohydrates, proteins, acids, minerals, or alcoholic solvents to create the aqueous phase utilized to create food-grade NEs. The physicochemical characteristics of the created n-emulsion are greatly influenced by choice of the aqueous phase [47].

5.1.3 Stabilizers

Stabilizers impact emulsions long-term stability; as a result, choosing the right stabilizer is one of the most crucial elements to consider for the effective creation of n-emulsions. To increase the long-term stability of n-emulsions, several stabilizers are added. A list of these stabilizers is provided in (Table 2). Examples of stabilizers are emulsifiers, ripening inhibitors, texture modifiers, and weighing agents. The most popular stabilisers used in n-emulsions are emulsifiers. Different emulsifiers may be added, such as phospholipids, small-molecule surfactants, polysaccharides, and proteins [47].

The three main components of a n-emulsion are oil, water, and surfactant [19]. The emulsion's stability and characteristics are controlled by carefully mixing these components. The choice of surfactants during n-emulsion creation depends on how active the surface is, with ionic strength, temperature, pH, and stability of the n-emulsion system being particularly important [15]. Additionally, in n-emulsion systems, the stability and size of the particle depend on the water-to-oil ratio. The dispersive strength of n-emulsion is enhanced by Oil, water, thickening agents, weighting agents, emulsification, antioxidants, and polyunsaturated fats [18]. The quantity of water and its particular characteristics significantly impact the organoleptic aspects of foods. The water crystals significantly influence the emulsion's texture and flavour of food products. Through interfacial interaction, emulsifiers also prevent coalescence and flocculation in n-emulsions. Emulsifiers make it easier for droplets to separate, which creates smaller particles. The amount of

Table 2 Kinds of stabilizers added to improve long-term stability of n-emulsions

Stabilizers	Function	Examples	References
Emulsifiers	To stabilise emulsions by raising their kinetic stability, a single emulsifier or combination of emulsifiers is added	1. Amphiphilic proteins (caseinate or whey protein isolate) 2. Amphiphilic polysaccharides (gum Arabic or modified starch) 3. Small molecule surfactants	[33]
Weighting agents	substances that maintain a balance between the liquid densities in n-emulsions	1. Sucrose acetate isobutyrate 2. Brominated vegetable oil	[47]
Texture modifiers	Substances that increase the viscosity of n-emulsions	1. Polysaccharides (carrageenan, xanthan, pectin, alginate) 2. Sugars (high-fructose corn syrup or sucrose) 3. Proteins (whey protein isolate, gelatin or soy protein isolate) 4. Polyols (sorbitol or glycerol)	[14]
Ripening retarders	Hydrophobic compounds that delay or prevent Ostwald ripening stabilise n-emulsions	1. Mineral oil 2. Long-chain triglyceride	[2]

biopolymer necessary to completely cover all oil–water interfaces and the rate of coating are used to determine the emulsifier's concentration. oils are used as the oil phase in n-emulsion systems and are more inclined to guard themselves against oxidative destruction [19].

6 Green Micro Emulsions and NEs

Focusing on Micro-emulsions (MEs) and NEs, colloidal dispersions are suspensions of tiny particles dispersed inside a liquid media. A "microemulsion" is a stable crystalline liquid system in which oil is distributed with the aid of a co-surfactant and surfactant. Depending on its content and the surrounding conditions, this mixture can produce several systems (in particular temperature). One or more distinct phases that are in balance with one another may arise. These phases can contain spherical structures (like reverse micelles or micelles), cylindrical (like reverse micelles or rod-micelles), sponge-like (e.g., bicontinuous) or flat (like lamellar structures) [52] (Fig. 1). It relies on how well the surfactant molecules are packaged and curved, as well as how much oil is used in the system [30]. A NE can be thought of as a typical emulsion with minor particles. It is a colloidal distribution of two incompatible liquids that are highly unstable, and one of the liquids is scattered as small droplets (100 nm) inside the other liquid. Therefore, the same ingredients used to produce a ME are typically used to prepare a NE: surfactant, water, oil, and perhaps a co-surfactant. Even while the same materials can be used to make both types of colloidal dispersion, their proportions vary between the two formulations. Surfactants are present in NEs at a lower concentration than in MEs (usually 5%–10% vs. 20% in MEs) [49].

Due to the interfacial area being reduced by the comparatively low particle radius and great interfacial tension in NEs, the droplets in these systems tend to be spherical [48]. Whereas the polar head groups of the surfactant molecules are pointed toward the nearby aqueous medium, their relatively nonpolar tails protrude into the lipophilic core [22]. The prefixes used to identify them, which are connected to the particle size, are the main cause of the misperception between MEs and NEs. The word "micro" typically refers to 10^{-6}, which in this context indicates that the drops have a micrometric dimension, or one-millionth of a metre or less. NE droplet size can cross over into ME. Both should contain particles between 20 and 200 nm in size [53]. The main distinction between NEs and MEs is the thermodynamic stability: NEs have unstable thermodynamic properties, whereas MEs have stable thermodynamic properties. A NE is thermodynamically unstable because the colloidal distribution (droplets in water) has higher free energy than the constituent phases (oil and water). NE may yet have rather high kinetic stability (Metastability), by creating an appropriately high energy barricade between the two states [10].

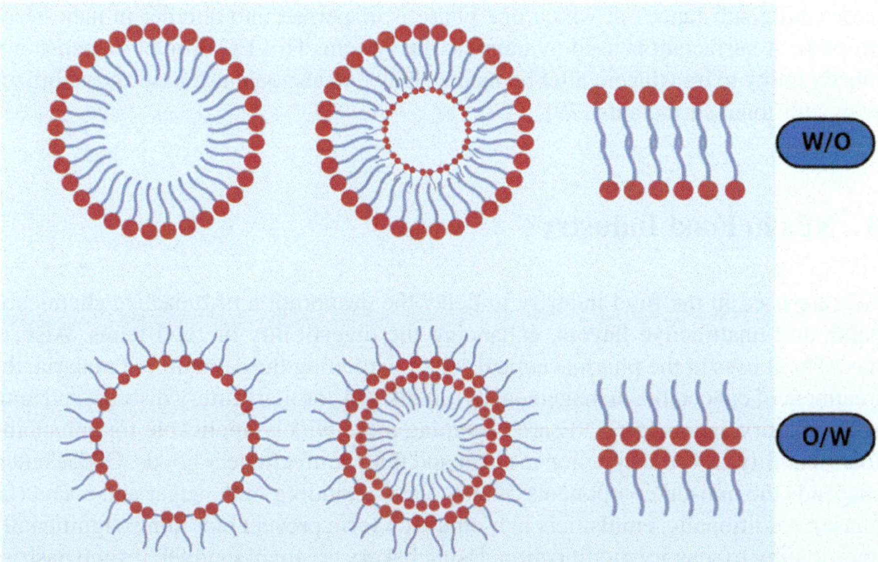

Fig. 1 Structure of oil–water (O/W) and water–oil (W/O) nanoemulsion

7 Nano Formulated Essential Oils as Novel Green Preservatives

The investigation and use of nanoparticles with at least one dimension up to 100 nm lie at the core of the emerging field of nanotechnology. Currently, nanotechnology is used in the production of numerous goods across the globe, including food, clothing, and cosmetics [6–9]. Although Eos and their active ingredients have a unique preservation potential, problems including rapid degradation under intense light, high temperatures, and high volatility bound their practical usage. Other significant obstacles to their broad use include their poor water solubility, adverse effects on food items organoleptic characteristics at higher doses, and considerable bioactive component change because of interaction with food components like fat, protein, and carbohydrate. Botanicals are frequently included in food-grade polymeric matrices that are biodegradable and non-reactive to food [29]. Starch, glycerides, carrageenan, whey proteins, pectin, gelatin, alginate, gum Arabic, phospholipids of natural origin, and chitosan are a few of the often utilized encapsulating substances [42]. Emulsification, liposomal approach, spray drying, nanoprecipitation, inclusion complexation, and coacervation are some common encapsulating techniques used in the food business to improve the bio-efficacy of green preservatives. It has been demonstrated that EOs and encapsulating agents can yield several morphologies, including nanosponge, nano-gel, nanofibers, nanoparticles, nanotubes, and NEs according to currently used methodologies. The most well-liked ones have been nanoparticles, nano-gels, and n-emulsions [11]. A NE is a colloidal system of hydrophobic and hydrophilic substances in which one phase is dispersed into another in nano-sized droplets. A surfactant is used to maintain the system. This EO droplet suspension is mostly milky to translucent after being distributed in an aqueous phase and stabilized with additional surfactants [20].

8 NEs in Food Industry

NEs are used in the food industry to delay the evaporation of bioactive chemicals' harsh and unattractive flavors, enhancing the digestibility of food items. Also, it has several uses in the pharmaceutical sector, including the detection of malaria, the treatment of cancer, the management of tumors and coronary artery disease, and anti-inflammatory agents [56]. NEsare becoming more widely applicable for enhancing foam stability and air dispersion in sugar and flour confectionery goods. Cocoa butter suspends the non-fat components of chocolate, including milk, sugar and cocoa (fat phase). Additionally, emulsifiers in n-emulsion form prevent blooming, significantly contributing to sensory modification. Using bakery cream in sandwiches and pastries creates enormous emulsions of market demands. Ice cream's texture and consistency have been effectively improved by Nes [55], and many more advantages of NEs are also shown in Fig. 2.

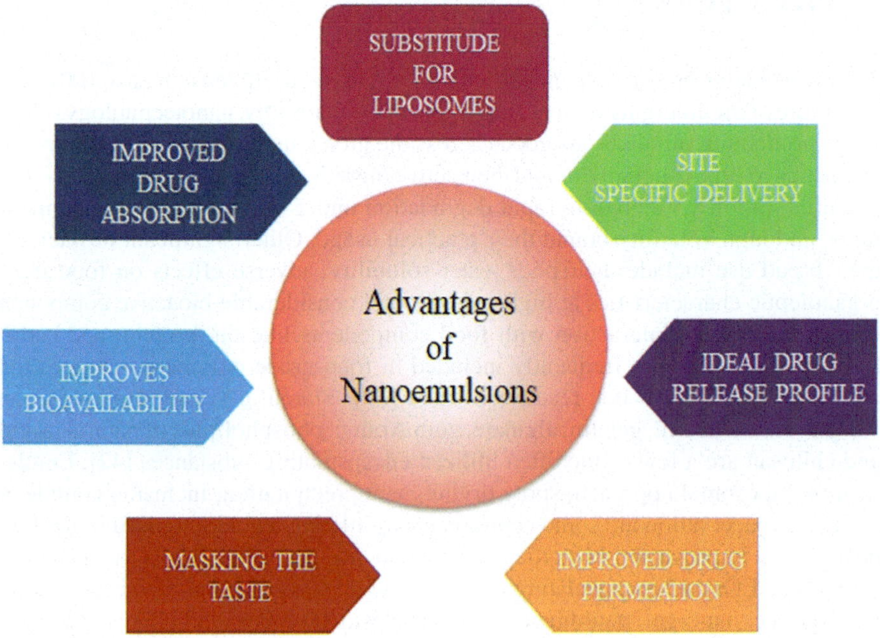

Fig. 2 Advantages of N-emulsions

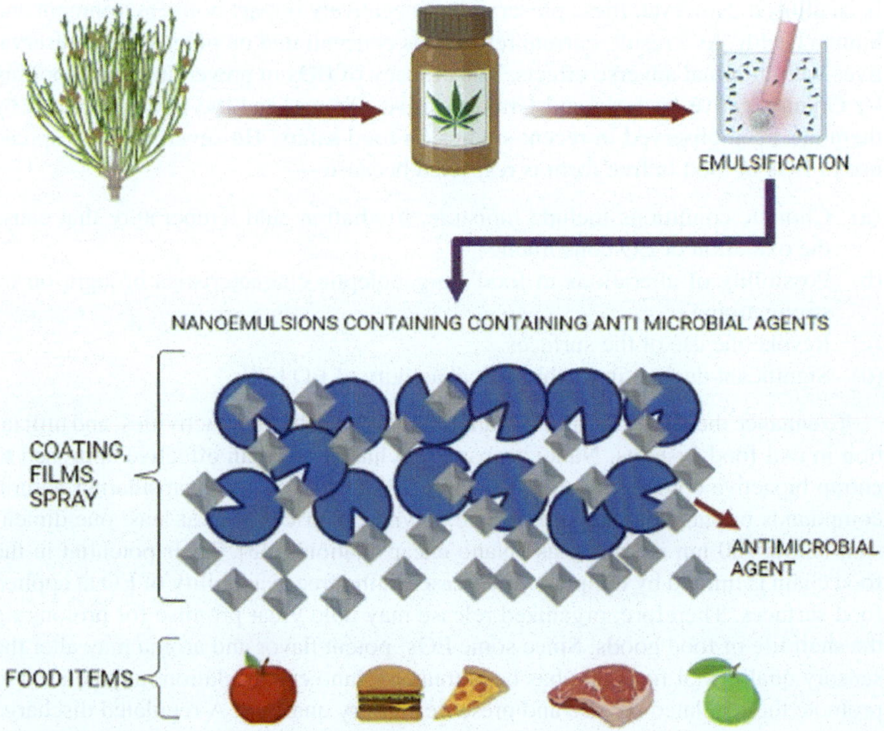

Fig. 3 Green n-emulsion controlling food spoilage

8.1 Food Edible Coatings

Green Nes controlling Food Spoilage shows in Fig. 3 that a liquid colloidal system known as FEC is created by evenly scattering natural food preservatives or their delivery Nanosystems into a solution containing consumable polysaccharides, proteins, and other biological macromolecules.

8.2 Essential Oils for Food Preservation in Nano Encapsulation

The aromatic and medicinal plants' flowers, buds, leaves, fruits, and stems are among the parts from which essential oils are extracted [45]. 300 of the nearly 3000 EOs isolated from the 2000 plant species have been used commercially. Alcohols, aldehydes, esters, phenylpropanoids, Terpenes, ketones, and terpenoids are only a few bioactive ingredients that makeup EOs. In order to prevent microbial contamination in food products that have been preserved, many synthetic preservatives have been utilised. However, these preservatives negatively impact both environment and human health. As a result, current research is concentrated on plant-based preservatives with minimal adverse effects. The efficacy of EOs in preventing food spoiling by Gram-negative bacteria and Gram-positive, fungus, and the toxins secreted by them has been observed in recent studies on food safety. However, the widespread use of EOs in food in free form is restricted because—

(a) Climatic conditions include moisture, irradiation, and temperature that cause the oxidation of EO constituents.
(b) Possibility of alterations in food's organoleptic characteristics brought on by strong aroma.
(c) Revoke the use of the surfaces.
(d) Significant decline in the biological activity of EO [39].

To enhance the microbial antioxidant capabilities, inhibitory activities, and utilization in real food systems, Nanoencapsulation has become an effective approach to entrap bioactive compounds and Eos [13]. The technique of encapsulating natural compounds or chemicals in appropriate polymer matrices with at least one dimension below 100 nm is known as "Nano encapsulation" [43]. Their potential in the food chain is limited by temporary decreases in the free availability of EO at applied food surfaces. Therefore, organized release may hold great promise for prolonging the shelf life of food goods. Since some EOs' potent flavor and aroma may alter the sensory qualities of food that has been treated, Nano encapsulation may be used to promote the regulated release and preserve sensory qualities. A regulated discharge of EOs loaded on zein nanoparticles has been recommended to preserve meat [57].

8.3 Functions of Essential Oils n-emulsion as Food Additives

In order to dodge the disadvantages of encapsulated essential oils for practical use with maximum compatibility and stability, consumers and current food businesses are concentrating on Nanoengineered EOs in the form of n-emulsion. Additionally, n-emulsions offer the most significant advantages related to the usage of EOs in food products [46], including:

- Higher dispersion on food surfaces, where bacteria typically multiply
- Decreased sensory effects
- N-emulsions with bioactive EO molecules have increased antibacterial activity
- The emulsion-based carriage organizations incorporating EOs may interact with microorganisms due to the many target locations in microbial cells, interfering with normal biological processes [46].

Microorganisms thrive in food and produce chemicals that alter flavor, color, and texture. These microorganisms include bacteria and fungi. The food will eventually lose its value for human consumption. Food is spoiled when it is kept with a hairy growth and becomes pulpy, producing unpleasant odors and encouraging the growth of moulds and yeasts. Mold and yeast-related spoilage include mold development on bread, the rotting of fruits and vegetables, and sour taste of milk. These organisms rarely harm humans, but bacterial contamination is frequently more severe since it may not necessarily seem awful even when food is seriously contaminated. When microbes exist in food, they utilize the nutrients already present and quickly proliferate. They alter the aroma of the meal and create fresh, potentially toxic substances. Food spoilage directly affects the food's color, flavor, aroma, consistency, and texture, and it may become unsafe to consume. Food that has a terrible odor or smell indicates that it might not be safe.

Reduced food rotting is the main contributor to food security, Food that has spoiled loses its ability to be edible and either becomes unfit for human consumption. Green MEs and NEs, pesticides made from n-emulsions and their constituents, nano-encapsulated essential oils for food preservation, the use of essential oils as food additives, and n-emulsion application are all examples. NEs are minuscule droplet-sized colloidal systems that are kinetically stable. They are more functionally effective than conventional emulsions. The structure and composition of the NEs can be organized to effectively encapsulate and distribute bioactive lipophilic substances. NEs can be used to deliver nutraceuticals, ingredients for colouring and flavouring, and antimicrobials in the food industry. Foods' quality, nutritional value, functionality, and shelf life can be increased by using biodegradable coverings and wrapping films made from active ingredient n-emulsion compounds.

9 Conclusion and Future Perspectives

Due to their distinctive features and superior steadiness to conservative emulsions, NEs offer enormous promise in various industries, cosmetics, pharma and the food sectors. The selection of the exact methodology and the enhancement of the conditions for the increased steadiness of NEs will enable the creation of a high amount of production and their various applications in the beverage, pharmaceutical, and food industries by their precise requirements. In order to increase the bioavailability, functionality, and solubility of nonpolar active chemicals, NEs are one of the most capable systems. This supports the use of NEs as drug delivery systems. Aside from their enormous potential, NEs are metastable, so further research is needed to produce stable NEs that can be used in manufacturing. In terms of the creation process, particularly their cost, as well as the characterization of the developed NEs and the food systems to which they will be deployed regarding product safety and acceptability, there are still issues with NE application that need to be resolved. The use of n-emulsions in food systems still presents difficulties that must be resolved in terms of their manufacture, particularly in terms of cost and their characterization in terms of the food systems to which they will be applied and the safety and acceptability of the final product.

Although they manufacture translucent beverages and meals, their increased physical stability and bioavailability are some of the possible benefits of n-emulsions over traditional ones. To enable the widespread use of n-emulsions, however, several regulatory issues must first be resolved. Most of the ingredients utilised in the preparation of NEs, whether by low-energy or high-energy methods, such as synthetic oils, synthetic polymers, organic solvents, or synthetic surfactants, are not generally acceptable usage in the food business. Therefore, food-grade substances that are ethically acceptable, affordable to use, and label-friendly, such as proteins, triglyceride oils, polysaccharides, and flavour oils, must be used to create food n-emulsions. Appropriate processing techniques should be used to produce affordable and reliable products to create food-grade n-emulsions on an industrial scale. Several proposed strategies established in research labs are inappropriate for commercial production since low-intensity strategies could not yet be validated in large-scale production. In the food business, only high-intensity methods are currently used to produce n-emulsions.

Finally, using very small lipid droplets in food is accompanied by a few safety concerns. For instance, a lipophilic component enclosed within nanometre-sized lipid droplets has very different absorption pathways, bioavailability, and possible toxicity than one that is disseminated within a bulk lipid phase. These factors make in-depth research in the field of n-emulsion safety necessary.

References

1. Ashaolu TJ (2021) Nanoemulsions for health, food, and cosmetics: a review. Environ Chem Lett 19(4):3381–3395. https://doi.org/10.1007/S10311-021-01216-9
2. Aswathanarayan JB, Vittal RR (2019) Nanoemulsions and their potential applications in food industry. Front Sustain Food Syst 3:95. https://doi.org/10.3389/FSUFS.2019.00095/BIBTEX
3. Atif M, Ilavenil S, Devanesan S, AlSalhi MS, Choi KC, Vijayaraghavan P, Alfuraydi AA, Alanazi NF (2020) Essential oils of two medicinal plants and protective properties of jack fruits against the spoilage bacteria and fungi. Ind Crops Prod 147:112239. https://doi.org/10.1016/J.INDCROP.2020.112239
4. Awuchi CG, Ondari EN, Ogbonna CU, Upadhyay AK, Baran K, Okpala COR, Korzeniowska M, Guiné RPF (2021) Mycotoxins affecting animals, foods, humans, and plants: types, occurrence, toxicities, action mechanisms, prevention, and detoxification strategies—a revisit. Foods 10(6). https://doi.org/10.3390/FOODS10061279
5. Azmi NAN, Elgharbawy AAM., Motlagh SR, Samsudin N, Salleh HM (2019) Nanoemulsions: factory for food, pharmaceutical and cosmetics. Processes 7(9):617. https://doi.org/10.3390/PR7090617
6. Bayda S, Adeel M, Tuccinardi T, Cordani M, Rizzolio F (2020) The history of nanoscience and nanotechnology: from chemical–physical applications to nanomedicine. Molecules 25(1). https://doi.org/10.3390/MOLECULES25010112
7. Bachheti A, Bachheti RK, Abate L, Husen A (2022) Current status of Aloe-based nanoparticle fabrication, characterization and their application in some cutting-edge areas. S Afr J Bot 147:1058–1069
8. Bachheti RK, Abate L, Bachheti A, Madhusudhan A, Husen A (2021) Algae-, fungi-, and yeast-mediated biological synthesis of nanoparticles and their various biomedical applications. In: Handbook of greener synthesis of nanomaterials and compounds. Elsevier, Amsterdam, pp 701–734. https://doi.org/10.1016/B978-0-12-821938-6.00022-0

9. Bachheti RK, Fikadu A, Bachheti A, Husen A (2020) Biogenic fabrication of nanomaterials from flower-based chemical compounds, characterization and their various applications: a review. Saudi J Biol Sci 27(10):2551–2562
10. Bhattacharjee K, Bhattacharjee K (2019) Importance of surface energy in nanoemulsion. Nanoemulsions Prop, Fabr Appl. https://doi.org/10.5772/INTECHOPEN.84201
11. Chaudhari AK, Dwivedy AK, Singh VK, Das S, Singh A, Dubey NK (2019) Essential oils and their bioactive compounds as green preservatives against fungal and mycotoxin contamination of food commodities with special reference to their nanoencapsulation. Environ Sci Pollut Res Int 26(25):25414–25431. https://doi.org/10.1007/S11356-019-05932-2
12. Chaudhari AK, Singh VK, Das S, Deepika, Prasad J, Dwivedy AK, Dubey NK (2020) Improvement of in vitro and in situ antifungal, AFB1 inhibitory and antioxidant activity of Origanum majorana L. essential oil through nanoemulsion and recommending as novel food preservative. Food Chem Toxicol: Int J Publ Br Ind Biol Res Assoc 143. https://doi.org/10.1016/J.FCT.2020.111536
13. Chaudhari AK, Singh VK, Das S, Dubey NK (2021) Nanoencapsulation of essential oils and their bioactive constituents: a novel strategy to control mycotoxin contamination in food system. Food Chem Toxicol Int J Publ Br Ind Biol Res Assoc 149:112019. https://doi.org/10.1016/J.FCT.2021.112019
14. Chen L (2015) Emulsifiers as food texture modifiers. Modif Food Texture: Novel Ingred Process Tech 27–49. https://doi.org/10.1016/B978-1-78242-333-1.00002-4
15. Choi SJ, McClements DJ (2020) Nanoemulsions as delivery systems for lipophilic nutraceuticals: strategies for improving their formulation, stability, functionality and bioavailability. Food Sci Biotechnol 29(2):149. https://doi.org/10.1007/S10068-019-00731-4
16. Das S, Singh VK, Dwivedy AK, Chaudhari AK, Deepika, Dubey NK (2021) Eugenol loaded chitosan nanoemulsion for food protection and inhibition of Aflatoxin B1 synthesizing genes based on molecular docking. CarbohydPolym 255. https://doi.org/10.1016/J.CARBPOL.2020.117339
17. Das S, Singh VK, Dwivedy AK, Chaudhari AK, Upadhyay N, Singh A, Deepika, Dubey NK (2020) Fabrication, characterization and practical efficacy of Myristica fragrans essential oil nanoemulsion delivery system against postharvest biodeterioration. Ecotoxicol Environ Saf 189. https://doi.org/10.1016/J.ECOENV.2019.110000
18. Dasgupta N, Ranjan S (2018) Ingredients and components of nanoemulsions 63–82. https://doi.org/10.1007/978-981-10-6986-4_4
19. Dasgupta N, Ranjan S, Gandhi M (2019) Nanoemulsion ingredients and components. Environ Chem Lett 17(2):917–928. https://doi.org/10.1007/S10311-018-00849-7
20. de Matos SP, Lucca LG, Koester LS (2019) Essential oils in nanostructured systems: challenges in preparation and analytical methods. Talanta 195:204–214. https://doi.org/10.1016/J.TALANTA.2018.11.029
21. Díaz-Montes E, Castro-Muñoz R (2021) Edible films and coatings as food-quality preservers: an overview. Foods 10(2). https://doi.org/10.3390/FOODS10020249
22. Esmaeili H, Mousavi SM, Hashemi SA, Lai CW, Chiang WH, Bahrani S (2021) Application of biosurfactants in the removal of oil from emulsion. Green Sustain Process Chem Environ Eng Sci: Biosurfactants Bioremed Polluted Environ 107–127. https://doi.org/10.1016/B978-0-12-822696-4.00008-5
23. Facciolà A, Riso R, Avventuroso E, Visalli G, Delia SA, Laganà P (2017) Campylobacter: from microbiology to prevention. J Prev Med Hyg 58(2):E79
24. Godeto YG, Ayele A, Ahmed IN, Husen A, Bachheti RK (2023) Medicinal plant-based metabolites in nanoparticles synthesis and their cutting-edge applications: an overview. Second Metab Med Plants 1–34
25. Gonçalves A, Nikmaram N, Roohinejad S, Estevinho BN, Rocha F, Greiner R, McClements DJ (2018) Production, properties, and applications of solid self-emulsifying delivery systems (S-SEDS) in the food and pharmaceutical industries. Colloids Surf, A 538:108–126. https://doi.org/10.1016/J.COLSURFA.2017.10.076

26. Gonfa YH, Tessema FB, Tadesse MG, Bachheti A, Bachheti RK (2023) Medicinally important plant roots and their role in nanoparticles synthesis and applications. In: Secondary metabolites from medicinal plants. CRC Press, pp 227–242
27. Guirin GF, Brusa V, Adriani CD, Leotta GA (2020) Prevalence of Campylobacter jejuni and Campylobacter coli from broilers at conventional and kosher abattoirs and retail stores. Rev Argent Microbiol 52(3):217–220. https://doi.org/10.1016/J.RAM.2019.07.002
28. Gupta A, Eral HB, Hatton TA, Doyle PS (2016) Nanoemulsions: formation, properties and applications. Soft Matter 12(11):2826–2841. https://doi.org/10.1039/C5SM02958A
29. Hasani S, Ojagh SM, Ghorbani M (2018) Nanoencapsulation of lemon essential oil in Chitosan-Hicap system. Part 1: Study on its physical and structural characteristics. Int J Biol Macromol 115:143–151. https://doi.org/10.1016/J.IJBIOMAC.2018.04.038
30. Hosseinpour S, Götz V, Peukert W (2021) Effect of surfactants on the molecular structure of the buried oil/water interface. Angewandte Chemie (International Ed. in English), 60(47):25143. https://doi.org/10.1002/ANIE.202110091
31. Hu F, Tu XF, Thakur K, Hu F, Li XL, Zhang YS, Zhang JG, Wei ZJ (2019) Comparison of antifungal activity of essential oils from different plants against three fungi. Food Chem Toxicol: Int J Publ Br Ind Biol Res Assoc 134. https://doi.org/10.1016/J.FCT.2019.110821
32. Jaiswal M, Dudhe R, Sharma PK (2015) Nanoemulsion: an advanced mode of drug delivery system. 3 Biotech 5(2):123. https://doi.org/10.1007/S13205-014-0214-0
33. Julian McClements D, Mahdi Jafari S (2017) Improving emulsion formation, stability and performance using mixed emulsifiers: a review
34. Kawacka I, Olejnik-Schmidt A, Schmidt M, Sip A (2020) Natural plant-derived chemical compounds as listeria monocytogenes inhibitors in vitro and in food model systems. Pathogens (Basel, Switzerland) 10(1):1–36. https://doi.org/10.3390/PATHOGENS10010012
35. Kumar Chaudhari A, Kumar Singh V, Das S Deepika, Kishore Dubey N (2022) Fabrication, characterization, and bioactivity assessment of chitosan nanoemulsion containing allspice essential oil to mitigate Aspergillus flavus contamination and aflatoxin B1 production in maize. Food Chem 372:131221. https://doi.org/10.1016/J.FOODCHEM.2021.131221
36. Kumar M, Bishnoi RS, Shukla AK, Jain CP (2019)Techniques for formulation of nanoemulsion drug delivery system: a review. Prev Nutr Food Sci 24(3):225. https://doi.org/10.3746/PNF.2019.24.3.225
37. Lages LZ, Radünz M, Gonçalves BT, Silva da Rosa R, Fouchy MV, de Cássia dos Santos da Conceição R, Gularte MA, Barboza Mendonça CR, Gandra EA (2021) Microbiological and sensory evaluation of meat sausage using thyme (Thymus vulgaris, L.) essential oil and powdered beet juice (Beta vulgaris L., Early Wonder cultivar). LWT, 148:111794. https://doi.org/10.1016/J.LWT.2021.111794
38. Liu Q, Huang H, Chen H, Lin J, Wang Q (2019) Food-grade nanoemulsions: preparation, stability and application in encapsulation of bioactive compounds. Molecules 24(23). https://doi.org/10.3390/MOLECULES24234242
39. Maurya A, Singh VK, Das S, Prasad J, Kedia A, Upadhyay N, Dubey NK, Dwivedy AK (2021) Essential oil nanoemulsion as eco-friendly and safe preservative: bioefficacy against microbial food deterioration and toxin secretion, mode of action, and future opportunities. Front Microbiol 12:3326. https://doi.org/10.3389/FMICB.2021.751062/BIBTEX
40. Mirza Alizadeh A, Hashempour-Baltork F, Mousavi Khaneghah A, Hosseini H (2021) New perspective approaches in controlling fungi and mycotoxins in food using emerging and green technologies. Curr Opin Food Sci 39:7–15. https://doi.org/10.1016/J.COFS.2020.12.006
41. Mohd Nizam NH, Mohammad Rawi NF, Mhd Ramle SF, Abd Aziz A, Abdullah CK, Rashedi A, Mohamad Kassim MH (2021) Physical, thermal, mechanical, antimicrobial and physicochemical properties of starch based film containing aloe vera: a review. J Market Res 15:1572–1589. https://doi.org/10.1016/J.JMRT.2021.08.138
42. Nasrollahzadeh M, Sajjadi M, Iravani S, Varma RS (2021) Starch, cellulose, pectin, gum, alginate, chitin and chitosan derived (nano) materials for sustainable water treatment: a review. Carbohyd Polym 251:116986. https://doi.org/10.1016/J.CARBPOL.2020.116986

43. Prakash A, Baskaran R, Paramasivam N, Vadivel V (2018) Essential oil based nanoemulsions to improve the microbial quality of minimally processed fruits and vegetables: a review. Food Res Int (Ottawa, Ont.) 111:509–523. https://doi.org/10.1016/J.FOODRES.2018.05.066
44. Radi M, Akhavan-Darabi S, Akhavan HR, Amiri S (2018) The use of orange peel essential oil microemulsion and nanoemulsion in pectin-based coating to extend the shelf life of fresh-cut orange. J Food Process Preserv 42(2):e13441. https://doi.org/10.1111/JFPP.13441
45. Ramos AVG, Eninger MC, Amrein AM, Araújo DL, de O, Zara RF, Cottica SM, Tiuman TS (2022) Effects of the combination of some essential oils with methylparaben and sodium metabisulfite against pathogenic foodborne bacteria. J Food Process Preserv 46(4). https://doi.org/10.1111/JFPP.16459
46. Salanță LC, Cropotova J (2022) An update on effectiveness and practicability of plant essential oils in the food industry. Plants 11(19). https://doi.org/10.3390/PLANTS11192488
47. Salem MA, Ezzat SM, Salem MA, Ezzat SM (2018) Nanoemulsions in food industry. Some New Asp Colloid Syst Foods. https://doi.org/10.5772/INTECHOPEN.79447
48. Schroën K, de Ruiter J, Berton-Carabin C (2020) The importance of interfacial tension in emulsification: connecting scaling relations used in large scale preparation with microfluidic measurement methods. ChemEngineering 4(4):1–22. https://doi.org/10.3390/CHEMENGINEERING4040063
49. Sjöblom J, Stenius P, Simon S, Grimes BA (2013) Emulsion stabilization. Encycl Colloid Interface Sci 415–454. https://doi.org/10.1007/978-3-642-20665-8_83
50. Souto EB, Cano A, Martins-Gomes C, Coutinho TE, Zielińska A, Silva AM (2022) Microemulsions and nanoemulsions in skin drug delivery. Bioengineering 9(4). https://doi.org/10.3390/BIOENGINEERING9040158
51. Sun X, Wang J, Dong M, Zhang H, Li L, Wang L (2022) Food spoilage, bioactive food fresh-keeping films and functional edible coatings: research status, existing problems and development trend. Trends Food Sci Technol 119:122–132. https://doi.org/10.1016/J.TIFS.2021.12.004
52. Tartaro G, Mateos H, Schirone D, Angelico R, Palazzo G (2020) Microemulsion microstructure(s): a tutorial review. Nanomaterials 10(9):1657. https://doi.org/10.3390/NANO10091657
53. Tenchov R, Bird R, Curtze AE, Zhou Q (2021) Lipid nanoparticles from liposomes to mRNA vaccine delivery, a landscape of research diversity and advancement. ACS Nano 15(11):16982–17015. https://doi.org/10.1021/ACSNANO.1C04996/ASSET/IMAGES/MEDIUM/NN1C04996_0026.GIF
54. Thomsen TB, Hunt CJ, Meyer AS (2022) Influence of substrate crystallinity and glass transition temperature on enzymatic degradation of polyethylene terephthalate (PET). New Biotechnol 69:28–35. https://doi.org/10.1016/J.NBT.2022.02.006
55. Tipu MMH, Baroi A, Rana J, Islam S, Jahan R, Miah MS, Asaduzzaman M, Tipu MMH, Baroi A, Rana J, Islam S, Jahan R, Miah MS, Asaduzzaman M (2021) Potential applications of nanotechnology in agriculture: a smart tool for sustainable agriculture. Agric Dev Asia—Potential Use of Nano-Mater Nano-Technol. https://doi.org/10.5772/INTECHOPEN.101142
56. Tripathi AD, Sharma R, Agarwal A, Haleem DR (2021) Nanoemulsions based edible coatings with potential food applications 3(1):112–125. https://doi.org/10.1080/24759651.2021.1875615; https://doi.org/10.1080/24759651.2021.1875615
57. Xavier LO, Sganzerla WG, Rosa GB, da Rosa CG, Agostinetto L, de Veeck AP, Bretanha LC, Micke GA, Dalla Costa M, Bertodi FC, Barreto PLM, Nunes MR (2021) Chitosan packaging functionalized with Cinnamodendron dinisii essential oil loaded zein: a proposal for meat conservation. Int J Biol Macromol 169:183–193. https://doi.org/10.1016/J.IJBIOMAC.2020.12.093
58. Zhang H, Wang J, Chang Z, Liu X, Chen W, Yu Y, Wang X, Dong Q, Ye Y, Zhang X (2021) Listeria monocytogenes contamination characteristics in two ready-to-eat meat plants from 2019 to 2020 in Shanghai. Front Microbiol 12:2326. https://doi.org/10.3389/FMICB.2021.729114/BIBTEX

Use of Nanoemulsions in Pesticide Formulation

Arshad Khan, Fadime Karabulut, Saba Fatima, and Mansoor Ahmad Siddiqui

Abstract The rapid reduction in agricultural production caused by various pests and diseases necessitates researching and designing novel, efficient, secure and eco-friendly formulations of pesticides. In agricultural farming, pesticides are used to efficiently manage pests and diseases, resulting in higher food production, reduced food waste, and increased farmer profit. Unfortunately, over 90% of pesticides drain into the environment and persist in agricultural products, causing environmental pollution, which are major drawbacks of a chemical-based traditional pesticide formulation. In the past few years, nanoemulsion-based pesticides have been developed to be effective, nontoxic, cost-effective, and possess excellent physico-chemical stability and greater biodegradability. For this purpose, there is a great interest in the design of efficient nanoemulsion-based pesticide formulations. Two normally immiscible liquids (water and oil) are combined in these formulations to produce an isotropic system. This chapter discusses various scientific concerns and approaches, including preparation method, composition, characterization, and biological properties pertaining to developing nanoemulsion-based pesticide formulations.

Keywords Nanoemulsion · Formulation · Pesticide · Eco-friendly · Biodegradability

1 Introduction

In the upcoming years, ensuring food security and feeding the growing world population will be a major challenge. Pesticides are essential for preventing crop losses and assuring food security [95, 149]. In the upcoming decades, pesticide use will be progressively crucial for resolving food security challenges related to ensuring a

A. Khan (✉) · S. Fatima · M. A. Siddiqui
Department of Botany, Aligarh Muslim University, Aligarh 202002, India
e-mail: arshad9758@gmail.com

F. Karabulut
Department of Biology, Firat University, Elazıg 23119, Turkey

A. Husen et al. (eds.), *Current Trends in Green Nano-emulsions*, Smart Nanomaterials Technology, https://doi.org/10.1007/978-981-99-5398-1_6

sufficient food supply for the expanding global population [32, 133]. Pesticide usage has risen significantly during the previous few decades. An estimated 5.2 billion pounds of pesticides are used annually worldwide [89]. The United Nations Population Division predicts the world population will rise by about 30% by 2050. The Food and Agriculture Organization of the United Nations (FAO) claims that 80% of the food required to maintain this development must rise from increased production of present crops [134]. In agriculture, synthetic pesticides, including organochlorides, organophosphates, carbamates, pyrethroids, and neonicotinoids, have been widely used since the 1950s due to their effectiveness and rapidity action, cheapness, and convenience of usage. Around 2.5 million tonnes of pesticides are expected to be employed on crops annually [39]. Current excessive usage of traditional synthetic pesticides poses several threats to human health and the environment [1]. There is a rising interest in developing innovative and effective pesticides with less detrimental ecological impacts than synthetic pesticides [116].

Pesticides are grouped into insecticides, fungicides, nematicides, herbicides, and plant growth regulators based on the types of living organisms they kill. They can be used in sustainable agriculture [121], forestry [113], and non-agricultural purposes [101]. It is well known that the majority of pesticide-active constituents are typically hydrophobic compounds that must be combined with a solvent, an emulsifier, and a dispersant in order to produce an appropriate water-based formulation, such as emulsifiable concentrates (ECs) and wettable powders (WPs), for fumigation [120, 152]. Pesticides are rarely utilized directly in their purified chemical form commercially. Therefore, commercial pesticide formulations can be categorized as solid (granules or powders), liquid (solutions or suspensions), and gaseous (vapors). In addition to solutions and emulsifiable concentrates, microemulsions, oil-in-water emulsions, and soluble concentrates are also considered liquid forms [137].

A nanoemulsion is a mixture of two or more liquids that cannot mix, in which tiny droplets of one liquid are dispersed throughout another liquid [57]. Most nanoemulsions have a milky, opaque, or transparent appearance due to the size of the droplets, which can be between 10–300 nm. Nanoemulsion is sometimes called a mini-emulsion, sub-micron emulsion, or ultrafine emulsion due to its size range of 20–500 nm [123]. A nanoemulsion has thermodynamic instability but dynamic stability. The formulation of a nanoemulsion involves both high-energy and low-energy processes. Hence, it uses the underlying physico-chemical attributes of the system's components and is sensitive to the arrangement in which chemicals are blended throughout the formulation [12]. The three primary components of nanoemulsions are oil, water, and surfactant. The system is formed subsequently when the surfactant layer separates the oil and aqueous phase. In a nanoemulsion system, the interfacial tension exerted by surfactants separates two immiscible phases,oil and water phase [33]. Conventional pesticide formulations are typically employed inefficiently due to loss and poor leaf adherence, resulting in significant resource waste and environmental pollution. Nanoemulsions possess tiny droplets, homogeneous distribution, high stability, convenient synthesis, and superior deposition, expansion, and penetration of the pest body and plant foliage surface [126, 152]. Therefore, developing more effective formulation methods for pesticides is being promoted globally

[9]. In order to develop efficient nanoemulsion-based pesticides, the appropriate formulation technologies must be developed.

2 Components of Nanoemulsion

The outlines of nanoemulsion components are depicted in Fig. 1.

2.1 *Oil Phase*

One of the essential parts of making a nanoemulsion is selecting the appropriate oil for the oily phase. This is because the oil affects how easily the active component dissolves and how easy it is to make a nanoemulsion for a specific purpose. If the pesticide is liquid at room temperature, it may be the only factor in the oil phase. Other suitable environmental solvents such as corn, coconut, olive, peanut, soybean, castor, corn, and sunflower oils could also be used. It may also be essential to combine a liquid pesticide that dissolves in water well with a solvent that does not dissolve in water. This is called a "ripening inhibitor," It slows the growth of droplets caused by Ostwald ripening, which is the movement of pesticide molecules from small droplets to large droplets when the Laplace pressure is different [22]. Pesticide particles can be suspended in an oil phase, but most suspension concentrates are usually dispersed in water [100]. The low solubility of oil in the solution improves nanoemulsion

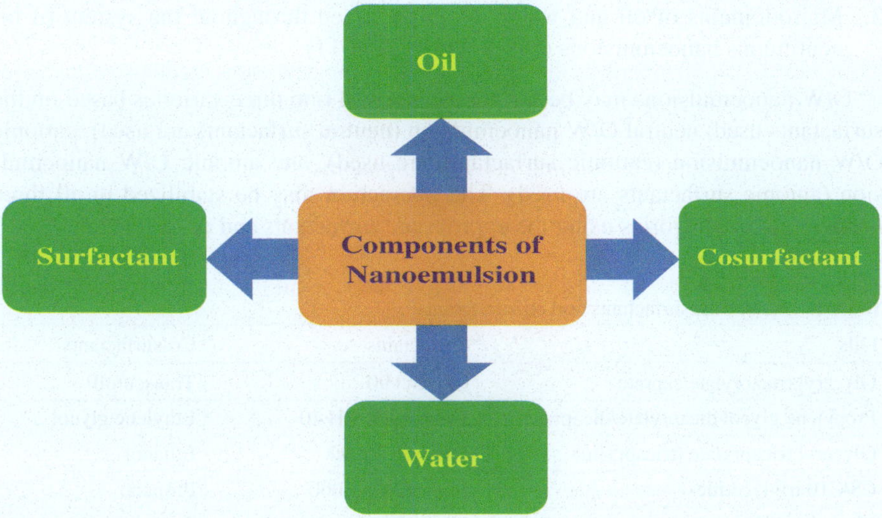

Fig. 1 Components of nanoemulsion

longevity by acting as a kinetic shield to Ostwald ripening. Ostwald ripening is the net movement of oil from tiny droplets to bigger droplets by continuous phase transport [141]. Because oil has many biological properties, dissolving it in a nanoemulsion could increase its cytotoxicity, genotoxicity, and antimicrobial properties against pathogenic microbes. The biological responses of essential oils to infections will be affected by the varying phytochemical compositions of oils [21, 48].

2.2 *Surfactants*

To facilitate the dispersion of all components, the surfactant must reduce the interfacial tension nearest to zero. The classification of surfactants is based on: a hydrophilic head and a hydrophobic tail. In preparing W/O nanoemulsion, surfactants with hydrophilic-lipophilic balance (HLB) values between 3 and 6 are desirable, but for the production of O/W nanoemulsion, surfactants with HLB values between 8 and 18 are favourable. Surfactants with an HLB value greater than 20 serve as cosurfactants for reducing surfactant concentrations to an appropriate level and forming microemulsions. The examples of various oils, surfactants, and co-surfactants are listed in Table 1.

Three different types of nanoemulsions can be produced, and they are as follows: water in oil (W/O), oil in water (O/W), and bi-continuous [23] (Fig. 2).

1. W/O Nanoemulsions that consist of water droplets dispersed in a continuous oil environment
2. O/W Nanoemulsions that consist of oil droplets dispersed in a continuous aqueous environment
3. Microdomains of oil and water are interspersed throughout the system of bi-continuous nanoemulsions (Devarajan et al. 2011).

O/W nanoemulsions may be further categorized into three varieties based on the surfactants used: neutral O/W nanoemulsion (neutral surfactants are used), cationic O/W nanoemulsion (cationic surfactants are used), and anionic O/W nanoemulsion (anionic surfactants are used). The interaction may be stabilized in all three nanoemulsion categories using the appropriate surfactants and co-surfactants.

Table 1 List of oils, surfactants and co-surfactants

Oils	Surfactants	Co-surfactants
Glyceryl tricaorylate/caprate	Caproyl 90	TranscutolP
Propylene glycol dicaprylate/dicaprate	Cremophor RH 40	Ethylene glycol
Glyceryl tricaprylate (tricaprylin)	Lauroglycol 90	Ethanol
C8/C10 triglycerides	PEG MW > 4000	Propano
Myristic acid isopropyl ester	Plurol Oleique CC 497	Propylene glycol
90:10% w/w C12 glyceride tridiesters	Poloxamer 124 and 188	Glycerin

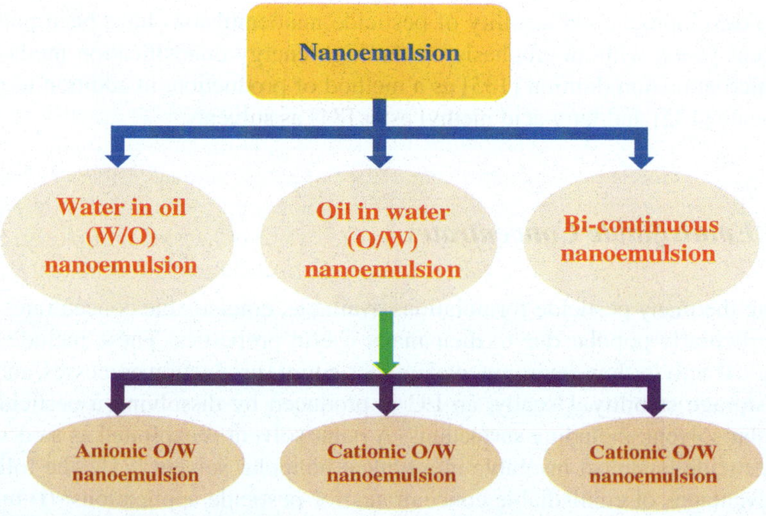

Fig. 2 Various types of nanoemulsion

3 Types of Nanoemulsion

3.1 Microemulsion

Microemulsions are thermodynamically stable systems that develop spontaneously when combining particular kinds and concentrations of surfactant, oil, and water. Microemulsions generally develop via thermodynamic self-assembly; however, nanoemulsions require external shear to separate the droplets. Microemulsions have numerous benefits, including thermodynamic stability, the convenience of preparation, transparency, low viscosity, homogeneity, and high solubility [19, 28]. Microemulsion formulations may provide a variety of benefits for pesticide encapsulation and application: (i) Preparation is simple because no specialist equipment is required (ii) Low environmental pollution (iii) Minimal production expenses because organic solvents are not used (iv) Using water as a solvent reduces the possibility of flammability and explosion and (v) Production of tiny oil droplets (10–100 nm in size) with increased permeability and absorption rate [25]. In comparison to nanoemulsion, microemulsion requires a more significant proportion of surfactant.

Pesticide microemulsion, a revolutionary water-based pesticide formulation, provides considerable development potential [49]. Microemulsions are defined by the tiny, homogenous droplets that make up the dispersed phase (less than 100 nm, [153]) which enhance the ability of pesticides to reach their target pests or plants [28]. Although the concentration of the active ingredient in microemulsions is low, the solubilization of the active ingredient and the presence of a large amount of surfactant may elevate the biological activity of the microemulsion. Considerable research on the development and stability of pesticide nanoemulsions have been published in recent years, with an emphasis on the high-energy emulsification method [85] and microemulsion dilution [135] as a method of production, in addition to methyl decanoate [135] and fatty acid methyl ester [84] as solvents.

3.2 Emulsifiable Concentrates

Among the many pesticide formulations available, emulsifiable concentrates (ECs) are particularly popular due to their many useful properties. These include strong biological activity, low environmental impact, simple production processes, and long-term storage stability. Usually, an EC is produced by dissolving a pesticide in a nonpolar solvent, including surfactants. A polar solvent is exploited as a co-solvent if a pesticide is almost insoluble in a typical nonpolar solvent [71]. The following are advantages of emulsifiable concentrates for pesticide applications: (i) minimal apparatus demands, (ii) simple processing demands, (iii) straightforward application, (iv) good storage stability, (v) high amounts of active components, and (vi) high-biological activity [143]. The emulsifiable concentrate formulation is completely biodegradable and somewhat harmless for animals and the environment [140].

4 Developments of Nanoemulsions-Based Pesticide

Several researchers have focused on developing more efficient pesticide formulations using nanoemulsions to avoid synthetic pesticides' shortcomings. The smaller drop sizes of nanoemulsions over traditional emulsions provide the system with enhanced wettability, spreadability, and biocompatibility. Nanoemulsion-based formulations may enhance pesticide characteristics and behaviors such as stability, solubility, mobility, dispersion, and targeted delivery [150]. Consequently, it has the potential to substantially enhance the feasibility, safeness, and economic effects of conventional pesticides by continuing to increase efficacy, extending effect duration, lowering dose demanded, facilitating the managed release of the active ingredients, and enhancing payload consistency in the environment, thereby minimizing run-off and environmental residuals.

4.1 Preparation Method of Nanoemulsions-Based Pesticides

Nanoemulsions are metastable systems, and how they are prepared influences the degree to which they are stable [135, 136]. Nanoemulsions lack thermodynamic stability but are kinetically stable. In addition, emulsions are understood to be nanoemulsions if the droplet sizes are measured in the nanometer range. The several nanoemulsion preparation processes include the following two methods: (1) low-energy emulsification and (2) high-energy emulsification. The high-energy emulsification process consists of high-energy stirring, high-pressure homogenization, ultrasonic emulsification, membrane emulsification, and microfluidization.

Low- and high-energy emulsification can be used to develop nanoemulsions. High-shear mixing, high-pressure homogenizers, high-energy emulsification techniques utilizing ultrasonic generators, and low-energy emulsification techniques utilizing the system's stored energy are all well known. The hydrophobic insect repellent N, N-diethyl phenylacetamide (DEPA) was emulsified using the low-energy phase inversion temperature (PIT) method [16]. The technique is based on nonionic polyethoxylated emulsifiers' capacity to alter their hydrophilic-lipophilic balance (HLB) values in response to temperature. The HLB decreases with increasing temperature. The ethoxylate unit's dehydration is the cause of this decline. Therefore, changing the temperature while keeping the composition constant can change the relative affinity of the surfactant for the different phases [13]. The polyethoxylated surfactant hydrates at low temperatures, increasing its affinity for the watery medium.

As a consequence of this, it fosters the development of an O/W macroemulsion that is stable. Dehydration of the polyoxyethylene chain makes the surfactant more lipophilic, which reduces the stability of the O/W emulsion as the temperature rises. This is because of the emulsion's interaction with the polymer. Surfactant stabilizes W/O emulsions at extremely high temperatures. The nonionic surfactant possesses exceptionally powerful affinities for both immiscible phases at either the PIT or the HLB temperature. The instability of both oil-in-water and water-in-oil macroemulsions in PIT results in a bi-continuous microemulsion structure or layered liquid crystal structure composed of oil, water, and surfactant [13]. As a result, nanoemulsions can be created, supporting the formation of only a few nanometers across emulsions. In order to make nanoemulsions that are kinetically stable through a transitional inversion, the emulsions must first be prepared in a region close to the PIT. Then they must be rapidly cooled or heated.

A W/O emulsion can change into an O/W emulsion due to the rapid cooling after an emulsion is formed in the PIT zone. During a process known as reversed phase inversion, an O/W emulsion can be transformed into a W/O emulsion by rapidly raising the temperature. Fast temperature increases, or decreases of about 25–30 °C are required to create kinetically stable W/O or O/W nanoemulsions, respectively. If this is not done, the resulting emulsions will be lumpy and difficult to work with [8]. The emulsion inversion point (EIP) method is another low-energy technique for creating nanoemulsions [117]. In this instance, unlike the PIT method, a change in composition rather than a change in temperature initiates the phase reversal. Adding water to an oil-in-water emulsion creates a W/O emulsion, which is stabilized by a hydrophilic surfactant with a greater affinity for the water phase than the oil phase. This method's primary goal is to produce a continuous oil phase. Emulsions stabilized with a surfactant with a higher affinity for the continuous phase are called "normal emulsions." When water is first added to an emulsion, it only causes more water droplets to form in the oil-continuous phase of the emulsion. It is possible to permanently transform a W/O emulsion into an O/W emulsion by adding more water than is required to slow the rate at which the oil droplets coalesce [93]. This phase inversion process is called catastrophic phase inversion to differentiate it from the transitional phase reversal between two regular emulsions (CPI). The EIP procedure aims to produce pyrethrin-containing O/W nanoemulsions that are kinetically stable. They only detail the creation of micro- and nanoemulsions as encapsulation media using materials that are risk-free and naturally toxic [66]. When conducting an emulsification process, it is possible to reduce the amount of damage done to the environment by employing "green" nonionic surfactants and an oil phase derived from natural sources. The oil phase contained a wide range of fatty acid methyl esters (FAMEs), including hexanoate, octanoate, decanoate, and laureate (methyl hexanoate/octanoate/decanoate/laurate = 3.2/50.8/44.0/2.0 (w/w)).

Nonionic surfactants such as 3-(3-hydroxypropyl)-heptamethyltrisiloxane (organosilicon), long-chain alkyl polyglycosides (LAPG), and short-chain alkyl polyglucosides (SAPG) are all eco-friendly options. Pseudo-ternary systems involving FAMEs, water, and surfactants containing 41% (w/w) glyphosate isopropylamine salt were analyzed using phase diagrams to determine which systems had the potential to form microemulsions. These phase diagrams were applied in various ratios and for each surfactant. Microemulsion-based formulations were created using the study's findings. Droplets smaller than 200 nm in size were achieved in O/W nanoemulsion glyphosate formulations by diluting the latter with much water (1:200) and mixing at a slow speed (200 rpm for 5 min). By contrasting the prepared glyphosate-loaded nanoemulsions with Roundup, a commercially available glyphosate formulation, their effectiveness in suppressing the narrow-leaf herb *Eleusine indica* was estimated. The optimal results were obtained with a nano vehicle consisting of 3.27% FAMEs oil, 74.58% water, and 22.15% 8:2 LAPG: organosilicon surfactant matter. A phytotoxicity study was conducted on weed 7–8 days old. Herbicide was sprayed on the plants. Nine distinct glyphosate concentrations were used. The probit transformation was used to create sigmoidal dose–response curves, linearize them, and determine the effective doses (ED50) for the various treatments. Commercial product doses of 0.25, 0.50, 0.75, and 1.00 kg herbicide/ha were primarily used to determine average weed control rates corresponding to the nanoformulation. The ED50 for the nanoemulsion formulation was less than that of Roundup (0.48 kg herbicide/ha) at 0.40 kg. Nanoemulsion formulations also had lower surface tensions, allowing pesticides to penetrate leaves more efficiently and increased wettability compared to commercial products. Oil-in-water (O/W) nanoemulsions of neem oil-based nano pesticides stabilized with the hydrophilic surfactant Tween 20 were developed and characterized. In experiments, droplets as small as 31.03 nm were produced in nanoemulsions with an oil-to-surfactant ratio of 1:3.The LC50 was calculated to be 11.75 mg/L at a droplet size of 31.03 nm, indicating a decreasing trend as droplet size was decreased [10].

4.2 Chemical Composition of Pesticide Nanoemulsions

A review of the creation and formulation of pesticide delivery systems based on nanoemulsions is presented. Nanoemulsions of oil in water are examples of colloidal dispersions characterized by their thermodynamic instability. They typically consist of oil droplets emulsified and suspended in water. They are similar to oil-in-water emulsions in this respect, which are also common. Droplets larger than 200 nm are considered part of a conventional emulsion, while droplets smaller than 200 nm are considered part of a nanoemulsion [94]. Nanoemulsions have some benefits over traditional emulsions because of the tiny size of the droplets in them. These benefits include improved droplet stability, enhanced optical clarity, and an apparent rise in bioactivity. Droplet aggregation and gravitational separation are the other benefits [94]. Thermodynamic stability, not particle size characteristics, distinguishes nanoemulsions from microemulsions. Unlike their microemulsion counterparts, nanoemulsions are notoriously unstable from a thermodynamic standpoint. The kinetic stability of nanoemulsions, however, can be improved over time.

If the oil phase is liquid at room temperature, it might only contain pesticide. However, solid pesticides at room temperature or with high water solubility may need oil phase components added. A suitable solvent can be used to dissolve crystalline pesticides prior to homogenization. The ideal solvent should have a low water solubility, high pesticide solubility, wide availability, good safety characteristics, a reasonable price tag, and low toxicity, among other characteristics. It should also be environmentally friendly and widely accessible. Toluene and xylene are two examples of traditional aromatic solvents [85]. However, more eco-friendly options exist, such as methyl oleate [42]. Other examples of these solvents include methyl decanoate and methyl oleate. In theory, other environmentally friendly solvents like food-grade triglyceride oils could also be utilized. Inhibitors, which are insoluble in water, can be added to liquid pesticides to prevent droplet maturation and the spread of the pesticide molecules [22].

Oil-in-water emulsions are made using various kinds of natural and synthetic emulsifiers. All necessary conditions must be met, including low toxicity and environmental compatibility. These conditions include compatibility with pesticides, solvents, and other ingredients and stability against environmental conditions like temperature changes. Any emulsifier should be able to produce excellent emulsifying performance, which entails tiny oil droplets at low dosage levels. In pesticide nanoemulsions, the primary function of emulsifiers is to rapidly adsorb the oil produced during homogenization onto the droplet surfaces, lowering the interfacial tension and making it easier for further droplet fragmentation to take place [92]. Due to the small droplet size, nanoemulsions form an interfacial layer with a highly curved surface. Since the pressure inside the droplet is greater than the pressure outside, there is a pressure difference between the inner and outer sides of the interface. This differential pressure is called the Laplace pressure (P) [129]. Droplet stabilization is yet another function of emulsifiers in pesticide nanoemulsions. For an emulsifier to be effective during homogenization, it must first form a protective coating around the oil droplets [64]. Emulsifying molecules can accomplish this by creating a significant electrostatic or steric attraction between the oil droplets in the mixture.

Additionally, the colliding droplets can create an impermeable interfacial film that is challenging to separate. For this reason, the stability of nanoemulsions depends significantly on the mechanical strength of the interfacial film [83]. Typically, to prevent oil droplets from coalescing, emulsifier molecules should almost completely encapsulate the oil droplets. Nanoemulsion stability can be improved using small molecular alcohols or mixed emulsifiers to form a composite interfacial film [148].

The most widely used co-emulsifiers are alcohols with a hydrocarbon chain, a small polar head group, and a shorter carbon chain. Using a wide range of physicochemical mechanisms, these molecules can alter the interfacial layer's optimal curvature, reduce interfacial tension, increase interfacial layer flexibility, and enhance emulsion formation and stability.

Understanding the scientific principles governing nanoemulsion behavior as a function of composition is aided by a discussion of the various nanoemulsion components. For example, an aqueous phase and an oil phase form the basis of a typical nanoemulsion, which is then stabilized by an emulsifier or a surfactant. These phases are separated from one another by an emulsifier. The formation of the nanoemulsion, as well as its stability and the functional characteristics it possesses, are all influenced by the properties of the oleic component. These properties include density, viscosity, the value of the interfacial tension, and the refractive index. Typically, water acts as the polar phase of nanoemulsions, dictating their rheological behavior, interfacial tension, polarity, and ionic strength. The coalescence of oil droplets, flocculation, the gradual increase in average droplet size (Ostwald maturation), and gravity separation and sedimentation are the causes of the nanoemulsion's breakdown into oil and water [77, 78]. As a result of electrostatic repulsion or steric hindrance, a surfactant that adsorbs oil droplets can prevent oil droplets from coalescing and colliding. By decreasing the IFT between the oil and water phases, oil droplets are stabilized against coalescence and collision [74, 75].

Co-emulsifiers can also be used interchangeably with emulsifiers. Coemulsifiers are molecules of surfactant that can adhere to the surfaces of droplets; however, when used on their own, coemulsifiers cannot stabilize emulsions. In some cases, the formation of an emulsion or its stability can be enhanced by adding a co-emulsifier.

4.3 Optimization and Characterization of Nanoemulsion-Based Pesticides

Dynamic light scattering (DLS), small angle neutron scattering (SANS), viscosity, and atomic force microscopy are some of the analytical methods used to characterize nanoemulsions: nanoparticles, extreme transparency, tunable viscosity, robust stability, and other desirable traits (AFM). Using dynamic light scattering (DLS), the size of the droplets in a nanoemulsion can be determined. The physicochemical stability of the nanoemulsion is assessed by measuring the droplet size as a function of time with DLS. The reports state that the DLS method determines the Ostwald ripening rate [142]. The SANS technique is helpful for determining the nanoemulsions' bulk structure because neutron wavelengths can investigate nanoscale structures [50]. The results from techniques like measuring the conductivity, dielectric properties, and nanoemulsion viscosity are appropriate at the macroscopic level. When the water content is raised, the nanoemulsion appears to have a lower viscosity. When the surfactant and co-surfactant content is reduced, it appears to have a higher viscosity. The viscosity variation aids in assessing the stability of nanoemulsion systems. Conductivity tests monitor the phase inversion phenomenon and establish whether the newly formed nanoemulsions are y/w or w/o in composition [27]. The dielectric properties are measured to assess nanoemulsions' structural and dynamic properties. Atomic force microscopy can determine the nanoemulsion system's droplet shape. Nanoemulsion systems are evaluated for their stability through battery stability tests, including centrifugation, freeze–thaw cycles, heating–cooling tests, and tests conducted at room temperature. The duration of stability that nanoemulsion systems can maintain without exhibiting Ostwald maturation, coalescence, creaming, or precipitation is examined in these tests. Consequently, the prepared system's shelf life is predicted [29].

Preformulation and parametric optimization are two crucial steps in creating agrochemicals based on nanoemulsions. To choose the optimization process's starting point, these parameters are crucial. After they pass the initial screening, the samples are chosen for chemical and biological characterizations. Establishing the critical micelle concentration (CMC) or the surfactants' hydrophobicity-limiting concentration (HLB) value is crucial in preformulation. The HLB value indicates the type of nanoemulsions produced, whether O/W, W/O, or some other type. When self-assembly begins, and the addition of surfactant no longer reduces the surface tension, the critical micelle concentration (CMC) value is reached. Fluorescent probe [26], surface plasmon resonance, dye solubility, tensiometry, viscosimetry, calorimetry, and electrical conductivity are just a few of the methods that have been used to determine the CMC value. According to previous studies, the pyrene fluorescent probe is one of the easiest and most accurate ways to determine CMC value. The peak difference method measures the absorbance of dissolved fleas in the systems. According to Lu et al. [88]'s graph of concentration versus absorbance, the CMC value is the intersection of the two line segments [45]. The choice of probe molecules impacts CMC measurement because of their charge and functional groups [62]. Several nanoemulsions with various parameters are created during the optimization process and put through the initial sieving process, which includes a centrifugation experiment, freeze–thaw cycle, heating cooling test, and 25 °C room temperature nanoemulsion stability test. The most stable nanoemulsion, which does not experience phase separation for a predetermined time, will be chosen for further characterization. In determining the thermodynamic stability studies, this step is crucial.

Characterizations such as dynamic light scattering (DLS), viscosity, zeta, pH measurement, morphology, biological properties, retention, and contact angle measurement are essential for nanoemulsion-based agrochemical formulations. The most popular method for figuring out the size and distribution of the nanoemulsion is DLS. To achieve maximum effectiveness, the nanoemulsion's small size is preferred. There were a few factors thought to influence nanoemulsion size. Electrostatic interaction-induced aggregation can lead to unwanted multiple scattering effects, so samples are diluted with deionized water before analysis. The polydispersity index (PDI), which gauges system stability, is another area of interest during measurement. Picking a PDI value of less than 0.5 is considered to have good particle diameter uniformity, making it suitable for agricultural use. This is because this value indicates that the particles have a similar size. To characterize samples using DLS measurement, samples with higher PDI will be discarded due to their polydisperse properties [37, 34].

Measurement of viscosity, zeta, and pH; Using a Zetasizer apparatus, the electrophoretic characteristics of the nanoemulsion, also referred to as the zeta potential, are analyzed. The particle's pH and the surrounding surface properties establish its zeta, a measure of the nanoemulsion's stability. As the repulsive forces created by a negative zeta value are greater than the attractive forces between the droplets, coagulation, and coalescence are prevented in a dispersed emulsion. Because the attractive forces are weaker than the repulsive ones, this is the case. Increasing the amount of oil in a system may impact the nanoemulsion's ability to maintain its stable structure [58]. Most pesticide nanoemulsions exhibit alkaline properties and have a pH ranging from 5 to 6 [15]. The nanoemulsion's viscosity was determined with an Ostwald viscometer. Several variables, such as the surfactants employed, the organic phase components, and the oil, can impact the measured viscosity. Since pesticide nanoemulsion is highly water-loaded and, therefore, of the O/W type, its viscosity is relatively low. However, the concentration of the surfactant can alter the nanoemulsion's viscosity [7].

Nanoemulsion morphology and structure are studied using several types of microscopes like atomic force microscopes (AFMs), transmission electron microscopes (TEMs), and cryogenic field emission scanning electron microscopes (Cryo-FESEM). Clusters of nano micelles formed during preparation led to the formation of spherical [14] or core–shell-like structures [131] shapes for pesticide nanoemulsions have frequently been reported. When the sample is stored, the study is carried out by either monitoring the appearance of the sample or performing periodic measurements of its physicochemical properties. These measurements may include zeta potential or particle size. A stable system is one in which the sample does not undergo visual changes like phase separation, creaming, flocculation, fusion, or precipitation. Before and after storage, the nanoemulsion's zeta potential and particle size are measured and compared [52].

Measurements of the attachment and contact angles of the leaves are taken to determine how much liquid pesticide should be applied to each leaf surface. The pesticide's effectiveness can be improved by improving the nanoemulsion's adherence to the leaves. Because the active ingredient has low interfacial tension, it allows for effective pesticide dilution on the plant surface. The fact that the nanoemulsion's contact angle drops as its agrochemical content rises is evidence of this [119]. The effectiveness of pesticide nanoemulsions in combating pathogenic organisms can be investigated through biological studies. Researchers have focused on these plant-damaging organisms (bacteria, fungi, and insects) to eradicate or at least bring them under better control [118]. A nanoemulsion based on peppermint oil can be used as a pesticide because it reduces the biomass of the fungus *Alternaria solani*, which is responsible for the early blight that affects tomato plants [108]. The garlic oil nanoemulsion had a minimum inhibitory concentration (MIC) value of 0.23%, which was lower than the value for pure garlic oil. This meant that it was more effective at destroying the proteins in *Penicillin italicum.*

Insect morphological and histological damage following nanoemulsion treatment has been documented in other studies. They tested the nanoemulsions' ability to irritate and seriously harm insect pests by penetrating their cuticle. *According to SEM, tribolium castaneum* exhibits necrosis and blackening due to *Pimpinella anisum* essential oil nanoemulsions [107]. Since they can significantly impact plant seedlings when used as a growth promoter or seed treatment, nanoemulsions are also regarded as growth promoters. It has been discovered that a nanoemulsion made of saponin and thymol essential oil can inhibit bacterial growth while fostering the growth of soybean plants [79]. The cottonseed plant's germination process and resistance to fusarium wilt disease were improved by eugol oil nanoemulsion [2]. As a seed stimulant, methylcellulose nanoemulsion caused an 18 and 33% increase in the length of the shoots and roots of maize seedlings [131]. Nanoemulsions of turmeric improved germination and growth when applied to watermelon seeds [6]. Pectin and neem oil nanoemulsion both show signs of stimulating growth in soybean seeds [125]. Many simulation studies that guide nanoemulsion EOR simulation studies using various simulation tools can be found throughout the emulsion EOR literature. Research and experiments conducted in the last few years are summarized in Table 2.

4.4 Accumulation, Dispersion, and Evaporation of Pesticide Nanoemulsions

Surfactant molecules are concentrated at the interface between the pesticide and the water in a nanoemulsion of a pesticide. This results in the pesticide being dispersed in the water as nanoscale droplets. Depending on the amount and kind of surfactant used, nanoemulsions can be put into either kinetically stable or thermodynamically stable. Nanoemulsions with high thermodynamic stability can be made when the concentration of a strong repellent surfactant in the aqueous phase is greater than the critical micelle concentration, and a non-polar pesticide is at least partially soluble in water. These nanoemulsions can be used to combat a variety of pests (CMC). Under these conditions, micelles are globular clusters formed when the nonpolar parts of the surfactant molecules arrange themselves inward and coalesce with one

Table 2 Recent research on nanoemulsions as an effective pesticide

Emulsifiers	Droplet size/types of nano emulsions	Oil phase	Processing method	Function and applications	Key referecnes
Tween 40	10–15 nm O/W	n-heptaneoil (a 7 carbon alkane)	ultrasonication technique	Used CMG-STARSTM to create a compositional reservoir model that accurately matched the results of CMOST's laboratory flooding tests on the surfactant/ NPs nanoemulsions	[77]
Tween 40	O/W	n-heptane oil	ultrasonication technique	The increase in nanoemulsion stability, improvement in wettability modification, and the wedging action were attributed to an improvement in oil RF of 8.54% by nanoemulsion + SiO_2 NPs	[74]
Tween 80	<40 nm O/W	Toluene	Ultrasonic emulsification	Reduced the pesticide load on the environment and human populations by ensuring that deltamethrin can still effectively fight the mosquito population at lower concentrations	[17]

(continued)

Table 2 (continued)

Emulsifiers	Droplet size/types of nano emulsions	Oil phase	Processing method	Function and applications	Key referecnes
Calcium alkyl aromatic sulfonate, polyoxyethylene aliphatic¼2:3	31–300 nm O/W	Solvent naphtha, xylene¼4:1	Low-energy emulsification	With the contact angle being smaller and the adhesion being higher, the plant leaves would be more easily absorbed into the liquid medicine, increasing the medicine's efficacy	[151]
Polyoxyethylene (40) castor oil ether (EL-40) and polyoxyethylene (20) castor oil ether (EL-20)	<200 nm O/W	Methyl oleate	Low-energy emulsification	The low-emulsifying process for highly stable pesticide nanoemulsions was developed, and methyl oleate was successfully used as an environmentally friendly alternative solvent	[42]
Polyoxyethylene ether	O/W	Cyclohexane	High energy (Ultrasonication)	By carrying out parallel core flood tests, the study sought to determine whether nanoemulsions could be used to enhance flow back rate and liquid capability of gas injection	[138]

another. The polar parts of the molecules extend outward into the aqueous solution. Micelles can be distinguished by their spherical shape. This aqueous solution's free energy decreases when a non-polar pesticide is added, resulting in the molecules of the pesticide being directed to diffuse into the micelles. Micelles swell and create tiny droplets in this situation. The surfactant and pesticide content primarily influences these droplets' size in each droplet and chemical composition [51]. This type of nanoemulsion is believed to be thermodynamically stable because the state of swollen micelles possesses the lowest amount of free energy. As a result, as long as the initial conditions are not altered, these nanoemulsions are stable indefinitely. This also means that emulsions form naturally when the surfactant, pesticide, and water components combine. Kinetic energy barriers or slow mass transport processes may require mechanical agitation or heating in practice [51]. These nanoemulsions are highly beneficial for pesticide formulations due to their high stability and straightforward preparation.

Consequently, there is currently a great deal of different commercial formulations available [132]. However, there are several drawbacks to this kind of nanoemulsion as well. For instance, they are expensive due to the sizeable amount of surfactant needed (typically 20% by weight of the formulation), and only small amounts of pesticide are added to the micelles [81]. A high surfactant concentration can expose plants to unfavorable phytotoxicity, which is undesirable [67]. In addition, because of their propensity to form micelles, the only appropriate surfactant systems are those with a high level of repulsion. Formulation development is a time-consuming process because the ideal combination of surfactants and co-surfactants varies depending on the pesticide that needs to be emulsified. This causes the ideal combination to change from formulation to formulation [94].

Unlike thermodynamically stable nanoemulsions, kinetically stable nanoemulsions form in environments where surfactant molecule aggregation into micelles is energetically unfavorable (such as with lower concentrations or weakly repellent surfactants). Kinetically stable nanoemulsions are almost entirely insoluble in pesticides. The water, pesticide, and surfactant combine to form a biphasic system at the outset, with the surfactant molecules preferentially accumulating at the pesticide-water interface. The system is subjected to shear (such as mixing), which blocks mixing due to interfacial repulsion. As a result, the pesticide phase becomes fragmented into droplets, and the surfactant coats each droplet. Larger pesticide droplets are stretched as this emulsion degrades further, becoming capillary unstable and fragmenting into smaller droplets. This procedure occurs until nanoscale pesticide droplets form [128]. Due to the nanoemulsion's higher free energy than the stirred state, these nanoemulsions are thermodynamically unstable.

Consequently, the Ostwald ripening process plays a significant part in separating the droplets from the aqueous phase. Surfactant on the droplet surfaces, however, creates an energy barrier that prevents mixing and slows down these processes. As a result of the fact that the pesticide droplets in the nanoemulsion in this scenario will continue to be dispersed for a sizeable amount of time, it has been determined that they are kinetically stable. This stability can last several years with considerable energy barriers (greater than 20 kT), which is sufficient for most practical applications [53]. Due to eliminating the need for micelles, kinetically stable nanoemulsions can be formulated with significantly less surfactant (5–10% by weight of the formulation) than thermodynamically stable nanoemulsions. This is because kinetically stable nanoemulsions do not form micelles. In addition, a broader selection of surfactant systems is available for use [94].

Many processes, such as ultrasonication, high-pressure homogenization, phase transition, and phase inversion temperature, fall into high-energy mixing methods. The coarse emulsion is mixed using the ultrasonication method, which involves using high power and ultrasonic frequencies to stir a vibrating solid surface. Due to cavitations, which are caused when tiny liquid-free bubbles form and burst, the emulsion droplets are broken up by powerful shock waves and fast-moving liquid jets. Because the sound field that is emitted is typically not homogeneous, it is frequently necessary to recirculate the emulsion. This aims to give each droplet the maximum shear rate possible [91]. Even though this method is typically used in laboratories, it is challenging to scale it up to create pesticide nanoemulsions in commercially viable quantities [135, 136]. By utilizing a high-pressure positive displacement pump capable of operating at up to 2000 atmospheres, the emulsion of water, surfactant, and pesticide that has been previously prepared is pushed through rigid microchannels with a width of approximately 100 μm. The emulsion droplets produce very small droplets in the submicron range due to the excessive collisions, hydraulic shear, cavitations, and intense turbulence resulting from this. A homogeneous nanoemulsion is then produced by filtering the bulk emulsion to remove large droplets. It is recommended that the emulsion be prepared using high-volume pesticide fractions and then diluted to boost the emulsification process's efficacy. Despite the possibility of coupling, more surfactant is added to stop it [87]. This technique is used in the lab and industry to create nanoemulsions [91]. Research in recent years has aimed to develop several low-energy emulsification techniques for dispersing non-polar liquids as nano-sized droplets to avoid the drawbacks associated with mixing with excessive shear. The primary goal of these strategies is to modify the hydrophilic-lipophilic balance (HLB) or other conditions that affect the solubility of surfactants, such as temperature and composition. To that end, these methods employ various means to promote the formation of droplets on the nanoscale. Compared to high-energy methods, these tend to have lower costs,however, their development is significantly more difficult [11].

Using the temperature-dependent solubility of nonionic surfactants, phase inversion temperature is a low-energy method for reducing the size of pesticide droplets in an emulsion. The emulsion's phase is altered (especially that of polyethoxylated surfactants) to achieve this. This procedure begins with creating a coarse O/W emulsion with a nonionic surfactant. This step is followed by Stage 2. Heating the emulsion increases the solubility of the nonionic surfactant in the pesticide phase and decreases solubility in the aqueous phase (reducing HLB). The pesticide droplets' size is also diminished simultaneously (to create more surface area to accommodate the surfactant). At the phase inversion temperature, an O/W emulsion changes into a W/O emulsion, and the surfactant preferentially becomes soluble in the pesticide phase (also abbreviated as PIT). If the temperature rises, the water droplets' size will continue to grow. The smallest droplet size is reached at temperatures just below the PIT. These minuscule droplets, though, are unstable and coalesce very quickly. Stabilizing the resulting nanoemulsion by rapidly cooling (to temperatures at least 20 degrees below PIT) increases fluid viscosity and decreases the rate at which pesticides diffuse through water. This slows the rate of aggregation and Ostwald ripening [61].

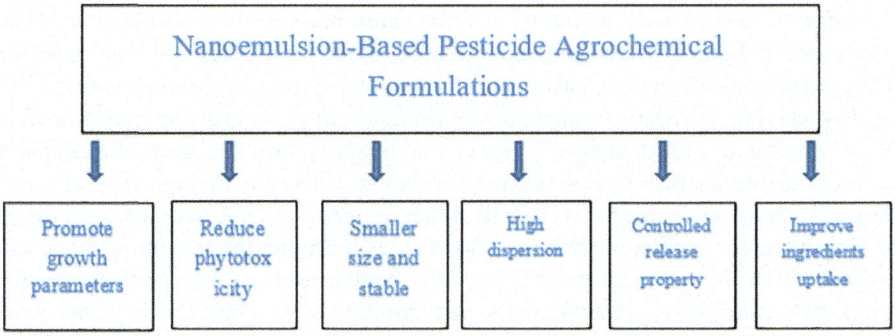

Fig. 3 Diagram of the production process for pesticide formulations based on nanoemulsions that have enhanced their physicochemical and biologic activities [102]

The phase transition method alters the environment to encourage the formation of pesticide droplets just below the temperature of PIT. A surfactant and pesticide are combined first. Then, water is added slowly to dilute the surfactant concentration and increase the ratio of pesticide to water, resulting in a practical solution [44].

The PIT for the emulsion shifts as a result of the gradual shift in composition that occurs while the temperature remains the same in this scenario.

To use the emulsification method (a water-soluble chemical and a pesticide), a mixture of pesticide, surfactant, and co-solvent must be prepared. After contact with water, an O/W nanoemulsion will automatically form independently. While intense fundamental research is still being done on the specific mechanisms these nanoemulsions produce, the "diffusion and envelopment" mechanism is currently the most widely accepted for self-emulsification [96]. There is mixing when this mechanism is set in motion, and the cosolvent diffuses into the aqueous phase, where it takes up some of the pesticides (forming a three-component phase near the interface). This is because the co-solvent continues to diffuse and dilute into the aqueous phase, rendering it incapable of retaining the pesticide in its dissolved state within the tricomponent phase. Thus, this pesticide causes the surfactant to supersaturate and separate from the water as nano-sized droplets, which are then quickly stabilized by the surfactant at the interface [80]. As shown in Fig. 3, nanoformulations' superior physicochemical and biological activities cause the properties mentioned above [102].

4.5 Advantageous of Nanoemulsion-Based Pesticide Formulations

The following benefits have increased interest in the use of nanoemulsion-based pesticides in agricultural farming:

1. Protects against hydrolysis and oxidation in the oil phase of O/W, and the nanoemulsion is impervious to water and air.
2. A high surfactant concentration is required to keep the microdroplets stable in producing a nanoemulsion.
3. Enhance the absorption rate in plant foliage.
4. Increases biodegradability.
5. Using nanoemulsions to develop lipophilic active-loaded formulations for pest control has significant promise.
6. The effectiveness of antimicrobial activity against microbial pathogens was changed by putting pesticides into nanoemulsions.
7. The nanoencapsulation method has the potential to enhance physicochemical qualities and stability by allowing water dispersibility, decreasing volatility, and shielding the substance from the environment.
8. The composition and concentration of the components used to generate nanoemulsion systems determine the range of flexibility in particle size and stability that may be produced with these agrochemicals.
9. Most research findings have demonstrated that pesticidal activities perform effectively in nanoemulsion-based formulations.
10. They have shown low toxicity against humans, mammals, and the environment.

5 Biological Properties of Nanoemulsion-Based Pesticides

Using synthetic pesticides is the most favorable method to protect the economically important crop from pests, weeds, plant parasitic nematodes, and fungi. Pesticides include fungicides, insecticides, nematicides, bactericides, and herbicides. Nonetheless, the use of synthetic chemicals is not suitable for the environment and human health. Biopesticides, including plant source-based and biocontrol microorganism-based, are highly desirable and more beneficial for crop protection. Nanomaterials have novel physical, mechanical, and chemical action [73]. Nanoparticles that come in contact with biological substances can receive a protein coating, which is called the corona effect, that results to enhance the interactivity of the nanoparticles with the cells modifying their mechanism of action in biological substances and having the capability to create toxicity [36, 82].

Much research has been put into establishing nano-based non-synthetic pesticides to benefit economic production by raising yield [65].

Nanoemulsions in pesticide formulation prevent plant disease from affecting economically significant crop yield and production. Nanoemulsions act as an agent that carries and transport biologically active metabolites and agrochemicals toward the target pathogen or phyto-pests [43]. The attractive physiochemical property of nanoemulsion is that a tunable nano size has developed a big surface area, enabling the functional mixture's emergence, uptake, and accumulation to be highly effective than their counterparts [43]. To obtain better kinetic stability [56], slight surface tension, and better wettability for improved leafage adhesion [145] and enhance the dissolution and solubility of imperfectly water-soluble agrochemicals the active ingredients must be incorporated into the nanoemulsion. It can also work as a coating sheet for pesticides, providing ample protection from photodecay [105]. It was determined that antimicrobic nanoemulsions are many sturdy oil-in-water emulsions with nanometer-sized droplets (positively charged) consisting of extensive properties against fungi, bacteria, and viruses [103, 135, 136]. To effectively reduce plant diseases, nanoemulsion should be used for biologically active pesticides. The common pesticides nanoemulsion formulations are functional chemicals that can kill fungi (Fungicides), destroy insects like snails and slugs (insecticides), kill nematodes (nematicides), and kill weeds (herbicides) [102].

5.1 Nanoemulsion as Fungicidal Agents

Diaz-Blancas et al. [34] determined the use of the fungicide TBZ (Tebuconazole) as a nanoemulsion formulation for agricultural applications. TBZ consists of average leaching potency into groundwater [34]. It is a long-lasting pesticide with an adsorption constant K_{oc} (for organic carbon) range of about 803 and 1251 ml/g [98]. The systemic action mechanism of TBZ permits it to be consumed by plant roots and leaves to prevent the formation of ergosterol. This kind of impediment averted the synthesis of pathogens on the cell wall and reduced the germination of germ tubes [90]. [2] characterized and formed Eugenol nanoemulsion antifungal activities using Tween 20 (non-ionic surfactant) with water as a continuous aqueous phase on cotton. In vitro, antifungal assay demarcated that Eugenol nanoemulsion inhibits the growth of *Fusarium oxysporum* f. sp. vasinfectum (which causes *Fusarium* wilt on important crops). Eugenol can inactivate important enzymes, which act on the cell membrane or interfere with nucleic acid activity [68]. Eugenol also blocks the ergosterol synthesis [30]. Nanoemulsion of Eugenol impedes radial growth and pigmentation, and sporulation of the pathogen [2]. Its mechanism of action is involved in the distortion of fungal mycelia. It has antimicrobial action against variable food-borne microbes [47], fungi toxigenic fungi [97], and phytopathogens [139]. Lemongrass and clove oil-based non-ionic nanoemulsion suppressed the growth and development of the vascular wilt-causing fungal agent *Fusarium oxysporum* f. sp. *lycopersici* [124]. Da Silva et al. [125] demonstrated that Eugenol nanoemulsion could be a secure formulation for use in farming and not cause a depletion of cell viability, DNA damage, and reactive oxygen species production. Also, they did not present ecotoxicity to *Folsomia candida.*

5.2 Nanoemulsion as Nematicidal Agents

Plant parasitic nematodes damage various economically important crops worldwide. It is very difficult to control nematodes; using insecticides is a good tool against them, but it has several negative impacts on living organisms and the environment. Developing an alternative pest management strategy that is equally effective as a synthetic pesticide and remains safe for the environment is essential. The greatest energy technique is characterized by its speed and efficiency in producing nanoemulsion with nano droplet diameters and small size dispersion [46]. Much research demonstrated that monoterpenes (the primary component of essential oils) are potent antibacterial agents in nanoemulsions [4, 5]. The nematicidal activity of a few nanoemulsions of monoterpenes, namely cinnamaldehyde, citral, geraniol, (R)-carvone, and pulegone, highly reduced egg hatching and increased juvenile mortality of root-knot nematode, *Meloidogyne javanica* [3]. The nematicidal activity of nanoemulsion dramatically impedes root galls, egg masses, and the population of *M. javanica.* The mode of action of monoterpenes and essential oils on nematodes is unknown (Abdelrasoul and Habashy 2021). However, many volatile oils have been found to exhibit carcinogenic activity in *Drosophila melanogaster* by activating octopaminergic receptors and destroying GABA receptors [38, 70]. Monoterpenes nanoemulsion enhances the mechanism of action of defense-related enzymes like polyphenol oxidase and peroxidase in tomato plants inoculated with *M. javanica* (Abdelrasoul and Habashy 2021). Therefore, monoterpenes nanoemulsion provides resistance against phytonematodes. Nanoemulsion obtained from Citronella oil potentially suppressed the development of the nematode population as effectively as a synthetic carbofuran-killed nematode [35].

5.3 Nanoemulsion as Insecticidal Agents

To control pest and insects, the use of pesticides effectively eradicate insects pest but harmful or toxic to the environment. Ecofriendly substances attract much attention as an alternative to pesticides, and plant-derived products can produce defensive action against insecticides. Nanoemulsion obtained from the crude extract of fruits of *Manikara subsericea* effectively controls the cotton pest known as *Dysdercus peruvianus.* Essential oils significantly impact several stored grain insects by contact [20] and reacting by ingestion [40]. Nanoemulsion positively affects essential oils by improving their physical stability, bioactivity and water diffusion. The bioactivity and persistence of volatile oil nanoemulsion from *Pimpinella anisum* provide an innovative strategy for managing *Tribolium castaneum*, an economically important stored grain pest, and for managing many insects attacking stored products [55]. The fruits of *Pimpinella anisum* is a rich source of essential oils (Anisi aetheroleum) and proved to have an anti-convulsant, neuroprotective, anti-ulcer and anti-inflammatory response [60]. The chemical composition shows the presence of phenylpropanoids, with (E)-anethole (1-methoxy-4-(prop-1-enyl) benzene), p-anisaldehyde, chavicol and (E)-pseudoisoeugenyl 2-methylbutyrate [59]. The essential oil of *P. anisum* and (E)-anethole were found to be very effective against larva as well as mature insects [114, 115] and showed fumigant impact on larvae of *Lycoriella ingenua* [111]. They displayed great efficacy on the Filariasis vector *Culex quinquefasciatus* [18], and displayed strong contact noxious impact against both sexes of *Blattella germanica* adults [146]. Aniseed oil-based nanoemulsion of *P. anisum* provides insecticidal bioactivity. [99] reported that the increased toxicity of garlic oil nanoemulsion might be due to the increase in the emulsion droplet's surface area, which leads to increased biological action and become more acricidal than normal emulsion. Nanoemulsion of *Mentha longifolia* oil increases the toxicant effect against *Ephestia kuehniella* and boosts the oil persistence time [86]. *M. longifolia* composed pulegone, p-menthan-3-one-cis, 1,8-cineol, p-menthan-3-one-trans and β-pinene have great insecticidal properties. The ecofriendly nanoemulsion technique can be applied in storage houses to conserve foodstuff from insect damage [86].

5.4 Nanoemulsion as Herbicidal Agents

Herbicides are used to control undesirable plants (weeds). For example, Glyphosate [N-(phosphonomethyl) glycine] is a non-selective foliage-sprayed herbicide to manage monocotyledonous and dicotyledonous weeds species [122]. Glyphosate plays a major role in preventing the biosynthesis of phenylalanine, tryptophan and tyrosine (aromatic amino acids) by ceasing the shikimate pathway, deactivating 5-enolpyruvyl shikimate-3-phosphate synthase to slow down weed growth [130]. To minimize synthetic herbicides, the essential way is to help penetrate functional ingredients into leafage [63]. Nanoemulsion-based herbicides can increase the bioavailability of glyphosate, as the hydrated herbicide does not absorb, translocate and penetrate well into the waxy sheet of leaves [63]. They examined that green nanoemulsion formulation with water-soluble glyphosate effectively controls *Eleusine indica*. Natural herbicide nanoemulsion of *Satureja hortensis* actively eradicated two weed species: *Chenopodium album* and *Amaranthus retroflexus*. *S. hortensis* majorly contains chemical constituents such as γ-terpinene and carvacrol, which have herbicidal activity due to their phytotoxic effect [56] (Table 3).

6 Conclusion

Nanotechnology facilitates the production of active substances with nanoscale dimensions. Nano-pesticides offer various benefits beyond their mass equivalents, including (i) secure and straightforward management and less toxicity to non-target organisms, (ii) Higher pesticide effectiveness as a result of the regulated release and consistent surface distribution, (iii) reduced pollution and costs attributable to lower pesticide and toxic solvent concentrations. Nanopesticide formulations are regarded as an efficient method of pest control. Nanoemulsion is a feasible nanoformulation that can enhance pesticide formulations' consistency, water dispersion, and bioactivity because nanoemulsion formulations may promote efficacy and eliminate synthetic pesticides. In the past few years, nanoemulsions have gained widespread

Table 3 Biocidal activity of various phytonanoemulsions-based pesticides

Nanoemulsion source	Biocidal activity	Pathogen	Key references
Allium sativum	Acaricidal	Eriophydae (eriophyid olive mites)	[99]
Azadirachta indica	Insecticidal (larvicidal)	*Culex quinquefasciatus*	[10]
A. indica, A. sativum, Lippia sidoides	Insecticidal	*Rhopalosiphum padi* and *Myzus persicae*	[112]
Ba-har essential oil	Fungicidal	*Aspergillus niger, Colletotrichum gloeosporiodes, Penicilium chrysogenum*	[41]
Bio-silica and Citronella	Nematicidal	(*Meloidogyne* spp.)	[35]
Carum copitum	Insecticidal	*Sitophilus granaries* and *Tribolium confusum*	[155]
Chitosan coated nanoemulsion	Nematicidal	*Bursaphelenchus xylophilus*	[144]
Cleome viscosa	Fungicidal	*Candida albicans*	[72]
Cocos nucifera	Herbicidal	*Diodia ocimifolia*	[147]
Cymbopogon	Insecticidal	*Aedes aegypti*	[106]
Eucalyptus globulus	Insecticidal	*Tribolium castaneum*	[110]
Eugenol	Fungicidal	*Aspergillus niger, Penicillium* sp., *Rhizopus* sp.	[2]
Eugenol and Mancozeb	Fungicidal	*Glomerella cingulata*	[125]
Jatropha curcas	Insecticidal	*Periplaneta americana*	[109]
Mentha longifolia	Insecticidal	*Ephestia kuehniella*	[86]
Nanoemulsion (X8W60PC)	Fungicidal	*C. albicans, C. tropicalis, Microsporum gypseum, Trichophyton mentagrophytes, T. rubrum, Fusarium oxysporum, Aspergillus fumigatus*	[103]
Pimpinella anisum	Insecticidal	*Tribollium castaneum* and *Culex pipiens*	[69, 104]
Satureja hortensis	Herbicidal	*Amaranthus retroflexus* and *Chenopodium album*	[56]
Thymus and *Mentha spicata*	Nematicidal and Fungicidal	*Meloidogyne javanica*	[54]

popularity as a water-insoluble pesticide delivery technology due to their excellent efficiency, eco-friendliness, and safety.

References

1. Abbassy MA, Marei AESM, Al-Ashkar MAM, Mossa ATH (2014) Adverse biochemical effects of various pesticides on sprayers of cotton fields in El-Behira Governorate, Egypt. Biomed Aging Pathol 4(3):251–256
2. Abd-Elsalam KA, Khokhlov AR (2015) Eugenol oil nanoemulsion: antifungal activity against *Fusarium oxysporum* f. sp. *vasinfectum* and phytotoxicity on cottonseeds. App Nanosci 5(2):255–265
3. Abdel Rasoul MA, El-Habashy DE (2021) Nematicidal activity of some nanoemulsions of monoterpenes on Tomato root-knot nematodes (*Meloidogyne javanica*). J Plant Prot Path 12(9):655–661
4. Abdelrasoul MA, Amer GM (2018) Comparative fungicidal activity of some pure and nanoemulsion monoterpenes effects on soil borne plant pathogenic fungi. J Plant Prot Path 9(10):707–713
5. Abdelrasoul MA, Eid AR, Badawy ME (2020) Preparation, characterizations and antibacterial activity of different nanoemulsions incorporating monoterpenes: in vitro and in vivo studies. Arch Phytopath Plant Prot 53(7–8):310–334
6. Acharya P, Jayaprakasha GK, Crosby KM, Jifon JL, Patil BS (2020) Nanoparticle-mediated seed priming improves germination, growth, yield, and quality of watermelons (*Citrullus lanatus*) at multi-locations in Texas. Sci Rep 10:1–17
7. Al-Assiuty BA, Gomah A, Ageba ME (2019) Chemical profile, characterization and acaricidal activity of essential oils of three plant species and their nanoemulsions against *Tyrophagus putrescentiae*, a stored-food mite Basma. Exp Appl Acarol 79:359–376
8. Allouche J, Tyrode E, Sadtler V, Choplin L, Salager J-L (2004) Simultaneous conductivity and viscosity measurements as a technique to track emulsion inversion by the phase-inversion-temperature method. Langmuir 20:2134–2140
9. Alonso ML, Laza JM, Alonso RM, Jiménez RM, Vilas JL, Fañanás R (2014) Pesticides microencapsulation. A safe and sustainable industrial process. J Chem Tech Biotechnol 89(7), 1077–1085
10. Anjali CH, Sharma Y, Mukherjee A, Chandrasekaran N (2012) Neem oil (Azadirachta indica) nanoemulsion—a potent larvicidal agent against *Culex quinquefasciatus*. Pest Manag Sci 68(2):158–163
11. Anton N, Benoit JP, Saulnier P (2008) Design and production of nanoparticles formulated from nano-emulsion templates: a review. J Control Release 128:185–199
12. Anton N, Vandamme TF (2011) Nano-emulsions and micro-emulsions: clarifications of the critical differences. Pharm Res 28(5):978–985
13. Anton N, Gayet P, Benoit J-P, Saulnier P (2007) Nano-emulsions and nanocapsules by the PIT method: an investigation on the role of the temperature cycling on the emulsion phase inversion. Int J Pharm 344:44–52 (2 times)
14. Badawy MEI, Abdelgaleil SAM, Mahmoud NF, Marei AESM (2018) Preparation and characterizations of essential oil and monoterpene nanoemulsions and acaricidal activity against two-spotted spider mite (*Tetranychus urticae* Koch). Int J Acarol 44:330–340
15. Badawy MEI, Saad A-FSA, Tayeb E-SHM, Mohammed SA, Abd-Elnabi AD (2019) Development and characterization of nanoemulsions of some insecticides by high energy technique for targeting delivery. J Agric Res 57:15–23
16. Balaji APB, Mishra P, Suresh Kumar RS, Mukherjee A, Chandrasekaran N (2015) Nanoformulation of poly (ethylene glycol) polymerized organic insect repellent by PIT emulsification method and its application for Japanese encephalitis vector control. Colloid Surf B 128:370–378

17. Balaji APB, Sastry TP, Manigandan S, Mukherjee A, Chandrasekaran N (2017) Environmental benignity of a pesticide in soft colloidal hydrodispersive nanometric form with improved toxic precision towards the target organisms than non-target organisms. Sci Total Environ 579:190–201
18. Benelli G, Pavela R, Iannarelli R, Petrelli R, Cappellacci L, Cianfaglione K, Maggi F (2017) Synergized mixtures of Apiaceae essential oils and related plant-borne compounds: larvicidal effectiveness on the filariasis vector *Culex quinquefasciatus* Say. Ind Crops Prod 96:186–195
19. Bera A, Mandal A, Kumar T (2014) Physicochemical characterization of anionic and cationic microemulsions: water solubilization, particle size distribution, surface tension, and structural parameters. J Chem Eng Data 59(8):2490–2498
20. Bossou AD, Ahoussi E, Ruysbergh E, Adams A, Smagghe G, De Kimpe N, Mangelinckx S (2015) Characterization of volatile compounds from three *Cymbopogon* species and *Eucalyptus citriodora* from Benin and their insecticidal activities against *Tribolium castaneum*. Ind Crops Prod 76:306–317
21. Caputo L, Smeriglio A, Trombetta D, Cornara L, Trevena G, Valussi M, Nazzaro F (2020) Chemical composition and biological activities of the essential oils of *Leptospermum petersonii* and *Eucalyptus gunnii*. Front Microbiol 11:409
22. Chang T, McLandsborough L, McClements DJ (2012) Physical properties and antimicrobial efficacy of thyme oil nanoemulsions: influence of ripening inhibitors. J Agric Food Chem 60:12056–12063
23. Chavda VP, Shah D (2017) A review on novel emulsification technique: a nanoemulsion. J Pharmacol Toxicol Stud 5(1):32–33
24. Chen Y, Li G, Wu X, Chen Z, Hang J, Qin B, Wang R (2008) Self-microemulsifying drug delivery system (SMEDDS) of vinpocetine: formulation development and in vivo assessment. Biol Pharm Bull 31(1):118–125
25. Chen J, Tian B, Yin X, Zhang Y, Hu D, Hu Z, Zhang Y (2007) Preparation, characterization and transfection efficiency of cationic PEGylated PLA nanoparticles as gene delivery systems. J Biotechnol 130(2):107–113
26. Chen T, Tu L, Wang G, Qi N, Wu W, Zhang W, Feng J (2020) Multi-functional chitosan polymeric micelles as oral paclitaxel delivery systems for enhanced bioavailability and anti-tumor efficacy. Int J Pharm 578
27. Chiesa M, Garg J, Kang YT, Chen G (2008) Thermal conductivity and viscosity of water-in-oil nanoemulsions. Colloids Surf A 326(1–2):67–72
28. Chin CP, Lan CW, Wu HS (2012) Study on the performance of lambda cyhalothrin microemulsion with biodiesel as an alternative solvent. Ind Eng Chem Res 51(12):4710–4718
29. Craig DQM, Barker SA, Banning D, Booth SW (1995) An investigation into the mechanisms of self-emulsification using particle size analysis and low frequency dielectric spectroscopy. Int J Pharm
30. De-Oliveira Pereira F, Mendes JM, de-Oliveira Lima, E, (2013) Investigation on mechanism of antifungal activity of eugenol against *Trichophyton rubrum*. Medical Mycol 51(5):507–513
31. De Castro e Silva P, Pereira LAS, Lago AMT, Valquíria M, de Rezende ÉM, Carvalho GR, Oliveira JE, Marconcini JM (2019) Physical-mechanical and antifungal properties of pectin nanocomposites/neem oil nanoemulsion for seed coating. Food Biophys 14:456–466
32. Delcour I, Spanoghe P, Uyttendaele M (2015) Literature review: Impact of climate change on pesticide use. Int Food Res 68:7–15
33. Devarajan V, Ravichandran V (2011) Nanoemulsions: as modified drug delivery tool. Int J Compr Pharm 4(01):1–6
34. Diaz-Blancas V, Medina DI, Padilla-Ortega E, Bortolini-Zavala R, Olvera-Romero M, Luna-Barcenas G (2016) Nanoemulsion formulations of fungicide tebuconazole for agricultural applications. Molecules 21(10):1271
35. Djiwanti SR, Noveriza R, Nurhayati H, Pasaribu G (2022) Application of bio-silica and citronella oil nanoemulsion, neem oil, and endophytic bacteria formula for the java tea root-knot nematode population suppression. IOP Conf Ser: Earth Environ Sci 974(1):012072

36. Docter D, Bantz C, Westmeier D, Galla HJ, Wang Q, Kirkpatrick JC, Stauber RH (2014) The protein corona protects against size-and dose-dependent toxicity of amorphous silica nanoparticles. Beilstein J Nanotechnol 5(1):1380–1392
37. Díaz-Blancas V, Medina DI, Padilla-Ortega E, Bortolini-Zavala R, Olvera-Romero M, Luna-Bárcenas G (2016) Nanoemulsion formulations of fungicide tebuconazole for agricultural applications. Molecules 21:1271
38. Enan E (2001) Insecticidal activity of essential oils: octopaminergic sites of action. Comp Biochem Physiol Part C: Toxicol Pharmacol 130(3):325–337
39. FAO (2012) Current world fertilizer trends and outlook to 2016. Rome: Food and Agriculture Organization of the United Nations (FAO)
40. Fabres A, de Campos Macedo da Silva J, Fernandes KV, Xavier-Filho J, Rezende GL, Oliveira AEA, (2014) Comparative performance of the red flour beetle *Tribolium castaneum* (Coleoptera: Tenebrionidae) on different plant diets. J Pest Sci 87(3):495–506
41. Felizardo MGAC, Pereira de-Lima TP, de-Sousa TL, Guterres CVF, Oliveira JPM, Everton GO, APM (2021) Chemical profile, stability and fungicide activity of oil-in-water nanoemulsion (O/A) incorporated with Ba-har essential oil. Ciencia E Natura. 43:57–57
42. Feng JG, Shi YL, Yu QY, Sun CC, Yang GT (2016) Effect of emulsifying process on stability of pesticide nanoemulsions. Colloids Surf A Physicochem Eng Asp 497:286–292
43. Feng J, Zhang Q, Liu QI, Zhu Z, McClements DJ, Jafari SM (2018) Application of nanoemulsions in formulation of pesticides. In: Nanoemulsions. Academic Press, pp 379–413
44. Forgiarini A, Esquena J, Gonzalez C, Solans C (2001) Formation and stability of nano-emulsions in mixed nonionic surfactant systems. Prog Colloid Polym Sci 118:184–189
45. Fu J, Cai Z, Gong Y, O'Reilly SE, Hao X, Zhao D (2015) A new technique for determining critical micelle concentrations of surfactants and oil dispersants via UV absorbance of pyrene. Colloids Surf A Physicochem Eng Asp 484:1–8
46. Ghosh V, Mukherjee A, Chandrasekaran N (2013) Ultrasonic emulsification of food-grade nanoemulsion formulation and evaluation of its bactericidal activity. Ultra Sonochem 20(1):338–344
47. Ghosh V, Mukherjee A, Chandrasekaran N (2014) Eugenol-loaded antimicrobial nanoemulsion preserves fruit juice against, microbial spoilage. Colloids Surf, B 114:392–397
48. Gonfa YH, Tessema FB, Gelagle AA, Getnet SD, Tadesse MG, Bachheti A, Bachheti RK (2022) Chemical compositions of essential oil from aerial parts of Cyclospermum leptophyllum and its application as antibacterial activity against some food spoilage bacteria. J Chem. https://doi.org/10.1155/2022/5426050
49. Gong DL, Zhao F, Xia HY, Wu YQ (2011) Research the formulation of triazophos microemulsion. Appl Mech Mater Trans Tech Publ Ltd 79:111–116
50. Graves S, Meleson K, Wilking J, Lin MY, Mason TG (2005) Structure of concentrated nanoemulsions. J Chem Phys 10(1063/1):1874952
51. Green JM, Beestman GB (2007) Recently patented and commercialized formulation and adjuvant technology. Crop Prot 26(3):320–327
52. Guan W, Tang L, Wang Y, Cui H (2018) Fabrication of an Effective avermectin nanoemulsion using a cleavable succinic ester emulsifier. J Agric Food Chem 66:7568–7576
53. Gutierrez JM, Gonzalez C, Maestro A, Sole I, Pey CM, Nolla J (2008) Nano-emulsions: new applications and optimisation of their preparation. Curr Opin Colloid Interface Sci 13:245–251
54. Hammad AE, Hassanin MMH (2022) Antagonistic effect of nanoemulsions of some essential oils against Fusarium oxysporum and root-knot nematode *Meloidogyne javanica* on coleus plants. Pak J Nematol 40(1):35–48
55. Hashem AS, Awadalla SS, Zayed GM, Maggi F, Benelli G (2018) *Pimpinella anisum* essential oil nanoemulsions against *Tribolium castaneum*-insecticidal activity and mode of action. Environ Sci Poll Res 25(19):18802–18812
56. Hazrati H, Saharkhiz MJ, Niakousari M, Moein M (2017) Natural herbicide activity of *Satureja hortensis* L. essential oil nanoemulsion on the seed germination and morphophysiological features of two important weed species. Eco Environ Saf 142:423–430

57. Helgeson ME (2016) Colloidal behavior of nanoemulsions: interactions, structure, and rheology. Curr Opin Colloid In 25:39–50
58. Heydari M, Amirjani A, Bagheri M, Sharifian I, Sabahi Q (2020) Eco-friendly pesticide based on peppermint oil nanoemulsion: preparation, physicochemical properties, and its aphicidal activity against cotton aphid. Environ Sci Pollut Res 27:6667–6679
59. Iannarelli R, Caprioli G, Sut S, Dall Acqua S, Fiorini D, Vittori S, Maggi F (2017) Valorizing overlooked local crops in the era of globalization: the case of aniseed (Pimpinella anisum L.) from Castignano (central Italy). Ind Crops Prod 104:99–110
60. Iannarelli R, Marinelli O, Morelli MB, Santoni G, Amantini C, Nabissi M, Maggi F (2018) Aniseed (Pimpinella anisum L.) essential oil reduces pro-inflammatory cytokines and stimulates mucus secretion in primary airway bronchial and tracheal epithelial cell lines. Ind crops Prod 114:81–86
61. Izquierdo P, Feng J, Esquena J, Tadros TF, Dederen JC, Garcia MJ, Azemar N, Solans C (2005) The influence of surfactant mixing ratio on nano-emulsion formation by the pit method. J Colloid Interface Sci 285:388–394
62. Jaiswal S, Mondal R, Paul D, Mukherjee S (2016) Investigating the micellization of the triton-X surfactants: A non-invasive fluorometric and calorimetric approach. Chem Phys Lett 646:18–24
63. Jiang LC, Basri M, Omar D, Rahman MBA, Salleh AB, Rahman RNZRA, Selamat A (2012) Green nano-emulsion intervention for water-soluble glyphosate isopropylamine (IPA) formulations in controlling *Eleusine indica* (*E. indica*). Pest Biochem Physiol 102(1):19–29
64. Jin QW, Li XB, Cai ZX, Zhang F, Madhav PY, Zhang HB (2017) A comparison of corn fiber gum, hydrophobically modified starch, gum arabic and soybean soluble polysaccharide: interfacial dynamics, viscoelastic response at oil/water interfaces and emulsion stabilization mechanisms. Food Hydrocoll 70:329–344
65. Kah M, Weniger AK, Hofmann T (2016) Impacts of (nano) formulations on the fate of an insecticide in soil and consequences for environmental exposure assessment. Env Sci Technol. 50(20):10960–10967
66. Kalaitzaki A, Papanikolaou NE, Karamaouna F, Dourtoglou V, Xenakis A, Papadimitriou V (2015) Biocompatible colloidal dispersions as potential formulations of natural pyrethrins: a structural and efficacy study. Langmuir 31:5722–5730
67. Katagi T (2008) Surfactant effects on environmental behaviour of pesticides. Rev Environ Contam Toxicol 194:71–177
68. Kim SW, Kim KS, Lamsal K, Kim YJ, Kim SB, Jung MY, Lee YS (2009) An in vitro study of the antifungal effect of silver nanoparticles on oak wilt pathogen *Raffaelea* sp. J Microbiol Biotechnol 19(8):760–764
69. Kimbaris AC, Koliopoulos G, Michaelakis A, Konstantopoulou MA (2012) Bioactivity of *Dianthus caryophyllus Koliopoulos, Lepidium sativum, Pimpinella anisum* and *Illicium verum* essential oils and their major components against the West Nile vector *Culex pipiens*. Parasitol Res 111(6):2403–2410
70. Kostyukovsky M, Rafaeli A, Gileadi C, Demchenko N, Shaaya E (2002) Activation of octopaminergic receptors by essential oil constituents isolated from aromatic plants: possible mode of action against insect pests. Pest Manag Sci: For Pest Sci 58(11):1101–1106
71. Kozuki Y, Ohtsubo T (2009) A predictive solubility tool for pesticide emulsifiable concentrate formulations. In: 29th symposium on pesticide formulations and delivery systems. ASTM International
72. Krishnamoorthy R, Gassem MA, Athinarayanan J, Periyasamy VS, Prasad S, Alshatwi AA (2021) Antifungal activity of nanoemulsion from *Cleome viscosa* essential oil against food-borne pathogenic *Candida albicans*. Saudi J Biol Sci 28(1):286–293
73. Kumar P, Huo P, Zhang R, Liu B (2019) Antibacterial properties of graphene-based nanomaterials. Nanomaterials 9(5):737
74. Kumar N, Mandal A (2020) Experimental investigation of PEG 6000/tween 40/SiO2 NPs stabilized nanoemulsion properties: a versatile oil recovery approach. J Mol Liq 319:114087

75. Kumar N, Mandal A (2020) Wettability alteration of sandstone rock by surfactant stabilized nanoemulsion for enhanced oil recovery—a mechanistic study. Colloids Surf A 601:125043
76. Kumar S, Nehra M, Dilbaghi N, Marrazza G, Hassan AA, Kim KH (2019) Nano-based smart pesticide formulations: emerging opportunities for agriculture. J Cont Rel. 294:131–153
77. Kumar N, Pal N, Mandal A (2021) Nanoemulsion flooding for enhanced oil recovery: theoretical concepts, numerical simulation and history match. J Petrol Sci Eng 202:108579
78. Kumar N, Verma A, Mandal A (2021) Formation, characteristics and oil industry applications of nanoemulsions: a review. J Petrol Sci Eng 206:109042
79. Kumari S, Kumaraswamy RV, Choudhary RC, Sharma SS, Pal A, Raliya R, Biswas P, Saharan V (2018) Thymol nanoemulsion exhibits potential antibacterial activity against bacterial pustule disease and growth promotory effect on soybean. Sci Rep 8:1–12
80. Lamaallam S, Bataller H, Dicharry C, Lachaise J (2005) Formation and stability of miniemulsions produced by dispersion of water/oil/surfactants concentrates in a large amount of water. Colloids Surf A 270–271:44–51
81. Lawrence MJ, Warisnoicharoen W (2006) Recent advances in microemulsions as drug delivery vehicles. In: Torchilin VP (ed) Nanoparticles as drug carriers. Imperial College Press, London
82. Lee YK, Choi EJ, Webster TJ, Kim SH, Khang D (2015) Effect of the protein corona on nanoparticles for modulating cytotoxicity and immunotoxicity. Int J Nanomed 10:97–113
83. Liang H, Guo D, Ma LR, Luo JB (2017) Investigation of film formation mechanism of oil-inwater (O/W) emulsions at high speeds. Tribol Int 109:428–434
84. Lim CJ, Basri M, Omar D, Rahman MBA, Salleh AB, Rahman RNZRA (2012) Physicochemical characterization and formation of glyphosate-laden nano-emulsion for herbicide formulation. Ind Crops Prod 36(1):607–613
85. Liu Y, Wei FL, Wang YY, Zhu GN (2011) Studies on the formation of bifenthrin oil-in-water nano-emulsions prepared with mixed surfactants. Colloids Surf A Physicochem Eng Asp 389:90–96
86. Louni M, Shakarami J, Negahban M (2018) Insecticidal efficacy of nanoemulsion containing Mentha longifolia essential oil against Ephestia kuehniella (Lepidoptera: Pyralidae). J Crop Prot 7(2):171–182
87. Lovelyn C, Attama AA (2011) Current state of nanoemulsions in drug delivery. J Biomater Nanobiotechnol 2:626–639
88. Lu Y, Yue Z, Xie J, Wang W, Zhu H, Zhang E, Cao Z (2018) Micelles with ultralow critical micelle concentration as carriers for drug delivery. Nat Biomed Eng 2:318–325
89. Mahmood I, Imadi SR, Shazadi K, Gul A, Hakeem KR (2016) Effects of pesticides on environment. In Plant, soil and microbes. Springer, Cham, pp 253–269
90. Marrs CT, Ballantyne B (2004) Pesticides toxicology and international regulation. John Wiley and Sons, Ltd. pp. 554
91. Mason TG, Wilking JN, Meleson K, Chang CB, Graves SM (2006) Nanoemulsions: formation, structure, and physical properties. J Phys Condens. Matter 18:R635–R666
92. Matsumiya K, Nakanishi K, Matsumura Y (2011) Destabilization of protein-based emulsions by diglycerol esters of fatty acids. The importance of chain length similarity between dispersed oil molecules and fatty acid residues of the emulsifier. Food Hydrocoll 25:773–780
93. McClements DJ (2011) Edible nanoemulsions: fabrication, properties, and functional performance. Soft Matter 7:2297–2316
94. McClements DJ (2012) Nanoemulsions versus microemulsions: terminology, differences, and similarities. Soft Matter 8:1719–1729. 2 times
95. McNabb DE (2019) The population growth barrier. In: Global pathways to water sustainability. Springer, pp 67–81
96. Miller CA (2005) Spontaneous emulsification: recent developments with emphasis on self-emulsification. In: Sjoblom J (ed) Emulsions and emulsion stability. CRC Press, NY USA
97. Morcia C, Malnati M, Terzi V (2012) In vitro antifungal activity of terpinen-4-ol, eugenol, carvone, 1, 8-cineole (eucalyptol) and thymol against mycotoxigenic plant pathogens. Food Add Contamin: Part A 29(3):415–422

98. Morell I, Hernández F (2000) El agua en Castellón: un reto para el siglo XXI. Publicacions de la Universitat Jaume I, Servei de Comunicació i Publicacions.
99. Mossa ATH, Afia SI, Mohafrash SM, Abou-Awad BA (2018) Formulation and characterization of garlic (*Allium sativum* L.) essential oil nanoemulsion and its acaricidal activity on eriophyid olive mites (Acari: Eriophyidae). Environ Sci Poll Res 25(11):10526–10537
100. Mulqueen P (2003) Recent advances in agrochemical formulation. Adv colloid interface sci 106(1–3):83–107
101. Murugan K, Abd-Elsalam KA (2022) Sustainable nanoemulsions for agri-food applications: today and future trends. In: Bio-based nanoemulsions for agri-food applications. Elsevier, pp 1–11
102. Mustafa IF, Hussein MZ (2020) Synthesis and technology of nanoemulsion-based pesticide formulation. Nanomaterials 10(8):1608
103. Myc A, Vanhecke T, Landers JJ, Hamouda T, Baker JR (2002) The fungicidal activity of novel nanoemulsion (X8W60PC) against clinically important yeast and filamentous fungi. Mycopath 155(4):195–201
104. Nenaah GE, Ibrahim SI (2011) Chemical composition and the insecticidal activity of certain plants applied as powders and essential oils against two stored-products coleopteran beetles. J Pest Sci 84(3):393–402
105. Nguyen MH, Hwang IC, Park HJ (2013) Enhanced photoprotection for photo-labile compounds using double-layer coated corn oil-nanoemulsions with chitosan and lignosulfonate. J Photochem Photobiol B: Biol 125:194–201
106. Nuchuchua O, Sakulku U, Uawongyart N, Puttipipatkhachorn S, Soottitantawat A, Ruktanonchai U (2009) In vitro characterization and mosquito (*Aedes aegypti*) repellent activity of essential-oils-loaded nanoemulsions. Aaps Pharmscitechnol 10(4):1234–1242
107. Osman Mohamed Ali E, Shakil NA, Rana VS, Sarkar DJ, Majumder S, Kaushik P, Singh BB, Kumar J (2017) Antifungal activity of nano emulsions of neem and citronella oils against phytopathogenic fungi, *Rhizoctonia solani* and *Sclerotium rolfsii*. Ind Crops Prod 108:379–387
108. Pandey S, Giri VP, Tripathi A, Kumari M, Narayan S, Bhattacharya A, Srivastava S, Mishra A (2020) Early blight disease management by herbal nanoemulsion in *Solanum lycopersicum* with bio-protective manner. Ind Crops Prod 150:112421
109. Pant M, Kumar N, Dubey S, Patanjali PK (2017) Biodiesel waste based new generation formulation of permethrin for cockroach control. J Sci. Ind. Res 76:184–186
110. Pant M, Dubey S, Patanjali PK, Naik SN, Sharma S (2014) Insecticidal activity of eucalyptus oil nanoemulsion with karanja and jatropha aqueous filtrates. Int Biodeter Biodegrad 91:119–127
111. Park IK, Choi KS, Kim DH, Choi IH, Kim LS, Bak WC, Shin SC (2006) Fumigant activity of plant essential oils and components from horseradish (*Armoracia rusticana*), anise (*Pimpinella anisum*) and garlic (*Allium sativum*) oils against *Lycoriella ingenua* (Diptera: Sciaridae). Pest Manag Sci for Pest Sci 62(8):723–728
112. Pascual-Villalobos MJ, Guirao P, Diaz-Banos FG, Canto-Tejero M, Villora G (2019) Oil in water nanoemulsion formulations of botanical active substances. In: Nano-biopesticides today and future perspectives. Academic Press, pp 223–247
113. Patir K (2022) Nanoemulsion and its application in pesticide formulation. In: Handbook of research on nanoemulsion applications in agriculture, food, health, and biomedical sciences. IGI Global, pp 401–424
114. Pavela R (2014) Insecticidal properties of *Pimpinella anisum* essential oils against the *Culex quinquefasciatus* and the non-target organism *Daphnia magna*. J Asia-Pac Entomol 17(3):287–293
115. Pavela R (2015) Acute toxicity and synergistic and antagonistic effects of the aromatic compounds of some essential oils against *Culex quinquefasciatus* Say larvae. Parasitol Res 114(10):3835–3853
116. Pavoni L, Benelli G, Maggi F, Bonacucina G (2019) Green nanoemulsion for biopesticide formulations. In: Nano-biopesticides today and future interventions perspectives. Academic Press, pp 133–160

117. Perazzo A, Preziosi V, Guido S (2015) Phase inversion emulsification: current understanding and applications. Adv Colloid Interface Sci 222:581–599
118. Pongsumpun P, Iwamoto S, Siripatrawan U (2020) Response surface methodology for optimization of cinnamon essential oil nanoemulsion with improved stability and antifungal activity. Ultrason Sonochem 60:104604
119. Qian K, Zhao H, He L (2018) Preparation of fenpropathrin nanoemulsion by phase inversion temperature method and its study on performance. Int J Nanosci 17:1–7
120. Qin H, Zhang H, Li L, Zhou X, Li J, Kan C (2017) Preparation and properties of lambda-cyhalothrin/polyurethane drug-loaded nanoemulsions. RSC Adv 7:52684–52693
121. Ravichandran M, Samiappan S C, Rangaraj S, Murugan K, Al-Dhabi NA, Karuppiah P (2022) Nanoemulsion formulations with plant growth promoting rhizobacteria (PGPR) for sustainable agriculture. In: Bio-based nanoemulsions for agri-food applications. Elsevier, pp 207–223
122. Santos LDT, Meira RMSA, Ferreira FA, Sant Anna-Santos BF, Ferreira LR (2007) Morphological responses of different eucalypt clones submitted to glyphosate drift. Environ Exp Bot 59(1):11–20
123. Sharma N, Bansal M, Visht S, Sharma PK, Kulkarni GT (2010) Nanoemulsion: a new concept of delivery system. Chron Young Sci 1(2):2–6
124. Sharma A, Sharma NK, Srivastava A, Kataria A, Dubey S, Sharma S, Kundu B (2018) Clove and lemongrass oil based non-ionic nanoemulsion for suppressing the growth of plant pathogenic *Fusarium oxysporum* f. sp. *lycopersici*. Ind C Prod 123:353–362
125. da Silva GS, Dos Reis TR, Copetti PM, Favarin FR, Sagrillo MR, da Silva AS, Ourique AF (2019) Evaluation of cytotoxicity, genotoxicity and ecotoxicity of nanoemulsions containing Mancozeb and Eugenol. Eco Environ Saf 169:207–215
126. Singh P, Dasgupta N, Singh V, Mishra NC, Singh H, Purohit SD, Mishra BN (2020) Inhibitory effect of clove oil nanoemulsion on fumonisin isolated from maize kernels. LWT, 134:110237
127. Sivastava VK, Kini G, Rout D (2006) Detergency in spontaneously formed emulsions. J Colloid Interface Sci 304:214–221
128. Song SL, Liu XH, Jiang JH, Qian YH, Zhang N, Wu QH (2009) Stability of triazophos in self-emulsifying pesticide delivery system. Colloids Surf A 350:57–62
129. Srivastava HM, Deniz S, Saad KM (2021) An efficient semi-analytical method for solving the generalized regularized long wave equations with a new fractional derivative operator. J King Saud Univer Sci 33(2):101345
130. Steinrucken HC, Amrhein N (1980) The herbicide glyphosate is a potent inhibitor of 5-enolpyruvylshikimic acid-3-phosphate synthase. Biochem Biophys Res Commun 94(4):1207–1212
131. Surendhiran M, Raja K, Jerlin R, Marimuthu S, Srivignesh S (2019) Nano emulsion seed invigoration for improved germination and seedling vigour in Maize. Int J Agric Sci Res 9:333–340
132. Tomlin CDS (2009) The pesticide manual, 15th edn. British Crop Protection Council, Surrey
133. Tyczewska A, Woźniak E, Gracz J, Kuczyński J, Twardowski T (2018) Towards food security: current state and future prospects of agrobiotechnology. Trends Biotechnol 36(12):1219–1229
134. WHO (2018) Pesticides residues in food. World Health Organization. www.who.int
135. Wang L, Li X, Zhang G, Dong J, Eastoe J (2007) Oil-in-water nanoemulsions for pesticide formulations. J Colloid Interface Sci 314:230–235
136. Wang L, Li X, Zhang G, Dong J, Eastoe J (2007) Oil-in-water nanoemulsions for pesticide formulations. J colloid Interface sci 314(1):230–235
137. Wang YY, Liu SR, Zong FL, Yestie JM (2004) Pcide formulations name and code instructions. Pestic Sci Adm 25:1–5
138. Wang J, Zhou FJ (2021) Cause analysis and solutions of water blocking damage in cracked/non-cracked tight sandstone gas reservoirs. Petrol Sci 18:219–233
139. Wang C, Zhang J, Chen H, Fan Y, Shi Z (2010) Antifungal activity of eugenol against *Botrytis cinerea*. Tropic P Path. 35:137–143

140. Wiwattanapatapee R, Sae-Yun A, Petcharat J, Ovatlarnporn C, Itharat A (2009) Development and evaluation of granule and emulsifiable concentrate formulations containing *Derris elliptica* extract for crop pest control. J Agric food Chem 57(23):11234–11241
141. Wooster TJ, Golding M, Sanguansri P (2008) Impact of oil type on nanoemulsion formation and Ostwald ripening stability. Langmuir 24(22):12758–12765
142. Wooster TJ, Golding M, Sanguansri P (2008) Ripening stability. Langmuir 24:12758–12765
143. Wu Y, Li YH, Gao XH, Chen HD (2013) The application of nanoemulsion in dermatology: an overview. J Drug Target 21(4):321–327
144. Yang E, Lee JW, Chang PS, Park IK (2021) Development of chitosan-coated nanoemulsions of two sulfides present in onion (Allium cepa) essential oil and their nematicidal activities against the pine wood nematode, Bursaphelenchus xylophilus. Env Sci Poll Res. 28(48):69200–69209
145. Yang Q, Xin X, Wang L, Lu H, Ren H, Tan Y, Xu G (2014) Modification of the stability of oil-in-water nano-emulsions by polymers with different structures. Coll Poly Sci 292(6):1297–1306
146. Yeom HJ, Jung CS, Kang J, Kim J, Lee JH, Kim DS, Park IK (2015) Insecticidal and acetylcholine esterase inhibition activity of Asteraceae plant essential oils and their constituents against adults of the German cockroach (Blattella germanica). J Agricul food chem. 63(8):2241–2248
147. Zainuddin NJ, Ashari SE, Salim N, Asib N, Omar D, Lian GEC (2019) Optimization and characterization of palm oil-based nanoemulsion loaded with *Parthenium hysterophorus* crude extract for natural herbicide formulation. J Oleo Sci 68(8):747–757
148. Zeeb B, Herz E, McClements DJ, Weiss J (2015) Reprint of: impact of alcohols on the formation and stability of protein-stabilized nanoemulsions. J Colloid Interface Sci 449:13–20
149. Zhang C, Guanming S, Shen J, Hu RF (2015) Productivity effect and overuse of pesticide in crop production in China. J Integr Agric 14:1903–1910
150. Zhao X, Cui H, Wang Y, Sun C, Cui B, Zeng Z (2017) Development strategies and prospects of nano-based smart pesticide formulation. J Agric food Chem 66(26):6504–6512
151. Zhao HK, Lan Y, Nan C, Hu Y, Rao P, Tian Y, Yan W, Qian K, He L (2016) Preparation of 8% fenpropathrin_cyflumetofen nano-emulsion and its performance. Sci Agric Sin 49:2700–2710
152. Zhao L, Liu F, Ge J, Ma L, Wu L, Xue X (2018) Changes in eleven pesticide residues in jujube (*Ziziphus jujuba* Mill.) during drying processing. Dry Technol 36(8):965–972
153. Zhu Y, An X, Li S, Yu S (2009) Nanoencapsulation of β-cypermethrin by complex coacervation in a microemulsion. J Surf Deter 12(4):305–311
154. Zhu X, He M, Zhang W, Wei H, Lyu X, Wang Q, Li L (2020) Formulation design of microemulsion collector based on gemini surfactant in coal flotation. J Clean Prod 257:120496
155. Ziaee M, Moharramipour S, Mohsenifar A (2014) Toxicity of C arum copticum essential oil-loaded nanogel against Sitophilus granarius and Tribolium confusum. J Appl Entomol 138(10):763–771

Plant Growth-Promoting Rhizobacteria Nanoemulsion and Their Applications

Shivani Tyagi, Pranchal Rajput, Atreyi Pramanik, Versha Dixit, Aashna Sinha, Anish Kumar Pal, Kundan Kumar Chaubey, B. T. Manjunath, Deepak Kumar Verma, Rakesh Kumar Bachheti, and Archana Bachheti

Abstract The toxicity of synthetic pesticides and herbicides to plants, soil, and the environment has raised consumer concerns; hence, innovative techniques are required. Plant growth-promoting rhizobacteria (PGPR) are particularly promising for using natural microbial-based biofertilizers to solve this enduring problem among many recognized systems. Moreover, PGPR can enhance plant productivity and protectivity through various direct and indirect methods, including the production of growth hormones, the development of siderophores, the solubilization of minerals, and the induction of systemic resistance. Based on its smart formulation process and advertised benefits for the environment and human health, nanotechnology is presently being used as a continuous and perpetual agricultural delivery system. The introduction of nanotechnology-based delivery technologies has improved PGPR. As nanoemulsions (NE) have a lower manufacturing cost and a simpler process than other delivery methods, it is the optimal carrier for a large-scale system. Because of its

S. Tyagi
Department of Botany, Dayalbagh Educational Institute, Uttar Pradesh, Agra 282005, India

P. Rajput · A. Pramanik · A. Sinha · A. K. Pal · K. K. Chaubey (✉)
Division of Research and Innovation, School of Applied and Life Sciences, Uttaranchal University, P.O. Chandanwari, Arcadia GrantPremnagar, Dehradun, Uttarakhand 248007, India
e-mail: kundan2006chaubey@gmail.com

V. Dixit · B. T. Manjunath
Department of Life Sciences, Christ University, Bangalore, Karnataka 560029, India

D. K. Verma
Department of Biotechnology, Sanskriti University, Mathura, UP, India

R. K. Bachheti
Centre of Excellence in Nanotechnology, Department of Industrial Chemistry, Addis Ababa Sciences and Technology University, P.O. Box-16417, Addis Ababa, Ethiopia

A. Bachheti
Department of Environment Science, Graphic Era University, Dehradun, Uttarakhand 248002, India

K. K. Chaubey
School of Basic and Applied Sciences, Sanskriti University, Mathura, Uttar Pradesh-281401, India

A. Husen et al. (eds.), *Current Trends in Green Nano-emulsions*, Smart Nanomaterials Technology, https://doi.org/10.1007/978-981-99-5398-1_7

enhanced dispersion, increased solubility, greater bioavailability, controlled release, and stability, the NE-based delivery technique was chosen in this instance. This makes PGPR more effective and can enhance agricultural operations and food production. This chapter focuses on the potential of the use of nanoemulsions PGPR formulation for improved agricultural and environmental sustainability. This technique can fully meet the future demands for food.

Keywords Plant · Rhizobacteria · Growth promoting · Nanoparticle · Nanoemulsion · Green chemistry

1 Introduction

Nanoemulsions (NEs) are dispersed particle aggregates utilized for delivery, clinical aids, and biotechnological use in medicines, toiletries, medication therapies, and diagnostics [2]. In order to improve the transport of active substances into a system, NEs, having a size of 20–200 nm, are utilized. With the help of emulsifiers such as surfactant and co-surfactant, two immiscible liquids are combined to produce a homogenized phase of Nanoemulsions [53]. Nanotechnology is being used in agriculture, which provides the potential role in increasing crop yield, plant protection, identification of plant diseases, and surveillance of growth of the plants. It also monitors the soil quality and pesticides in the soil [67]. Due to the limits of traditional agricultural technologies, i.e., global urbanization, deforestation, a lack of space, and water shortages, endangers the ecosystem. Like soil degradation, limited irrigation and the use of synthetic insecticides for agricultural purposes have significantly expanded in recent years, which is not a sustainable approach for soil and plants. In order to decrease the frequent use of chemical fertilizers and pesticides in farming without negatively influencing the environment, these bacteria are a great option. Natural microbial-based biofertilizers have become more widespread among the various tried-and-true techniques because of the significant rise in the usage of synthetic pesticides for agricultural purposes in recent years. When it comes to natural plant-based products utilization for the formation of fertilizers, Nanoemulsion is an excellent option for this. Rhizobacteria have long been known to stimulate plant growth, provide protection from diseases, and lessen the burden of repeated use of chemical fertilizers, insecticides, and herbicides. Plant growth-promoting microbial agents collaborating with Nanoemulsion is a prominent technique to resolve this ongoing concern (Fig. 1).

Bacteria that aid in plant growth by phosphates dissolution induce hormone production and assimilate nitrogen via using their metabolism can directly affect the metabolism of plants. Additionally, PGPR enhances plant nutrient and water uptake, enhancing root growth, boosting enzyme catalytic activity, and showing antimicrobial properties against various pathogens [59]. Nanotechnology-based delivery technologies have enhanced PGPR. Due to their improved distribution, increased solubility, greater bioavailability, controlled release, and stability, NE-based administration systems can improve agricultural operations and food production. This increases

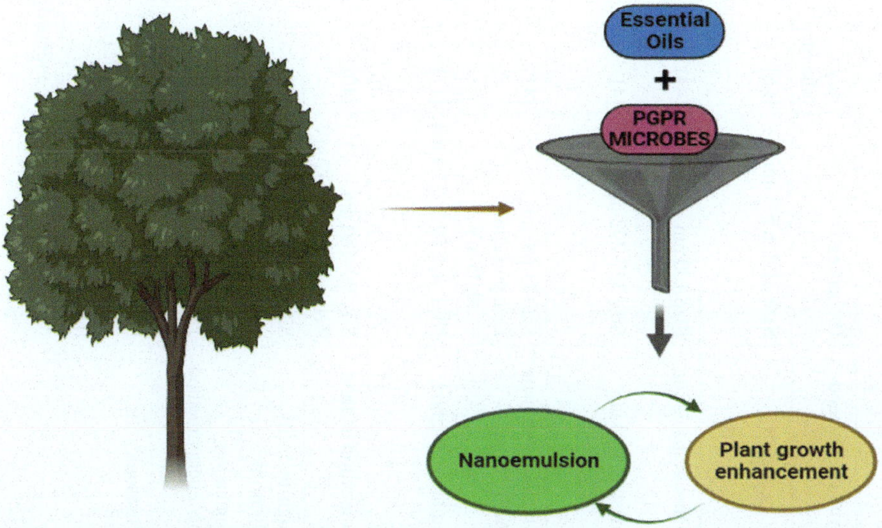

Fig. 1 Overview of NE synthesis from plants

the effectiveness of PGPR. Moreover, As NE has a lower manufacturing cost and a more straightforward distribution mechanism than other delivery systems, it is the optimum carrier for a large-scale system. The potential applications of NEs in developing PGPR for improved agricultural and environmental sustainability are the main topic of this chapter [54].

2 NE Synthesis

Top-down and bottom-up approaches are different methods utilized in the formulation of nanomaterials [3, 34]. According to its name, top-down methods create nanoscale materials by breaking them down into macro- or micro particles. In contrast, bottom-up methods assemble atomic particles to create nanoparticles by bringing the particles from the atomic level to the nanoscale level (Beer and Singh; [32, 33]. In terms of total size, both procedures result in the formation of 100 nm particles. Some of the advanced techniques of lithography that are being applied in top-down methods include optical, electron beam, scanning probe, soft nanoimprint, and copolymer lithography. At the same time, some other techniques characterized bottom-up are nano-layer aggregation, molecular self-assembly, sol–gel nano-formation, DNA-scaffolding for nanoelectronics, and vapor-phase deposition [3] (Table 1).

NEs are made by combining a liquid with an immiscible phase that has a typical size of 100 nm. Small size increases surface area, optical transparency, and kinetic stability [25]. A three-part NE system consists of an aqueous phase, a surfactant,

Table 1 Plant-based NEs and their characterization

Source of NE	Green ingredients	Application	Synthesis method	Properties	Key reference
Aloe vera	Erythromycin	Antifungal (*Escherichia coli* & *Staphylococcus aureus)*	–	Size 21.2 ± 5.7 nm	[66]
Cymbopogonmartinii	–	Antibacterial (*Enterococcus faecalis)*	Nanoencapsulation	Size 100 nm	[44]
Cymbopogonpendulus	Tween 20/80, SDS & ethanol	antibacterial, anti-inflammatory, & anti-diabetic	constant shaking	Size 327-392 nm	[4]
Kaempferiagalanga	ethyl-p-methoxycinnamate	Antifungal (*Aspergillusflavus)*	Ultrasonic acoustic energy	Size 71.68 nm Zeta potential − 20.05 mV	[41]
Essential oils- NET (Thymol), NEE (Eugenol),NEG (Geraniol), NEM (Menthone)	Thymol,eugenol, geraniol and menthone	Antibacterial (Citrus Canker)	Agitation method	NET size 59.8–73.9 nm	[11]
Quillaja tree	Thymol&saponin	Antibacterial (pustules disease)	Sonication method	Size 90–180 nm Zeta potential − 32 mV	[40]
Aniseed, peppermint, lemongrass	Trans-Anethole, caryophyllene	Pesticide	High-intensity ultrasonic	Size 100 nm	[43]
Pterodon emarginatus	Polysorbate 80/sorbitan (mono &trioleate)	Larvicidal activity against *Culexquinquefasciatus*	Sonication; low-energy method	Size approx. 151.0–160.7 nm	[49]
Vitexagnus-castus	Triacetin, labrasol&cremophor	Drug delivery	Ultrasonication	Size 11.82 ± 0.125 nm,PdI 0.117 ± 0.019	[52]

(continued)

Table 1 (continued)

Source of NE	Green ingredients	Application	Synthesis method	Properties	Key reference
Thymus daenensis	–	Antibacterial (*E. coli*)	High-intensity ultrasound	Size 143 nm	[24]
Eucalyptus oil	–	Antibacterial (*Listeria monocytogenes*)	Low-energy emulsification method	Size 50–100 nm polydispersity index (PDI) < 0.2	[13]
BoswelliaSerrata	Isopropyl Myristate, Tween 80 and Ethanol	Drug encapsulation	High-speed homogenizer	Size 11.25 nm, Zeta potential-0.223 mV	[64]
Phyllanthusemblica	Rhamnolipids, Ethanol	Delivery agent	High pressure homogenization	Size 191.63 ± 4.07 nm Zeta potential – 10.19 ± 0.54 mV	[42]

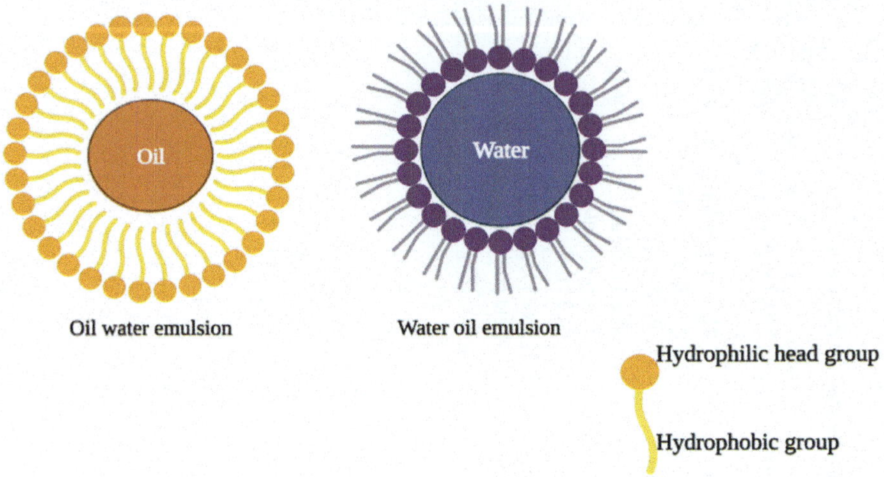

Fig. 2 Water in oil and oil in water NE illustration

and oil. Depending on how these components are arranged, there are several kinds of NEs, for example, oil in water, water dispersed in oil, and bi-continuous NE. Oil dispersion in water refers to the droplets of oil being spread in water, whereas "water in oil" refers to the opposite of it, in which droplets of water are spread into the regular oil phase [5] (Fig. 2). In bi-continuous NE, oil domain and water domain remain dispersed within a system.

There are two main methodologies for synthesizing NEs, one is low energy method and another is high-energy method. Various techniques can implement them, including solvent displacement, phase inversion temperature (PIT), high-pressure homogenization, micro fluidization, and ultrasonic emulsification [36].

The transformation of macro emulsions into NEs demands much energy in the high-energy method. The high-energy categories include micro fluidization, ultrasonication, and high-pressure homogenization.

When using inexpensive energy techniques, the system moves with less interfacial tension and needs less energy as an input. Phase inversion method, bubble bursting, evaporative ripening, and hydrogel methods are concluded in low energy methods [50].

3 Properties of Nanoemulsion

NEs are kinetically stable liquid-in-liquid dispersions with droplet diameters of around 100 nm. Because of their small size, they have beneficial qualities like a high surface area per unit volume, long-lasting stability, optical transparency, and tunable rheology. Biocompatibility, low toxicity, non-immunogenicity, drug entrapment, broad surface area, nanoscale size, long-term and restricted release, thermodynamic stability, and simple formulation mode are remarkable properties of Nanoemulsions, which make them feasible delivering agents (S., Ahmed, and Ramalingam 1AD). NE is used in several sectors, including medicines, food, cosmetics, drug delivery, and material development.

NEs exhibit several special physicochemical characteristics and functional characteristics. Compared to conventional emulsion compositions, they may have better physical stability, increased oral bio-occurrence, and distinct textural properties. The droplets that are the building blocks of NEs having their unique properties such as their constitution, morphology, accumulation state, and surface characteristics; apart from these properties, NEs themselves have some physicochemical properties like optical, rheological, and stability properties making them more sufficient for future applications in various fields. To maximize their potential, many researchers have been modifying these features [25]. Engineering Nano emulsions could enhance the bioavailability of bioactive compounds.

Furthermore, some researchers highlight the relationship between Nano emulsions' structure and function. The surfactant utilized affected the physical stability of the emulsions. For example, stability could be enhanced by Soy lecithin < Tween 20 < Whey protein isolate in increasing order, according to an article [21].

4 Applications

4.1 Application of Nanotechnology in Agriculture

The world's population is expanding and increasing the demand for food. Current agricultural practices must be improved to feed the growing population [9]. One of the most promising technologies is nanotechnology, which can potentially increase agricultural output with effective pesticides and herbicides, regulation of soil features, wastewater management, and disease identification. Enhanced food production with good market value, heightened nutritional and sensory properties, improved safety, and better antimicrobial protection are all advantages of industrial food processing. Using nanoparticles to extend the shelf life can also help decrease post-farming losses.

Due to their stability and cautious usage in the biological system, nanoparticles are produced by using green and microbial synthesis processes [27, 28, 29, 30, 62, 63, 31], Bachheti et al. (2022). Nano-persuade polysaccharide powder for condensation detention and soil assemblage, nano-seed stimulation, slow-moving and constant release of pesticides, the regulated delivery of nutrients from nanofertilisers, the development of nano condensed herbicides, and the regulated out of its active constituents, nano alleviation of soil and marine pollutants, and production of nano-films for long-term storage are some advanced research areas that primarily focus on nanoparticle synthesis. While nanotechnology goods and their manufacturing are not adequately regulated, this can lead to serious health and environmental risks [20].

A recent development in precision farming is nanotechnology. In addition to helping with the systematic use of natural resources like water, nutrients, and other essential substances, Nano sensors may be used to determine the need for fertilisers and pesticides in every area of agricultural land. It aids farmers in carrying out adequate agricultural land maintenance. Farmers and farm managers may remotely access agricultural pests, weeds, or any signs of biotic and abiotic stress using nanomaterials and GPS (Global positioning system) and manage the crop as necessary [46, 60, 38, 39, 47]. Agriculture mostly uses bio-nanosensors and electrical-nano biosensors. Certain nanosensors have been created to test the microcystin toxicity caused by cyanobacteria, which is harmful to human health and agriculture [60].

Conventional fertilizers and pesticides are extensively used for plant growth and disease control, but around 90% of applied fertilizers get wasted due to runoff and other activities. To overcome this problem, nano fertilizers are very effective as they are more environmentally safe and efficient even in small quantities [60]. Nano fertilizers or nano-encapsulated slow-release fertilizers are also used to save fertilizer exhaustion and decrease environmental pollution [39]. With the help of nanotechnology, modification of the genetic material of the crop plant is also possible, which can give another aspect for crop improvement [46].

Nanoformulations enhance solubility and stimulate the active components in a regulated manner. It enhances the functioning of nano pesticides, decreases the chemical dose, and simultaneously increases sustainability listed in (Table 2) [67]. These formulations are widely used in the pesticide and agrochemical industries to develop NE and nanocapsules containing active components. Some common metal oxide nanoparticles (ZnO, TiO_2, Cu, and SiO_2) are used in pesticides and fungicides to protect and develop plants [61].

4.2 Nanotechnology and PGPR

Plant growth-promoting rhizobacteria (PGPR) are well studied for their phyto stimulatory actions on seed and crop in combination with a physiological aspect such as; asymbiotic N_2 fixation, production of phytohormones [indole-3-acetic acid (IAA), cytokinin and gibberellins]. They are free-living, soil-born, and root-colonizing bacteria, playing a significant role in the conservation of soil quality, plant pathogen control, nutrient cycling, ecosystem functioning, and crop yield. With the advancement of nanotechnology, nanoparticles are instigated in the environment because of a number of human and anthropogenic activities. They are immensely resistant to degradation, and due to this, they accumulate in water bodies and soil. Developing new types of analytic tools using nanotubes, nanowires, fullerene derivatives, and quantum dots for biotechnology and agronomy has received extensive

Table 2 Nanoformulations and their application

Carrier system	Plant or micro-organism	Agent	Agricultural product	Application	Method	Key reference
Wheat gluten	Wheat	Ethofumesate and montmorillonites(MMT)	Pesticide	Controlled release of pesticides	Bi-vis extrusion	[17]
Xyloglucan	Tamarind seed	Tropocamide	–	A significant increase in corneal permeation across excised goat cornea	Nanoaggregates	[19]
Polyhydroxybutyrate-co-hydroxyvalerate microspheres (PHBV-MS)	*Lactuca sativa* plants	Atrazine	Herbicide	Biodegradable herbicide and decreased genotoxicity	Encapsulation	[16]
Chitosan and sodium tripolyphosphate (TPP)	Maize (*Zea mays*) and Mustard (*Brassica* sp.)	Paraquat	Herbicide	Decrease cytotoxicity and genotoxicity and also have the potential to control weed	Encapsulation	[23]
Organic–inorganic nanohybrid	–	2,4-dichlorophenoxyacetate (24D)	Herbicide	Controlled release	Self-assembly technique	[22]
Sodium alginate (Na-Alg)	Neem oil	Azadirachtin	NE-encapsulated beads	Controlled release of Aza-A and toxic in lymphophtic cells	Encapsulation	[35]

(continued)

Table 2 (continued)

Carrier system	Plant or micro-organism	Agent	Agricultural product	Application	Method	Key reference
Oil in water	*Eleusineindica*	Glyphosate isopropylamine (IPA)	Herbicide	Enhance bio-efficacy, and increase the negative effect of pesticide formulations on environment	NE	[15]
Sodium alginate	Bhindi (OKRA, *Abelmoschusesculentus)*	Imidacloprid	Pesticide	More effective pesticides with lower toxicity and decrease the risk of environmental pollution	Encapsulation	[37]
Poly citric acid and poly ethylene glycol	–	Imidacloprid	Pesticide	Decreased essential dose and environmental risk of pesticide	Encapsulation	[45]
Chitosan-saponin and Cu-chitosan	–	Chitosan, saponin and copper sulphate	Fungicide	Antifungal	Cross-linking	[57]

(continued)

Table 2 (continued)

Carrier system	Plant or micro-organism	Agent	Agricultural product	Application	Method	Key reference
Aliginate/chitosan	–	Paraquat	Herbicide	Change the release of herbicide and reduce negative impact on the environment	Pre-gelation of alginate then complexation between alginate and chitosan	[48]
Carboxymethylchitosan (CMCS)	–	Methomyl	Pesticide	Controlled release of encapsulate	Encapsulation	[48]
Chitosan	Probiotic bacteria	–	PGPR formulation for agriculture	Increase degradability, survival, and long-term performance	Encapsulation	[56]
Monoterpenes (carvone, cinnamaldehyde, citral, geraniol and pulegone)	Tomato knot-root nematode *M. javanica*	–	Nematicides	Decreased egg hitching, increase shoot and root growth of plant	NE	[1]

recognition among the scientific and industrial community. Although Ecotoxicological properties and associated risks of the nanoparticles are not clearly understood, some recent studies put forward the effect of nanoparticles on microorganisms. As many nanoparticles are described to have antimicrobial activities, they directly affect the soil microorganisms when entering the soil environment [58]. Titanium dioxide nanoparticles (TiO_2NPs) significantly inhibit PGPR of three groups: nitrogen fixers, phosphate solubilizers, and biofilm formers. NPs had more inhibitory effects on gram-positive bacteria, giving a clue that bacterial cell walls might have a significant role in the toxicity of TiO_2 NPs [11]. One more study was done on the same group of bacteria (nitrogen fixers, phosphate solubilizers, and biofilm formers) and showed the bactericidal effect of silver nanoparticles. Zinc oxide nanoparticles showed bacteriostatic effects [14] (Table 3).

Though the effect of nanoparticles on living entities remains, they produce reactive oxygen species (ROS) and oxidative stress, which can explain the toxicological effect of nanoparticles on living organisms. The introduction of nanotechnology in agriculture is still developing, so some guidelines and precautions are required for their use and the development of new green nanotechnologies and other future inventions. Understanding all the aspects of nanomaterials and their eco toxicological effect is required before their applications [58].

Table 3 PGPR used to alleviate stress and its application

PGPR	Abiotic stress	Plant	Application	Key reference
Azotobacter chroococcum	PGPR with slight drought stress	*Trachyspermumammi*	Enhance essential oil yield, increased total phenolic, and flavonoid content, and increased antioxidant activity	[26]
Bacillus mycoides PM35	PGPR with salt stress	*Zea mays* L	PGPR alleviate salt stress and enhance plant growth, including enhanced production of indole acetic acid, siderophore, ACC deaminase, and exo-polysaccharides	[6]
Enterobacter cloacae PM23	PGPR with salt stress	*Zea mays* L	Multi-stress resistance for both biotic and abiotic stresses, attenuate salt stress through up-regulation of stress related genes, increase plant growth, biomass and chlorophyll content	[6]

4.3 *Plant Membrane Absorption of Nanoparticles and Possible Consequences*

After they have penetrated the plant, nanoparticles can move across tissues through either the apoplast or the symplast. Apoplast transport occurs outside the plasma membrane through extracellular spaces, neighbouring cells' cell walls, and xylem vessels. In contrast, synplastic transport involves the passage of water and materials between the cytoplasm of cells through specialized structures called plasmodesmata and sieve plates. The apoplastic channel is crucial for radial movement inside plant tissues. It enables nanoparticles to enter the vascular tissues and root central cylinder for the continued ascent of the aerial portionNanoparticles, can follow the transpiration stream into the central cylinder, and then proceed through the xylem and into the aerial portion. To enter the xylem through the root, the Casparian strip, a barrier to the apoplastic route, must be passed in a symplastic fashion via endodermal cells. The Casparian strip can stagnate and collect some nanomaterials.

Another important symplastic transport is also feasible, allowing distribution to tissues and organs that are not photosynthetic using the sieve tube components in the phloem. In the case of foliar sprays, nanomaterials must get through the cuticle's barrier by either taking the hydrophilic or the lipophilic path. Although the lipophilic pathway necessitates diffusion via cuticular waxes, the hydrophilic pathway is carried out through polar aqueous pores found in the cuticle and stomata [8, 51]. Endocytosis is the mechanism that can cross the cell membrane barrier and can lead to the cellular uptake of small nanoparticles. Phagocytosis, pinocytosis, and receptor-mediated endocytosis help cellular uptake of the nanoparticles. The substance can also cross the membrane through diffusion, but nanoparticles are relatively larger for diffusion. Therefore, phagocytosis and pinocytosis are the commonly reported mechanisms for the cellular uptake of nanoparticles. For pharmacological purposes, we sometimes need to target the non-phagocytic cells. Therefore, for cellular uptake in non-phagocytic cells, particle size should be around 50 nm. Charges on particles (positive or negative charge) also enhance cellular uptake of cells. Positively charged particles are taken up more rapidly only when their absolute values are equivalent [18].

5 Conclusion and Future Prospects

Given that 46% of our country's population relies directly on agriculture, agriculture plays a significant part in the country's economy. Indian Agriculture's Economic Contribution in 2022 according to the most recent data for 2020–21, agriculture contributed 19.9% of India's GDP, a rise from 17.8% in the prior session, 2019–20. Nanotechnology is an amazing combination of science, engineering, and technology with many applications in many areas of human life, including agriculture. The synthesis of NEs using plant systems offers a feasible and environmentally friendly method. On the other hand, plant growth-promoting bacteria can directly alter plant metabolism, stimulate plant growth, and use their metabolism to solubilize phosphates, create hormones, and fix nitrogen.

When both these systems are merged for plant growth, they act as nanofertilizers and show a significant potential to operate as powerful antimicrobial agents to stimulate plant development. In order to limit the use of chemical fertilizers and pesticides without negatively affecting the environment or losing yields, farmers can greatly benefit from these NEs. Global concerns about potential adverse health effects, the emergence of antibiotic resistance, and environmental contamination have been raised by the usage of synthetic chemicals in agriculture around the world. It has become the most need for sustainable development in agriculture to use biological or eco-friendly alternatives to combat above mentioned global issues.

References

1. Rasoul A, Mona A, El-Habashy DE (2021) Nematicidal activity of some nanoemulsions of monoterpenes on tomato root-knot nematodes (Meloidogyne Javanica). J Plant Prot Pathol 12(9):655–661. https://doi.org/10.21608/JPPP.2021.203160
2. Abdelrasoul Mona A, El-Habashy DE (2021) Journal of Plant Protection and Pathology Nematicidal Activity of Some Nanoemulsions of Monoterpenes on Tomato Root-Knot Nematodes (Meloidogyne Javanica)." J Plant Protection and Pathology, Mansoura Univ 12(9):655–61. https://doi.org/10.21608/jppp.2021.203160.
3. Abid, Namra, Aqib Muhammad Khan, Sara Shujait, Kainat Chaudhary, Muhammad Ikram, Muhammad Imran, Junaid Haider, Maaz Khan, Qasim Khan, Muhammad Maqbool (2022) Synthesis of nanomaterials using various top-down and bottom-up approaches, influencing factors, advantages, and disadvantages: a review. Adv Colloid Int Sci 300. https://doi.org/10.1016/J.CIS.2021.102597
4. Agnish S, Sharma AD, Kaur I (2022) Nanoemulsions (O/W) containing Cymbopogon Pendulus essential oil: development, characterization, stability study, and evaluation of in vitro antibacterial, anti-inflammatory, anti-diabetic activities. BioNanoScience 12(2):540–554. https://doi.org/10.1007/S12668-022-00964-4/FIGURES/8
5. Akram Salman, Nicolas Anton, Ziad Omran, Thierry Vandamme (2021) Water-in-Oil Nano-Emulsions Prepared by Spontaneous Emulsification: New Insights on the Formulation Process. Pharmaceutics 13(7). https://doi.org/10.3390/PHARMACEUTICS13071030
6. Ali B, Wang X, Saleem MH, Azeem MA, Afridi MS, Nadeem M, Ghazal M, Batool T, Qayyum A, Alatawi A, Ali S (2022) Bacillus Mycoides PM35 Reinforces Photosynthetic Efficiency, Antioxidant Defense, Expression of Stress-Responsive Genes, and Ameliorates the Effects of Salinity Stress in Maize. Life 12(2):219. https://doi.org/10.3390/LIFE12020219/S1
7. Ali Baber, Xiukang Wang, Muhammad Hamzah Saleem, Sumaira, Aqsa Hafeez, Muhammad Siddique Afridi, Shahid Khan, Zaib-Un-nisa, Izhar Ullah, Antônio Teixeira Do Amaral Júnior, Aishah Alatawi, Shafaqat Ali (2022) PGPR-Mediated Salt Tolerance in Maize by Modulating Plant Physiology, Antioxidant Defense, Compatible Solutes Accumulation and Bio-Surfactant Producing Genes. Plants 11(3):345. https://doi.org/10.3390/PLANTS11030345.
8. Ali Shahid, Asif Mehmood, Naeem Khan (2021) Uptake, translocation, and consequences of nanomaterials on plant growth and stress adaptation. J Nanomater. https://doi.org/10.1155/2021/6677616
9. Ashish, Deepti Singh, Neeraj Gupta (2020) Impact of nanoparticles on PGPR and soil nutrient contents. Biogenic Nano-Particles and Their Use in Agro-Ecosystems 247–257. https://doi.org/10.1007/978-981-15-2985-6_14

10. Bachheti RK, Abate L, Deepti, Bachheti A, Madhusudhan A, Husen H (2021) Algae-, fungi-, and yeast-mediated biological synthesis of nanoparticles and their various biomedical applications. *In*: Handbook of Greener Synthesis of Nanomaterials and Compounds, Volume 1: Fundamental Principles and Methods (Eds. Kharisov B, Kharissova O) Elsevier Inc. 50 Hampshire St., 5th Floor, Cambridge, MA 02139, USA, pp. 701–734
11. Baranowska-Wójcik E (2021) Factors conditioning the potential effects TiO2 NPs exposure on human microbiota: a mini-review. Biol Trace Elem Res 199(12):4458. https://doi.org/10.1007/S12011-021-02578-5
12. Beer, Prof, Pal Singh (n.d) TOP-DOWN AND BOTTOM-UP APPROACHES FOR SYNTHESIS OF NANOMATERIALS
13. Chandrasekaran, Dr. Natarajan (2015) Nanoemulsion formation and characterization by spontaneous emulsification: investigation of its antibacterial effects on listeria monocytogenes. Asian J Pharm (AJP) 23–28. https://doi.org/10.22377/AJP.V9I1.427
14. Chavan Sangeeta, Vigneshwaran Nadanathangam (2019) Effects of nanoparticles on plant growth-promoting bacteria in Indian agricultural soil. Agronomy 9(3):140. https://doi.org/10.3390/AGRONOMY9030140
15. Jiang C, Lim MB, Omar D, Rahman MBA, Salleh AB, Rahman RNZRA, Selamat A (2012) Green nano-emulsion intervention for water-soluble glyphosate isopropylamine (IPA) formulations in controlling Eleusine Indica (E. Indica). Pestic Biochem Physiol 102(1):19–29. https://doi.org/10.1016/J.PESTBP.2011.10.004
16. Chen Xiao ting, Tongxin Wang (2019) Preparation and characterization of atrazine-loaded biodegradable PLGA nanospheres. J Integ Agricul 18(5):1035–1041. https://doi.org/10.1016/S2095-3119(19)62613-4
17. Chevillard A, Angellier-Coussy H, Guillard V, Gontard N, Gastaldi E (2012) Controlling Pesticide release via structuring agropolymer and nanoclays based materials. J Hazard Mater 205–206:32–39. https://doi.org/10.1016/J.JHAZMAT.2011.11.093
18. Di Jinwei, Xiang Gao, Yimeng Du, Hui Zhang, Jing Gao, Aiping Zheng (2021) Size, shape, charge and 'stealthy' surface: carrier properties affect the drug circulation time in Vivo. Asian J Pharmac Sci 16(4):444–458. https://doi.org/10.1016/J.AJPS.2020.07.005
19. Dilbaghi N, Kaur H, Ahuja M, Kumar S (2013) Evaluation of tropicamide-loaded tamarind seed xyloglucan nanoaggregates for ophthalmic delivery. Carbohyd Polym 94(1):286–291. https://doi.org/10.1016/J.CARBPOL.2013.01.054
20. Elizabath A, Babychan M, Mathew AM, Syriac GM, India S (2019) Application of nanotechnology in agriculture. Int J Pure App Biosci 7(2):131–139. https://doi.org/10.18782/2320-7051.6493
21. Fuentes Karen, Claudia Matamala, Nayaret Martínez, Rommy N Zúñiga, Elizabeth Troncoso (2021) Comparative study of physicochemical properties of nanoemulsions fabricated with natural and synthetic surfactants. Processes 9(11):2002. https://doi.org/10.3390/PR9112002
22. Grillo R, de Melo NFS, de Lima R, Lourenço RW, Rosa AH, Fraceto LF (2010) Characterization of Atrazine-Loaded Biodegradable Poly(Hydroxybutyrate-Co-Hydroxyvalerate) Microspheres. J Polym Environ 18(1):26–32. https://doi.org/10.1007/S10924-009-0153-8
23. Grillo R, Pereira AES, Nishisaka CS, De Lima R, Oehlke K, Greiner R, Fraceto LF (2014) Chitosan/Tripolyphosphate nanoparticles loaded with paraquat herbicide: an environmentally safer alternative for weed control. J Hazard Mater 278:163–171. https://doi.org/10.1016/J.JHAZMAT.2014.05.079
24. Guo M, Zhang L, He Q, Arabi SA, Zhao H, Chen W, Ye X, Liu D (2020) Synergistic antibacterial effects of ultrasound and thyme essential oils nanoemulsion against escherichia coli O157:H7. Ultrason Sonochem 66:104988. https://doi.org/10.1016/J.ULTSONCH.2020.104988
25. Gupta A, Burak Eral H, Alan Hatton T, Doyle PS (2016) Nanoemulsions: formation, properties and applications. Soft Matter 12(11):2826–2841. https://doi.org/10.1039/C5SM02958A
26. Hashemi M, Behboodian B, Karimi E, Oskoueian E (2022) Enhancing biosynthesis and bioactivity of trachyspermum ammi seed essential oil in response to drought and azotobacter chroococcum stimulation. Chem Biol Technol Agricul 9(1):1–10. https://doi.org/10.1186/S40538-022-00292-5/TABLES/8

27. Husen A (2019a) Natural product-based fabrication of zinc oxide nanoparticles and their application, *In*: N*anomaterials* and Plant Potential (Eds. Husen A, Iqbal M) Springer International Publishing AG, Gewerbestrasse 11, 6330 Cham, pp. 193–291. https://doi.org/10.1007/978-3-030-05569-1_7
28. Husen A (2019b) Medicinal plant-product based fabrication nanoparticles (Au and Ag) and their anticancer effect, *In*: Plants that Fight Cancer - Second Edition (Eds. Kintzios SE, Barberaki M, Flampouri) Taylor & Francis/CRC Press pp. 133–147
29. Husen A (2023a) Secondary metabolites based green synthesis of nanomaterials and their applications. Springer Nature Singapore Pte Ltd., 152 Beach Road, #21–01/04 Gateway East, Singapore 189721, Singapore
30. Husen A (2023b) Nanomaterials from agricultural and horticultural products. Springer Nature Singapore Pte Ltd., 152 Beach Road, #21–01/04 Gateway East, Singapore 189721, Singapore
31. Husen A, Siddiqi KS (2023) Advances in smart nanomaterials and their applications. Elsevier Inc., 50 Hampshire St., 5th Floor, Cambridge, MA 02139, USA
32. Husen A, Siddiqi KS (2014) Phytosynthesis of nanoparticles: concept, controversy and application. Nanoscale Res Lett 9(229):1–24
33. Husen A, Iqbal M (2019a) N*anomaterials*and plant potential: an overview. In: N*anomaterials*and Plant Potential (Eds. Husen A, Iqbal M) Springer International Publishing AG, Gewerbestrasse 11, 6330 Cham, pp. 3–29. https://doi.org/10.1007/978-3-030-05569-1_1
34. Husen A (2020) Introduction and techniques in nanomaterials formulation, *In*: Nanomaterials for Agriculture and Forestry Applications (Eds. Husen, A, Jawaid M) Elsevier Inc. 50 Hampshire St., 5th Floor, Cambridge, MA 02139, USA, pp. 1–14
35. Jerobin J, Sureshkumar RS, Anjali CH, Mukherjee A, Chandrasekaran N (2012) Biodegradable polymer based encapsulation of neem oil nanoemulsion for controlled release of Aza-A. Carbohyd Polym 90(4):1750–1756. https://doi.org/10.1016/J.CARBPOL.2012.07.064
36. Kumar M, Bishnoi RS, Shukla AK, Jain CP (2019) Techniques for formulation of nanoemulsion drug delivery system: a review. Preven Nut Food Sci 24(3):225. https://doi.org/10.3746/PNF.2019.24.3.225
37. Kumar Sandeep, Gaurav Bhanjana, Amit Sharma, Sidhu MC, Neeraj Dilbaghi (2014) Synthesis, characterization and on field evaluation of pesticide loaded sodium alginate nanoparticles. Carbohyd Poly 101(1):1061–1067. https://doi.org/10.1016/J.CARBPOL.2013.10.025
38. Kumar A, Choudhary A, Kaur H, Guha S, Mehta S, Husen A (2022) Potential applications of engineered nanoparticles in plant disease management: a critical update. Chemosphere 295:133798. https://doi.org/10.1016/j.chemosphere.2022.133798
39. Kumar A, Choudhary A, Kaur H, Mehta S, Husen A (2021) Smart nanomaterial and nanocomposite with advanced agrochemical activities. Nanoscale Res Lett 16(156):1–26. https://doi.org/10.1186/s11671-021-03612-0
40. Kumari Sarita, Kumaraswamy RV, Ram Chandra Choudhary, Sharma SS, Ajay Pal, Ramesh Raliya, Pratim Biswas, Vinod Saharan (2018) Thymol nanoemulsion exhibits potential antibacterial activity against bacterial pustule disease and growth promotory effect on soybean. Sci Reports 8(1):1–12. https://doi.org/10.1038/s41598-018-24871-5
41. Kundu A, Mandal A, Dutta A, Saha S, Raina AP, Kumar R, Ghosh A (2022) Nanoemulsification of Kaempferia Galanga essential oil: characterizations and molecular interactions explaining fungal growth suppression. Process Biochem 121:228–239. https://doi.org/10.1016/J.PROCBIO.2022.07.008
42. Madrid Rossana, Puttasiddaiah Rachitha, Krupashree Krishnaswamy, Renal Antoinette Lazar, Vijai Kumar Gupta, Baskaran Stephen Inbaraj, Vinay Basavegowda Raghavendra, Minaxi Sharma, Kandi Sridhar (2023) Attenuation of hyperlipidemia by medicinal formulations of emblica officinalis synergized with nanotechnological approaches. Bioengineering 10(1):64. https://doi.org/10.3390/BIOENGINEERING10010064
43. Mahran Hesham A (2022) Using nanoemulsions of the essential oils of a selection of medicinal plants from Jazan, Saudi Arabia, as a Green Larvicidal against Culex Pipiens. PLoS One 17(5). https://doi.org/10.1371/JOURNAL.PONE.0267150

44. Marinković J, Nikolić B, Marković T, Radunović M, Ilić J, Bošković M, Ćirić A, Marković D (2021) Cymbopogon citratus essential oil: an active principle of nanoemulsion against enterococcus faecalis root canal biofilm. Future Microbiol 16(12):907–918. https://doi.org/10.2217/FMB-2021-0081
45. Memarizadeh N, Ghadamyari M, Adeli M, Talebi K (2014) Preparation, characterization and efficiency of nanoencapsulated imidacloprid under laboratory conditions. Ecotoxicol Environ Saf 107:77–83. https://doi.org/10.1016/J.ECOENV.2014.05.009
46. Mendes Jorge, Tatiana M Pinho, Filipe Neves Dos Santos, Joaquim J Sousa, Emanuel Peres, José Boaventura-Cunha, Mário Cunha, Raul Morais (2020) Smartphone Applications targeting precision agriculture practices—a systematic review. Agronomy 10(6):855. https://doi.org/10.3390/AGRONOMY10060855
47. Mondal R, Dam P, Chakraborty J, Paret ML, Katı A, Altuntas S, Sarkar R, Ghorai S, Mandal AK, Husen A (2022) Potential of nanobiosensor in sustainable agriculture: the state-of-art. Heliyon. https://doi.org/10.2139/ssrn.4134873
48. Niculescu Adelina Gabriela, Alexandru Mihai Grumezescu (2022) Applications of chitosan-alginate-based nanoparticles—an up-to-date review. Nanomaterials 12(2). https://doi.org/10.3390/NANO12020186
49. Oliveira AEMFM, Duarte JL, Cruz RAS, Souto RNP, Ferreira RMA, Peniche T, Conceição EC, Oliveira LAR, Faustino SMM, Florentino AC, Carvalho JCT, Fernandes CP (2017) Pterodon Emarginatus oleoresin-based nanoemulsion as a promising tool for Culex Quinquefasciatus (Diptera: Culicidae) control. J Nanobiotech 15(1):1–11. https://doi.org/10.1186/S12951-016-0234-5/TABLES/4
50. Pandey Shipra, Aradhana Mishra, Ved Prakash Giri, Madhuree Kumari, Sumit Soni (2019) A green nano-synthesis to explore the plant microbe interactions. New and Future Developments in Microbial Biotechnology and Bioengineering: Microbial Biotechnology in Agro-Environmental Sustainability 85–105. https://doi.org/10.1016/B978-0-444-64191-5.00007-9
51. Pérez-de-Luque Alejandro (2017) Interaction of nanomaterials with plants: what do we need for real applications in agriculture? Front Environ Sci 5(APR):12. https://doi.org/10.3389/FENVS.2017.00012/BIBTEX
52. Piazzini Vieri, Elena Monteforte, Cristina Luceri, Elisabetta Bigagli, Anna Rita Bilia, Maria Camilla Bergonzi (2017) Nanoemulsion for improving solubility and permeability of vitex agnus-castus extract: formulation and in vitro evaluation using PAMPA and Caco-2 approaches 24(1):380–90. https://doi.org/10.1080/10717544.2016.1256002
53. Pour MM, Saberi-Riseh R, Mohammadinejad R, Hosseini A (2019) Nano-encapsulation of plant growth-promoting rhizobacteria and their metabolites using alginate-silica nanoparticles and carbon nanotube improves UCB1 pistachio micropropagation. J Microbiol Biotechnol 29(7):1096–1103. https://doi.org/10.4014/JMB.1903.03022
54. Ravichandran Mythili, Sumathi C. Samiappan, Suriyaprabha Rangaraj, Kasi Murugan, Naif Abdullah Al-Dhabi, Ponmurugan Karuppiah (2022) Nanoemulsion formulations with plant growth promoting rhizobacteria (PGPR) for sustainable agriculture. Bio-Based Nanoemul Agri-Food Appl 207–223. https://doi.org/10.1016/B978-0-323-89846-1.00017-6
55. Mohamed Akif S, Jahangir Ahmed, Karthikeyan Ramalingam. 1AD. The essential properties of nanoemulsions: basics of nanoemulsion. Https://Services.Igi-Global.Com/Resolvedoi/Resolve.Aspx? pp 1–23. https://doi.org/10.4018/978-1-7998-8378-4.CH001
56. Riseh S, Roohallah ET, Hajabdollahi N, Vatankhah M, Thakur VK, Skorik YA (2022) Chitosan microencapsulation of rhizobacteria for biological control of plant pests and diseases: recent advances and applications. Rhizosphere 23:100565. https://doi.org/10.1016/J.RHISPH.2022.100565
57. Saharan V, Mehrotra A, Khatik R, Pokhar Rawal SS, Sharma, and Ajay Pal. (2013) Synthesis of chitosan based nanoparticles and their in vitro evaluation against phytopathogenic fungi. Int J Biol Macromol 62:677–683. https://doi.org/10.1016/J.IJBIOMAC.2013.10.012
58. Sánchez-López Elena, Daniela Gomes, Gerard Esteruelas, Lorena Bonilla, Ana Laura Lopez-Machado, Ruth Galindo, Amanda Cano, Marta Espina, Miren Ettcheto, Antoni Camins, Amélia

M Silva, Alessandra Durazzo, Antonello Santini, Maria L Garcia, Eliana B Souto (2020) Metal-based nanoparticles as antimicrobial agents: an overview. Nanomaterials 10(2). https://doi.org/10.3390/NANO10020292
59. dos Santos R, Mendes PA, Diaz E, Lobo LLB, Rigobelo EC (2020) Use of plant growth-promoting rhizobacteria in maize and sugarcane: characteristics and applications. Front Sust Food Syst 4:136. https://doi.org/10.3389/FSUFS.2020.00136/BIBTEX
60. Sharma Pankaj, Vimal Pandey, Mayur Mukut Murlidhar Sharma, Anupam Patra, Baljinder Singh, Sahil Mehta, Azamal Husen (2021) A review on biosensors and nanosensors application in agroecosystems. Nanoscale Res Lett 16(1). https://doi.org/10.1186/S11671-021-03593-0
61. Sharma P, Pandey V, Sharma MMM, Patra A, Singh B, Mehta S, Husen A (2021) A review on biosensors and nanosensors application in agroecosystems. Nano Res Lett 16:136. https://doi.org/10.1186/s11671-021-03593-0
62. Siddiqi KS, Husen A (2016) Fabrication of metal nanoparticles from fungi and metal salts: scope and application. Nanoscale Res Lett 11(98):1–15
63. Siddiqi KS, Husen A (2017) Recent advances in plant-mediated engineered gold nanoparticles and their application in biological system. J Trace Elem Med Biol 40:10–23
64. Shehata Tamer M, Hanan M Elnahas, Heba S Elsewedy (2022) Development, characterization and optimization of the anti-inflammatory influence of meloxicam loaded into a eucalyptus oil-based nanoemulgel. Gels 8(5). https://doi.org/10.3390/GELS8050262
65. Slattery Matthew, Bryan Harper, Stacey Harper (2019) Pesticide encapsulation at the nanoscale drives changes to the hydrophobic partitioning and toxicity of an active ingredient. Nanomaterials 9(1). https://doi.org/10.3390/NANO9010081
66. Valizadeh Alireza, Mohammad Hasan Darvishi, Amir Amani, Ali Akbar Karimi Zarchi (2022) Design and development of novel formulation of aloe vera nanoemulsion gel contained erythromycin for topical antibacterial therapy: in Vitro and in Vivo assessment. J Drug Delivery Sci Technol 74:103519. https://doi.org/10.1016/J.JDDST.2022.103519
67. Vedamurthy AB, Sourav Bhattacharya, Arijit Das, Shruthi SD (2021) Exploring nanomaterials with rhizobacteria in current agricultural scenario. Advances in Nano-Fertilizers and Nano-Pesticides in Agriculture 487–503. https://doi.org/10.1016/B978-0-12-820092-6.00020-3
68. Yadav J, Jasrotia P, Kashyap PL, Bhardwaj AK, Kumar S, Singh M, Singh GP (2022) Nanopesticides: current status and scope for their application in agriculture. Plant Prot Sci 58(1):1–17. https://doi.org/10.17221/102/2020-PPS

Biomedical Application

Antimicrobial Activities of Nanoemulsion

Limenew Abate Worku, Archana Bachheti, Kundan Kumar Chaubey, Rakesh Kumar Bachheti, and Azamal Husen

Abstract According to the World Health Organization, bacterial, viral, fungal, and parasitic infections result in millions of fatalities worldwide. There needs to be more than existing treatments to treat many microbial infections. Nanoemulsions have demonstrated positive antibacterial effects on microorganisms such as bacteria, fungi, and viruses. When two immiscible liquids (water and oil) are combined to form a single phase using a suitable surfactant with a size distribution between 20 and 100 nm, a thermodynamically stable isotropic system is created called a nanoemulsion. Nanoemulsions can be synthesized from oil obtained from fish, sunflower, soybean, algae, flaxseed, olive, safflower, or corn because they are inexpensive and have good nutritional value. The antibacterial properties of nanoemulsions made from eugenol, thymol, geraniol, and menthone essential oil have been investigated for gram-positive and gram-negative bacteria species.Different mechanisms of action, such as rupturing cell membranes, lowering proton-motive force, or interfering with important metabolic processes, were examined to treat nanoemulsions' antibacterial properties. Research results showed that the nanosize characteristics of nanoemulsions provide great potential for antimicrobial applications to treat various fungi species such as *Cladosporium* sp., *Fusarium* sp., and *Penicillium* sp. as well as

L. A. Worku · R. K. Bachheti
Centre of Excellence in Nanotechnology, Addis Ababa Science and Technology University, Ethiopia

Department of Industrial Chemistry, College of Applied Sciences, Addis Ababa Science and Technology University, P.O. Box-16417, Addis Ababa, Ethiopia

A. Bachheti (✉)
Department of Environment Science, Graphic Era University, 248002, Dehradun, Uttarakhand, India

K. K. Chaubey
Division of Research and Innovation, School of Applied and Life Sciences, Uttaranchal University, Arcadia Grant, P.O. Chandanwari, Premnagar, Dehradun Uttarakhand-248007, India

School of Basic and Applied Sciences, Sanskriti University, Mathura, Uttar Pradesh 281401, India

A. Husen
Wolaita Sodo University, P.O. Box: 138, Wolaita, Ethiopia

A. Husen et al. (eds.), *Current Trends in Green Nano-emulsions*, Smart Nanomaterials Technology, https://doi.org/10.1007/978-981-99-5398-1_8

methicillin-resistant bacteria species such as *Staphylococcus aureus*, and *Streptococcus pyogenes*. Overall, this chapter provides a complete and up-to-date overview of how nanoemulsions are utilized to treat microbial infections and how they could be employed to develop novel antimicrobial drugs.

Keywords Antimicrobial activities · Nanoemulsion · Formulation of nanoemulsions

1 Introduction

A nanoemulsion is a heterogeneous system made up of two immiscible phases, one of which is an oil phase and the other an aqueous phase. The droplet size ranges from 5 to 200 nm. It is transparent, optically clear, and has a steady thermodynamic state [1]. High-energy processes, including high-pressure homogenizers, high-shear agitation, and ultrasound generators, create Nanoemulsions [2].

There are three different kinds of nanoemulsions, such as water in oil in water, oil in water, oil in water in oil, and water in oil, depending on their composition. The main distinction between emulsion and nanoemulsion is that the latter is thermodynamically and kinetically stable, whereas the former is unstable. In contrast to clear and translucent nanoemulsions, emulsions are hazy. Emulsions need a lot of energy input; however, nanoemulsions can form with or without a lot of energy input (sometimes spontaneously) [3]. Emulsions are less free-energy than nanoemulsions and have a smaller surface area to volume ratio. Compared to nanoemulsion, emulsions require a higher amount of surfactant to be prepared; for example, 20–25% surfactant is added, whereas 5–10% surfactant is added in the case of nanoemulsion [4]. Ostwald ripening is an important mechanism of nanoemulsion disintegration [5].

The physicochemical characteristics of nanoemulsions for use in practical applications are intriguing because of the long-term stability of nanoemulsions and the very smaller size of droplets. [6], used nanoemulsions drug delivery system in the agricultural sector for controlling pesticides, as well as for the protection of antimicrobial, antibiofilm, and medication delivery purposes in cosmetics and food industries (Fig. 1) [7]. In cosmetics as drug carriers for personal care or skin care products [8], as well as in pharmaceuticals as a matrix for encapsulating the bioactive chemicals desired for formulation without alcohol [9].

Nanoemulsions are widely employed for various purposes, including vaccines, DNA-encoded drugs, antibiotics, cosmetics, and topical pin oils. The primary distinction is reparations, which can be administered through various channels, including oral, pulmonary, ocular, and transdermal [10, 11]. There are several possible benefits of nanoemulsions over emulsions for encapsulating functional lipophilic components. Nanoemulsions' tiny droplet sizes considerably slowed down the rate at which destabilizing processes including gravity separation, flocculation, and coalescence took place. The fact that nanoemulsions are transparent or only mildly cloudy due

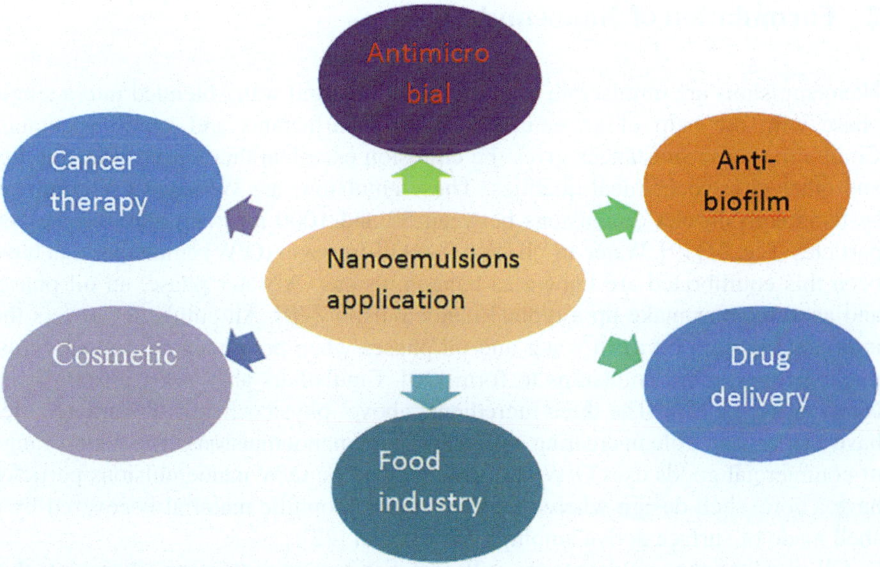

Fig. 1 Different applications of Nanoemulsions

to the small droplet size is another possible benefit [12]. The industries of pharmaceutical, food, and cosmetics may find nanoemulsions to be more advantageous than traditional emulsions in terms of their high bioavailability, low turbidity, sterilization by filtration, and strong physical stability. Pharmaceutical companies use nanoemulsions as delivery systems for drugs and other lipophilic bioactive substances, food companies use them for flavours and antimicrobials, agrochemical companies use them for water-insoluble pesticides, and the cosmetics industry uses them for skincare and other personal care products [13, 14]. Nanoemulsions are the most cutting-edge nanoparticle methods for the systemic delivery of active pharmaceuticals for regulated or sustained medication delivery and targeting [15].

Forming new antibacterial agents to combat drug-resistant bacteria is difficult but also crucial and essential. Researchers have attempted to produce nanoemulsions with desirable antibacterial characteristics and numerous potential targets or sophisticated processes. The bacterial cell membrane contains charged lipids with strong electrostatic attraction and excellent binding efficiency [16]. The nanoemulsions that comprise antibacterial nanodroplets have healthy-promoting bioactive chemicals in their core and charged lipids in their outer shell layers. Colloids of tiny droplets make up nanoemulsions, which exhibit persistent physicochemical characteristics over an extended period [17]. Herbal plant oil nanoemulsions exhibit extensive antibacterial properties. Because they damage the bacteria's outer membranes, they represent a new class of antimicrobials for treating Gram-positive and Gram-negative bacterial infections [18]. This book chapter's goals is to examine a nanoemulsion's antibacterial properties.

2 Formulation of Nanoemulsions

Nanoemulsions are immiscible liquids made of oil and water blended into a single phase with the help of an emulsifier like co-surfactants and such surfactants. Combining these substances gives the emulsion excellent thermodynamic stability and other physicochemical qualities. These emulsions are W/O or O/W emulsion systems with droplet dimensions between 50 and 1000 nm with nanometer-sized particles (Fig. 2) [19]. Water-in-oil (W/O) or Oil-in-water (O/W) emulsions that have been this equilibrated are known as nanoemulsions. A water phase, an oil phase, and an emulsifier make up a typical nanoemulsion [20]. An emulsifier lowers the interfacial tension between water and oil phases when present in modest amounts, making it easier for emulsions to form [20]. Emulsifiers also assist in stabilizing Nanoemulsions [21]. The three ingredients above' physicochemical characteristics have a significant role in creating and maintaining nanoemulsions. The widest range of commercial goods uses O/W nanoemulsions. The O/W nanoemulsions particles have a core–shell design where a core made of lipophilic material is covered by a shell made of surface-active amphiphilic material [22].

Oil-in-water nanoemulsions are colloidal dispersions containing oil droplets that have a diameter of 200 nm or less and are kinetically stable/thermodynamically unstable [23]. One of two procedures—low-energy or high-energy techniques—can be used to make them [24]. The system's intrinsic chemical energy is used in low-energy operations, either with or without mild agitation, to create tiny oil droplets

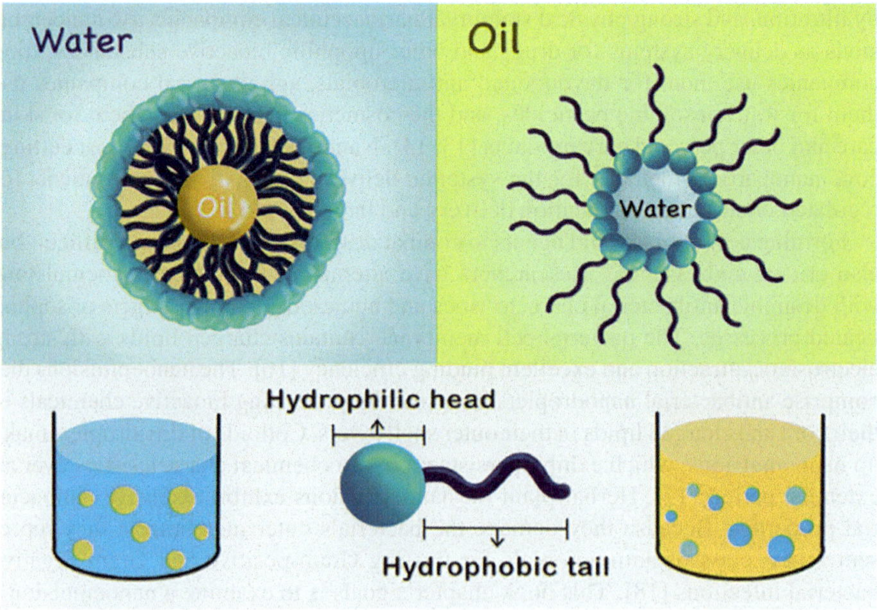

Fig. 2 Water-in-oil and oil-in-water emulsions [19]

[25]. On the other hand, sonicators, microfluidizers, and high-pressure homogenizers use powerful disruptive forces to separate the oil and water phases to produce tiny oil droplets [24, 25]. Although the low-energy approaches usually require a large amount of synthetic surfactants, they do not require any particular equipment [26].

2.1 Oil Phase

The chemical stability, interfacial tension, refractive index, polarity, density, water solubility, as well as viscosity, and of the oil phase have a considerable impact on the characteristics of Nanoemulsions [27–29]. Mineral oils, wax, triacylglycerols (TAG), diacylglycerols (DAG), monoacylglycerols (MAG), and free fatty acids (FFA), and various lipophilic nutraceuticals can all be used to create food-grade Nanoemulsions [22]. The most popular TAG oils used in nanoemulsions are those extracted from, safflower, fish, sunflower, algae, olive, flaxseed, corn, or soybeanor because they are inexpensive and have good nutritional value [22].

2.2 Aqueous Phase

Water can be combined with various alcoholic solvents, minerals, different acids, proteins, carbohydrates, and other polar molecules, which can then create food-grade Nanoemulsions [28]. The choice of the aqueous phase largely influences the physicochemical characteristics of the synthesized nanoemulsion.

2.3 Stabilizers

One of the most crucial things to consider for the efficient creation of nanoemulsions is the selection of the appropriate stabilizer since stabilizers impact Nanoemulsions' long-term stability [20, 28, 30]. Stabilizers include, among other things, emulsifiers, ripening inhibitors, and texture modifiers. Emulsifiers are the stabilizers in nanoemulsions that are most frequently used. Proteins, polysaccharides, small-molecule surfactants, and phospholipids are a few emulsifiers that can be added [31].

3 Antimicrobial Application of Nanoemulsions

3.1 Antibacterial Activities

As a result of improper and inappropriate usage of antibacterial drugs by patients and healthcare personnel, antibacterial resistance is quickly emerging in bacteria [32]. Novel antibacterial and antifungal drugs are urgently required to address the ineffectiveness of currently available pharmaceutical treatments, which is brought on by the rise of bacterial and fungal strains that are becoming more and more resistant. As a basic material for antibacterial activity, nanoemulsion is crucial. The antibacterial characteristics were determined using a well diffusion approach of *Cinnamomum Cassia* L. essential oil nanoemulsion, towards the gram-negative bacterium such as *K. pneumonia.* With a MIC of 0.025%, the nanoemulsion exhibited the highest level of inhibition against *K. pneumonia.* The remarkable antibacterial effects of *Cinnamomum Cassia* essential oil nanoemulsion are probably caused by their small size, low zeta potential (-5.00 mV), and the arrangement of plant bioactive compound [33]. These results confirmed that the concentration of *Cinnamomum Cassia* essential oil nanoemulsion and the size of the inhibitory zone that results from it have a linear connection [34]. Raghupathi and colleagues' findings suggest that using particles with small dimensions can considerably reduce the growth of bacteria [35]. According to research by [36], the proportion of bacteria destroyed by a specific particle rose in direct proportion to its size.

According to a study by [37], eugenol nanoemulsions made using a high-speed shearing process showed significant suppression of *Escherichia coli* and *Staphylococcus aureus.* SEM pictures of both bacteria treated by the nanoemulsion revealed severe deformation and membrane disruption. Thymol, an element of plant essential oils, was used to create an antibacterial nanoemulsion in the study [38]. Sonication was used to create the emulsion. Cryo-FESEM, TEM, FTIR spectroscopy, and DLS were used to analyze the nanoemulsion. Dilution stability, pH value, and the creaming index were also researched for practical use. The result showed that nanoscale thymol could be a potential antimicrobial agent. In a different investigation, the antibacterial effectiveness of thymol nanoemulsion and thymol towards *C. perfringens, E. coli,* and *S. aureus* were examined over the course of four weeks on a sausage product. The zeta potential of the droplets formed from thymol nanoemulsions -0.86 mV, the smaller size of 86.39 nm. Thymol's MBC and MIC values were roughly twice as high as those of thymol-nanoemulsion [30]. In the research of [39], nanoemulsions were formed from eugenol oil for antibacterial activities, with lecithin and gum arabic serving as food-grade natural emulsifiers. Their research compared the effects of freeze-drying and spray-drying on the dispersibility and morphology of nanoemulsion powders. Investigations were done into the best production technique, stability, antibacterial activity, and physicochemical and structural characterization. As a co-surfactant, ethanol was added to the aqueous phase mixture containing 0.5% gum arabic, 0.5% lecithin, and 1.25 percent eugenol oil to form the nanoemulsions, which had a particle size of 103.6 nm (Fig. 3).

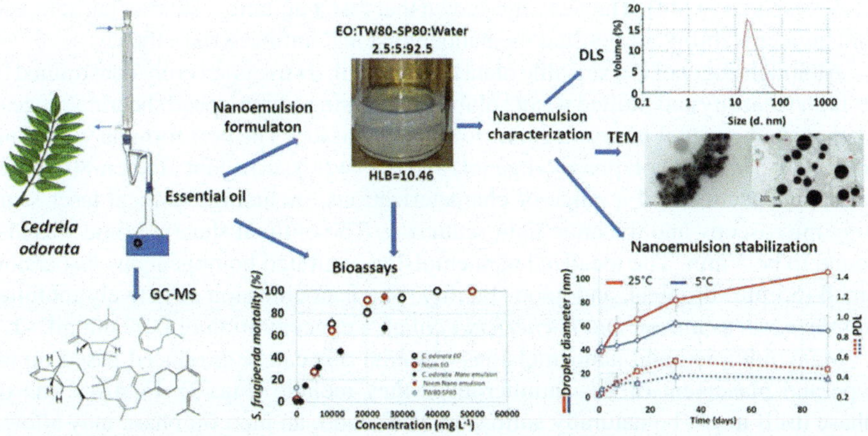

Fig. 3 Formulation of *Cedrela odorata* essential oil and its larvicidal effect against *Spodoptera frugiperda* [40]

The ability of Bergamot essential oil to preserve food is widely documented. Unfortunately, the low bioaccessibility and low solubility ofwater-made bergamot essential oil have restricted its use in other contexts. The conversion of Bergamot essential oil into nanoemulsions was examined to improve its water dispersibility and promote its antibacterial activity against *Saccharomyces cerevisiae*, *Escherichia coli,* and *Lactobacillus delbrueckii.* Mung bean, alfa seed, broccoli, and radish seeds contaminated with Salmonella enteritis and *Escherichia coli* were demonstrated to be efficiently inhibited by carvacrol nanoemulsions made using this technique [41]. Interestingly, the antibacterial activity of all the studied oils was greatly increased by applying Bergamot essential oil Nanoemulsions [42]. One study used medium-chain triglyceride and Tween 80 as the surfactant and co-surfactant to generate Geraniol nanoemulsions through spontaneous emulsification. Analysis was done on the droplet's physical and chemical characteristics, such as stability, viscosity, PDI, appearance morphology, zeta potential, pH, contact angle, and average particle size. The outcomes demonstrated that nanoemulsions' zeta potential was -17.95 mV, with a particle size of 90.33 $\pm$ 5.23 nm, and PDI value of 0.058 having encapsulation efficiency greater than 90%. The produced Geraniol nanoemulsions are fairly stable in long-term storage at 4∘C at a pH of 5. Moreover, Geraniol nanoemulsions had a larger inhibition property on *Salmonella typhimurium*, *Staphylococcus aureus*, *Escherichia coli*, and *Listeria monocytogenes* [43].

Citrus canker is a bacterial disease that affects citrus crops brought on by Xanthomonas microorganisms, significantly harming citrus production. In order to control citrus canker, nano emulsification can be used instead of utilizing the antibacterial activity of natural substances. The antibacterial properties of nanoemulsions based on the main components of essential oils eugenol, thymol, geraniol, and menthone were examined against two strains of Xanthomonas that cause citrus canker in the research of [44], The outcome shown that faster agitation, smaller particle size, and greater stability all boosted the nanoemulsions' antibacterial efficacy.

Although eugenol is a versatile plant essential oil, its use is severely constrained by its high volatility and limited water solubility. By using a high-speed shearing process, [19] created eugenol nanoemulsions to solve this issue. The best formula was found to contain 5% (w/w) oil phase (eugenol) and 8% (w/w) surfactant (Tween-80), based on eye inspection and a range of characterizations, including confocal laser scanning microscopy and dynamic light scattering. The optimal shearing time was also found to be 5 min. The idealized nanoemulsion exhibited homogeneous dispersion, small (85 nm) droplets, and good stability. The nanoemulsion effectively inhibited *Staphylococcus aureus* and *Escherichia coli*at, a concentration of 0.02 mg mL -1.

Many antimicrobial nanoemulsions essential oils can be produced based on the type and placement of the antimicrobials they include (Fig. 3). First off, the oil phase itself might be naturally antibacterial. Second, an inert oil phase may allow a polyphenol or other lipophilic antimicrobial to degrade, for example, corn oil. Fourth, a nanoemulsion might incorporate various antimicrobials in various places, resulting in a highly effective formulation. Lastly, an antimicrobial emulsifier may create and maintain the nanoemulsions. Various methods, such as short molecular diffusion or collisions, can transfer antimicrobial substances from nanoparticle surfaces to microbial ones [45] (Fig. 4).

3.2 *Antifungal Activities*

Synthetic fungicides are thought to be the most efficient treatment for various fungal illnesses. Their prolonged use, nonetheless, has brought about unfavorable pathogen resistance. Also, their residues in soil and food harm both the environment and humans [46, 47]. As a result, researchers are working to create alternative, eco-friendly, and safe antifungal substances, including nanoparticles made of particular vegetable oils or plants [48]. An investigation into the antifungal properties of a nanoemulsion made from *Croton cajucara* essential oil was done. The aqueous phase of the *Croton cajucara* nanoemulsion was created using an ultrasonic processor. The best *Croton cajucara* nanoemulsion antifungal activity was discovered against *Candida albicans* and *Mucor ramosissimus*, with MIC result of 12.2 and 25.6 g/mL, respectively. At pH 3 and pH 4, respectively, *Mucor ramosissimus* and *Candida albicans* showed the highest extracellular protease activity (Azevedo et al., 2021b). The study deals with formulating and characterizing a bio-based oil-in-water nanoemulsion [49]. DLC, stability tests, TEM, and TLC, were used to characterize the synthesized eugenol oil nanoemulsion. The Z-average diameter of the nanoemulsion droplets was determined to be 80 nm, and a TEM investigation confirms the spherical shape of the eugenol oil nanoemulsion (EON).In vitro and post-application on Minas Padro cheese [50] assess the antifungal efficacy of nanoemulsions containing oregano essential oil (Origanum vulgare). The phase inversion temperature approach was used to produce nanodispersions. The genera *Cladosporium, Fusarium,* and *Penicillium*

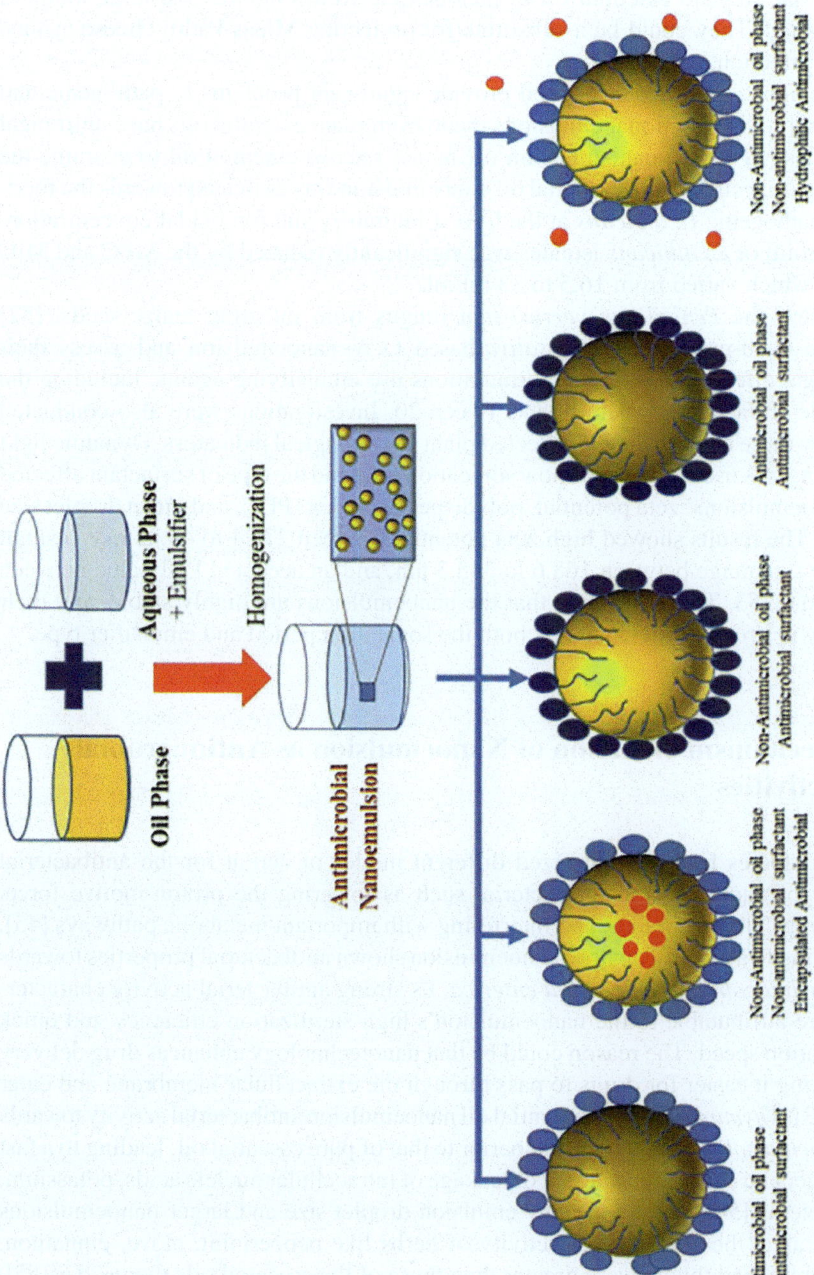

Fig. 4 Antimicrobial nanoemulsion classification depending on the location of the antibacterial ingredients [45]

were isolated and tested for antifungal activity from cheese samples. It was determined that oregano essential oil in nanoparticle form inhibited the three kinds of fungi tested. They could be a substitute for protecting Minas Padro cheese against fungus infection.

Infections associated with food provide significant problems to pathogenic and spoilage fungi. Traditional emulsions made from plant essential oils have antifungal properties [51] use a nanoemulsion of*Cleome viscosa* essential oil to examine the antifungal spectrum. The essential oil nanoemulsion was evaluated towards the food-borne pathogenic *C. albicans* at the lowest inhibitory and fungicidal concentration. The biofilm of *C. albicans* isolates was significantly reduced by the MFC and MIC values, which varied from 16.5 to 33 ml/ml.

To stop the *Penicillium verrucosum* fungus from infecting maize seeds, [52] produce Cold-pressed *Nigella sativa*-based O/W nanoemulsion and assess their antifungal effectiveness. These formulations use emulsifying agents, including the nonionic surfactants Tween 80 and Tween 20. Investigations were also conducted into how these nanoemulsions affected plant physiological indicators. Dynamic light scattering was used to examine how sonication time and the type of surfactant affected the nanoemulsions' zeta potential, polydispersity index (PDI), and mean droplet size (DLS). The results showed high zeta potential between 27.24 to 48.82 mV, a small particle size range between 168.6 to 345.3 nm, and an accepted PDI value between 0.181 to 0.353. This confirmed that the nanoemulsions are highly stable, and their stability is strongly influenced by both the sonication period and emulsifier type.

4 Mechanism of Action of Nanoemulsion as Antimicrobial Activities

Several studies have hypothesized different modes of action for the antibacterial activity of nanoemulsion on bacteria, such as lowering the proton-motive force, disrupting cell membranes, and interfering with important metabolic pathways [45]. According to a report, Linalool nanoemulsion shown antibacterial properties towards *Escherichia coli* and *Salmonella enterica.* Its strong antibacterial activity characteristics are attributable to the nanoemulsion's high sterilization efficiency and quick sterilization speed. The reason could be that nanotechnology enhances drug delivery by making it easier for drugs to pass through the extracellular membrane and enter cells [53]. *Thymus daenensis* essential oil nanoemulsion antibacterial activity towards *Escherichia coli* was noticeably superior to that of pure essential oil, leading to a fast and larger and exceed the amount of leakage of intracellular nucleic acids, potassium, and proteins [54]. Due to the tiny emulsion droplet size and larger nanoemulsions surface area, the antibacterial activity of herbs like peppermint, clove, cinnamon, lemongrass, and thyme were greater than those of the essential oils themselves[55]. The antibacterial effect of a medicine cannot always be enhanced through nanoemulsion production, but also associated with rupturing the bacterial species' outer membrane. For instance, in the research of [56], it was found that Linalool nanoemulsion was highly detrimental to the outer membrane of *A. hydrophila*, and TEM observed the leaking of sizeable amounts of internal contents. This cell membrane rupture harms the bacteria since it prevents the bacterial membrane's regular function of transporting materials [57]. Based on the results of SEM and TEM, their theory proposed that when the cell wall and membrane were disrupted by Linalool nanoemulsion, the intracellular material was initially ejected outwards, leading to the cells' collapse.

Subsequently, the exterior water started to seep into the cells, causing them to expand, and eventually, the loss of water caused the cells to become entirely distorted. Because of this antibacterial mechanism, linalool nanoemulsion should be employed as a promising antibacterial agent.

5 Conclusion and Future Prospects

The chapter on nanoemulsion's antibacterial activity emphasises the technology's potential for limiting a variety of microbial diseases. Numerous microorganisms, such as bacteria, fungi, viruses, and protozoa, have been proven to be resistant to nanoemulsions. The small droplet size, high surface area to volume ratio, and higher solubility of antimicrobial agents in nanoemulsion are some of its distinctive features that contribute to its enhanced antibacterial activity. Research findings indicated that nano-size properties of Nanoemulsions gave an excellent potential for antimicrobial applications to treat methicillin-resistant bacteria species such as *Staphylococcus aureus* (MRSA), *Streptococcus pyogenes* and different fungi species such as *Cladosporium* sp., *Fusarium* sp., and *Penicillium* sp. the nanoemulsion antimicrobial action can be affected by the type, concentration and drop size of nanoemulsions. The antimicrobial activities of nanoemulsion are due to several nanoemulsion treatment mechanisms such as cell membrane damage, a reduction in proton-motive force, and interference with important metabolic pathways. Nanoemulsion's potential for usage in antibacterial applications is promising for other application such as food, medicine, and cosmetics. The usage of nanoemulsion in antimicrobial applications is expected to increase in the upcoming years as scientists continue to experiment with new formulations and use for this technology.

References

1. Shah P, Bhalodia D, Shelat P (2010) Nanoemulsion: a pharmaceutical review. Syst Rev Pharm 1(1)
2. Azevedo MM, Almeida CA, Chaves FC, Ricci-Júnior E, Garcia AR, Rodrigues IA, Alviano CS, Alviano DS (2021) Croton cajucara essential oil nanoemulsion and its antifungal activities. Processes 9(11):1872

3. Devarajan V, Ravichandran V (2011) Nanoemulsions: as modified drug delivery tool. Int J Compr Pharm 2(4):1–6
4. Hamouda T, Hayes MM, Cao Z, Tonda R, Johnson K, Wright DC, Brisker J, Baker JR Jr (1999) A novel surfactant nanoemulsion with broad-spectrum sporicidal activity against Bacillus species. J Infect Dis 180(6):1939–1949
5. Subhaswaraj P, Siddhardha B (2020) Nanoemulsions for antimicrobial and anti-biofilm applications. Nanostructures for Antimicrobial Antibiofilm Applications, 347–373
6. Feng J, Shi Y, Yu Q, Sun C, Yang G (2016) Effect of emulsifying process on stability of pesticide nanoemulsions. Colloids Surf Physicochem Eng Aspects 497:286–292
7. Sonneville-Aubrun O, Simonnet J-T, L'alloret F (2004) Nanoemulsions: a new vehicle for skincare products. Adv Colloid Int Sci 108:145–149
8. Somayeh S, Sedigheh A, Mohsen R, Hamid-Reza A (2021) Antimicrobial activity of thymol and thymol-nanoemulsion against three food-borne pathogens inoculated in a sausage model. Food Bioprocess Technol. https://doi.org/10.1007/s11947-021-02689-w
9. Campani V, Biondi M, Mayol L, Cilurzo F, Pitaro M, De Rosa G (2016) Development of nanoemulsions for topical delivery of vitamin K1. Int J Pharm 511(1):170–177
10. Sharma N, Bansal M, Visht S, Sharma P, Kulkarni G (2010) Nanoemulsion: a new concept of delivery system. Chronicles of Young Sci 1(2):2–6
11. Taylor P (1998) Ostwald ripening in emulsions. Adv Colloid Int Sci 75(2):107–163
12. Chang Y, Mclandsborough L, Mcclements DJ (2012) Physical properties and antimicrobial efficacy of thyme oil nanoemulsions: Influence of ripening inhibitors. J Agricul Food Chem 60(48):12056–12063
13. Ghosh V, Mukherjee A, Chandrasekaran N (2013) Ultrasonic emulsification of food-grade nanoemulsion formulation and evaluation of its bactericidal activity. Ultrason Sonochem 20(1):338–344
14. Rodrigues IA, Ramos ADS, Falcão DQ, Ferreira JLP, Basso SL, Silva JRDA, Amaral ACF (2018) Development of nanoemulsions to enhance the antileishmanial activity of copaifera paupera oleoresins. BioMed Research International
15. Savardekar P, Bajaj A (2016) Nanoemulsions-a review. Inter J Res Pharm Chem 6:312–322
16. Matsuzaki K (1999) Why and how are peptide–lipid interactions utilized for self-defense? Magainins and tachyplesins as archetypes. Biochimica et Biophysica Acta -Biomembranes 1462(1–2):1–10
17. Balasubramani S, Rajendhiran T, Moola AK, Diana RKB (2017) Development of nanoemulsion from Vitex negundo L. essential oil and their efficacy of antioxidant, antimicrobial and larvicidal activities (Aedes aegypti L.). Environ Sci Poll Res 24:15125–15133
18. Hamouda T, Baker J (2000) Antimicrobial mechanism of action of surfactant lipid preparations in enteric Gram-negative bacilli. J Appl Microbiol 89(3):397–403
19. Fu, Xuan, Yuan G, Yan W, Ziluo Z, Sarker S, Yin Y, Qi, Liu, Feng J, Jing C (2022) Preparation of eugenol nanoemulsions for antibacterial activities. RSC Adv 12:3180–3190
20. Gupta A, Eral HB, Hatton TA, Doyle PS (2016) Nanoemulsions: formation, properties and applications. Soft Matter 12(11):2826–2841
21. Mason TG, Wilking JN, Meleson K, Chang CB, Graves SM (2006) Nanoemulsions: formation, structure, and physical properties. J Phys Condens Matter 18(41):R635
22. Mcclements DJ, Rao J (2011) Food-grade nanoemulsions: formulation, fabrication, properties, performance, biological fate, and potential toxicity. Critical Rev Food Sci Nut 51(4):285–330
23. Salvia-Trujillo L, Martín-Belloso O, Mcclements DJ (2016) Excipient nanoemulsions for improving oral bioavailability of bioactives. Nanomaterials 6(1):17
24. Jiang T, Liao W, Charcosset C (2020) Recent advances in encapsulation of curcumin in nanoemulsions: a review of encapsulation technologies, bioaccessibility and applications. Food Res Int 132:109035
25. Ren G, Sun Z, Wang Z, Zheng X, Xu Z, Sun D (2019) Nanoemulsion formation by the phase inversion temperature method using polyoxypropylene surfactants. J Colloid Interf Sci 540:177–184

26. Borrin TR, Georges EL, Moraes IC, Pinho SC (2016) Curcumin-loaded nanoemulsions produced by the emulsion inversion point (EIP) method: An evaluation of process parameters and physico-chemical stability. J Food Eng 169:1–9
27. Silva HD, Cerqueira MÂ, Vicente AA (2012) Nanoemulsions for food applications: development and characterization. Food Bioproc Technol 5:854–867
28. Mcclements DJ (2011) Edible nanoemulsions: fabrication, properties, and functional performance. Soft Matter 7(6):2297–2316
29. Gupta A, Eral HB, Hatton TA, Doyle PSJSM (2016b) Nanoemulsions: formation, properties and applications 12(11):2826–2841
30. Silva HD, Cerqueira MÂ, Vicente AA (2012b) Nanoemulsions for food applications: development and characterization. Food Bioproc Technol 5:854-867
31. Salem MA, Ezzat SM (2019) Nanoemulsions in food industry. Some New Aspects Of Colloidal Syst Foods 2:238–267
32. Sahoo KC, Tamhankar AJ, Johansson E, Lundborg CS (2010) Antibiotic use, resistance development and environmental factors: a qualitative study among healthcare professionals in Orissa, India. BMC Public Health 10:1–10
33. Aftab A, Mohammad J, Mohammed H, Mohammad A, Mohammad R (2023) Antioxidant, antibacterial, and anticancer activity of ultrasonic nanoemulsion of cinnamomum cassia L. Essential Oil. Plants 12(834):1–14
34. Gharari Z, Hanachi P, Sadeghinia H, Walker T (2022) Cichorium intybus bio-callus synthesized silver nanoparticles: a promising antioxidant, antibacterial and anticancer compound. Int J Pharm 625:122062
35. Raghupathi K, Koodali R, Manna A (2011) Size-dependent bacterial growth inhibition and mechanism of antibacterial activity of zinc oxide nanoparticles. Langmuir 27:4020–4028
36. Jeong Y, Lim DW, Choi J (2014) Assessment of size-dependent antimicrobial and cytotoxic properties of silver nanoparticles. Adv Mater Sci Eng C 2014:1–6
37. Xuan F, Yuan G, Weiyao Y, Ziluo Z, Shovra S, Yinyan Y, Qi L, Jianguo F, Jing C (2022) Preparation of eugenol nanoemulsions for antibacterial activities. RSC Adv 12:3180–3190
38. Sarita K, Kumaraswamy RV, Ram C, Sharma S, Ajay P, Ramesh R, Pratim B Vinod S (2018) Thymol nanoemulsion exhibits potential antibacterial activity against bacterial pustule disease and growth promotory effect on soybean
39. Hu Q, Gerhard H, Upadhyaya I, Venkitanarayanan K, Luo Y (2016) Antimicrobial eugenol nanoemulsion prepared by gum arabic and lecithin and evaluation of drying technologies. Int J Biol Macromol 87:130–140
40. Lemus de la Cruz AS, Barrera-Cortés J, Lina-García LP, Ramos-Valdivia AC, Santillán R (2022) Nanoemulsified formulation of *Cedrela odorata* essential oil and its larvicidal effect against *Spodoptera frugiperda* (JE Smith). Molecules 27(9):2975
41. Landry K, Chang Y, Mcclements D, Mclandsborough L (2014) Effectiveness of a novel spontaneous carvacrol nanoemulsion against Salmonella enterica Enteritidis and Escherichia coli O157:H7 on contaminated mung bean and alfalfa seeds. Int J Food Microbiol 187:15–21
42. Marchese E, D'onofrio N, Balestrieri ML, Castaldo D, Ferrari G, Donsì F (2020) Bergamot essential oil nanoemulsions: antimicrobial and cytotoxic activity. Zeitschrift für Naturforschung C 75(7–8):279–290
43. Feng X, Feng K, Zheng Q, Tan W, Zhong W, Liao C, Liu Y, Li S, Hu W (2022) Preparation and characterization of geraniol nanoemulsions and its antibacterial activity. Front Microbiol 13
44. Da Cruz Silva G, De Oliveira Filho JG, De Mori Morselli Ribeiro M, De Souza CWO, Ferreira MD (2021) Antibacterial activity of nanoemulsions based on essential oils compounds against species of Xanthomonas that cause citrus canker. Biointerface Res Appl Chem 12:1835-1846
45. Mcclements D, Arun K, Pubali D, Pramod K, Niloy C (2021) Nanoemulsion-based technologies for delivering natural plant-based antimicrobials in foods. Nat Antimicrobial Nanoem 5:1–23
46. Alghuthaymi MA, El-Sersy ZR, Tayel AA, Alsieni MA, Abd El Maksoud AI (2021) Anticandidal potentiality of biosynthesized and decorated nanometals with fucoidan. Green Proc Syn 10(1):811–823

47. Mustafa IF, Hussein MZ (2020) Synthesis and technology of nanoemulsion-based pesticide formulation. Nanomaterials 10(8):1608
48. Thiagarajan P (2011) Nanoemulsions for drug delivery through different routes. Research in Biotechnology
49. Abd-Elsalam KA, Khokhlov AR (2015) Eugenol oil nanoemulsion: antifungal activity against Fusarium oxysporum f. sp. vasinfectum and phytotoxicity on cottonseeds. Appl Nanosci 5:255–265
50. Bedoya-Serna CM, Dacanal GC, Fernandes AM, Pinho SC (2018) Antifungal activity of nanoemulsions encapsulating oregano (Origanum vulgare) essential oil: in vitro study and application in Minas Padrão cheese. Braz J Microbiol 49:929–935
51. Krishnamoorthy R, Gassem MA, Athinarayanan J, Periyasamy VS, Prasad S, Alshatwi AA (2021) Antifungal activity of nanoemulsion from Cleome viscosa essential oil against food-borne pathogenic Candida albicans. Saudi J Biol Sci 28(1):286–293
52. Mosa MA, Youssef K, Hamed SF, Hashim AF (2023) Antifungal activity of eco-safe nanoemulsions based on Nigella sativa oil against Penicillium verrucosum infecting maize seeds: biochemical and physiological traits. Front Microbiol. https://doi.org/10.3389/fmicb.2022.1108733
53. Ahmed K, Li Y, Mcclements DJ, Xiao H (2012) Nanoemulsion-and emulsion-based delivery systems for curcumin: encapsulation and release properties. Food Chem 132(2):799–807
54. Moghimi R, Ghaderi L, Rafati H, Aliahmadi A, Mcclements DJ (2016) Superior antibacterial activity of nanoemulsion of Thymus daenensis essential oil against E. coli. Food Chem 194:410–415
55. Vickers NJ (2017) Animal communication: when I'm calling you, will you answer too? Curr Biol 27(14):R713–R715
56. Yukuyama M, Ghisleni DDM, Pinto TDJA, Bou-Chacra NA (2016) Nanoemulsion: process selection and application in cosmetics–a review. Int J Cosmet Sci 38(1):13–24
57. Daood U, Matinlinna JP, Pichika MR, Mak K-K, Nagendrababu V, Fawzy AS (2020) A quaternary ammonium silane antimicrobial triggers bacterial membrane and biofilm destruction. Sci Rep 10(1):1–14
58. Wan J, Zhong S, Schwarz P, Chen B, Rao J (2019) Physical properties, antifungal and mycotoxin inhibitory activities of five essential oil nanoemulsions: impact of oil compositions and processing parameters. Food Chem 291:199–206
59. Zhong W, Tang P, Liu T, Zhao T, Guo J, Gao Z (2021) Linalool nanoemulsion preparation, characterization and antimicrobial activity against Aeromonas hydrophila. Int J Mol Sci 22(20):11003

Role of Nanoemulsion in Lung Cancer Treatment

Aashna Sinha, Chetan Shrivastava, Anish Kumar Pal, Atreyi Pramanik, Pranchal Rajput, Kundan Kumar Chaubey, Sujata Hariharan, Ashok Kumar, Rajesh Prasad Jayaswal, Rakesh Kumar Bachheti, and Archana Bachheti

Abstract Lung cancer, a collection of dangerous malignant tumours, is one of the major contributors to new cancer cases and cancer-related mortality. Lung cancer has a bad outcome and a relatively low life expectancy due to late detection and the lack of effectiveness of traditional therapy. Thus, drug delivery techniques that might prevent harm to healthy cells, increase treatment effectiveness, and serve as imaging instruments are receiving much interest. Unique drugs based on nanoscales have been created due to advances in material science, giving patients with lung cancer fresh hope. Because of their outstanding capacities to dissolve polar anti-cancer drugs, high thermal stability, bioactivity, and target ability, nanoemulsions (NEs) represent a viable platform for cancer treatment. In the continuous phase, NEs may be able to do various tasks with their submicron dispersed colloids, which helps overcome biological barriers that prevent traditional chemotherapeutic drug delivery from reaching the target area of lung cancer. By modifying it to target the cancer cells' surroundings better, surface-engineered NE, combined with chemotherapy, can

A. Sinha · C. Shrivastava · A. K. Pal · A. Pramanik · P. Rajput · K. K. Chaubey (✉) · S. Hariharan
Division of Research and Innovation, School of Applied and Life Sciences, Uttaranchal University, Prem Nagar, P.O. Chandanwari, Arcadia Grant, Dehradun 248007, Uttarakhand, India
e-mail: kundan2006chaubey@gmail.com

A. Kumar
Animal Health Division, Central Institute for Research On Goats, Makhdoom, Farah, Mathura 281122, Uttar Pradesh, India

R. K. Bachheti
Centre of Excellence in Nanotechnology, Department of Industrial Chemistry, Addis Ababa Sciences and Technology University, P.O. Box-16417, Addis Ababa, Ethiopia

A. Bachheti
Department of Environment Science, Graphic Era University, Dehradun 248002, Uttarakhand, India

R. P. Jayaswal
School of Medical and Allied Sciences, Sanskriti University, Mathura 281401, Uttar Pradesh, India

K. K. Chaubey
School of Basic and Applied Sciences, Sanskriti University, Mathura 281401, Uttar Pradesh, India

A. Husen et al. (eds.), *Current Trends in Green Nano-emulsions*, Smart Nanomaterials Technology, https://doi.org/10.1007/978-981-99-5398-1_9

transform the way lung cancer is treated. This chapter aims to give an overview of the role of nanoemulsion in lung cancer treatment.

Keywords Lung cancer · Nanoemulsions in treatment · Traditional treatment · Clinical aspects · Commercialisation

1 Introduction

People in both industrialized and developing nations are affected by cancer, which has emerged as one of the most important and significant health challenges worldwide [11, 22]. Lung cancer (LC) is one of the scariest diseases malignancies, has a substantial death rate, and is the main factor in fatalities due to tumours.LC, the greatest killer in men, is the second-highest mortal threat for women after breast cancer [19]. Genomic instability comes from lung damage brought on by an overwhelming chronic inflammation and the absence of repair mechanisms for injured cells, which in turn leads to LC [28]. It may be divided into primary and secondary cancers. Small-cell lung cancer (SCLC), non-small-cell lung carcinoma (NSCLC), mesothelioma, sarcoma, and carcinoid are the diverse types of lung carcinoma based on histological analysis. The most prevalent LCs are SCLC and NSCLC, which account for about 90% of all cases, while other types have been uncommon [41]. The four more NSCLC subtypes are squamous cell carcinoma, Pancoast tumour, adenocarcinoma, and large-cell undifferentiated carcinoma [29, 32]. An aggressively expanding tumour known as SCLC is classified as a combination of small-cell and "oat cell" cancer. The main pathogenesis-related variables for LC include smoking behaviours, heredity, urbanism, and external conditions (such as exposure to arsenic, toxins, and asbestos) [16, 31]. The World Health Organization (WHO) reported that 2.21 million new instances of LC were detected in 2020 alone, resulting in 180 million deaths(*Cancer*, n.d.). By 2035, it is anticipated that up to 3 million people will have died from LC [1].

The cornerstone of current treatment plans has been using traditional methods such as chemotherapy, radiation, and surgical resection, independently or in combination [14, 34]. However, with such conventional therapeutic approaches, serious adverse effects are frequently recorded. Furthermore, radiation harms the surrounding healthy cells; as a result, It should not be used as the first course of therapy for individuals with already severely compromised pulmonary systems since this might result in a loss of lung function [13, 26]. The three chemotherapy drugs for LC that are most frequently prescribed are carboplatin, paclitaxel, and gemcitabine. Because of their severe side effects on the patient, these medications only have a limited therapeutic efficacy [19, 25]. The focused and effective drug delivery toward the tumour site while preventing adverse systemic reactions is another significant problem in treating LC [33]. Using a more effective treatment strategy and an improved security profile, it is possible to administer chemotherapeutics directly to cancer cells; nanotechnology has introduced various unique approaches. Many nano-carriers have been investigated and created better to treat various disorders, including cancer therapy, including solid-lipid nanoparticles, solid-lipid carriers, polymeric micelles, nanoemulsions (NEs), nano-emulgels, liposomes, and nanoparticles [8]. Consequently, this chapter focuses on using NEs for more accurate and successful targeting in the care of LC.

2 Lung Cancer

In the US and worldwide, LC continues to be the leading cause of cancer-related death in both men and women. Smoking and using tobacco products account for around 90% of LC incidences. Nevertheless, other elements, including asbestos, radon gas, exposure to air pollution, and chronic infections, may also play a role in the development of LC. In addition, both acquired and inherited routes of vulnerability to LC have been postulated. Small-cell lung carcinomas (SCLC) and non-small-cell lung carcinomas are LC's two main histologic subtypes. Each has a distinct pattern of development and dissemination (NSCLC). Surgery, radiation, chemotherapy, and targeted therapy are all options for treating LC. Among many other things, the kind and stage of cancer affect the indicated therapeutic techniques. The regulatory networks that control normal cell growth and homeostasis are compromised in LC cells. Various genetic and epigenetic changes are believed to cause a benign LC phenotype to become cancerous, which then progresses to invasive cancer through clonal proliferation [30]. The procedures of penetration, metastasis, and susceptibility to anticancer therapy are all impacted by the ongoing accumulation of genetic and epigenetic aberrations accumulated throughout clonal proliferation following the start of the initial malignancy [30]. These molecular alterations must be identified and characterized for improved disease prevention, early detection, and treatment. Understanding the traits of the tumour and the patient's genetic composition will considerably improve each patient's individual prognosis and recommended course of treatment. Despite advances in diagnosis and therapy over the past 25 years, the outlook for those with LC remains grim. Even for the most firmly attached tumours, conventional treatment is now ineffective. However, a deeper comprehension of the biology behind these complex tumours may create more effective and individualized therapies [23].

2.1 Traditional Treatment for Lung Cancer

1. Non-small cell lung cancer (NSCLC), which represents 85% of all subtypes of the condition, must be handled seriously since it is the most common kind of LC and the major cause of cancer-related fatalities that pose a severe danger to public health globally [3]. The standard of care for NSCLC has produced

outstanding clinical outcomes based on, among other modalities, surgery, radiation, chemotherapy, targeted therapy, and immunotherapy[40] (Fig. 1). Traditional Chinese Medicine (TCM) may only be regarded as adjuvant therapy in the usual treatment of NSCLC due to the absence of significant clinical research that adheres to the concepts of evidence-based medicine and the limits of statistics. While treating NSCLC, using both Chinese and Western therapies has yielded good results. Moreover, it can lessen side effects while increasing the sensitivity of chemotherapy and radiation (such as bone marrow suppression, nausea, and vomiting). Combining chemotherapy with a Chinese herbal combination reduces the toxicity of adjuvant treatment [38]. When used with platinum chemotherapy as opposed to platinum chemotherapy alone, astragalus-based Chinese medicine lessens the side effects of platinum chemotherapy (such as neutropenia, nausea, and vomiting). It is essential to remember that herbal astragalus therapy based on symptom distinction is superior to astragalus oral medicine in terms of effectiveness [39]. Eastern medicine Combining kangai injection with platinum-based chemotherapy has the added benefits of improving clinical efficacy, reducing side events, and regulating tumour immunity [24]. Additional substances like FGS pods, Sun-Bai-Pi extract (SBPE) from traditional Chinese medicine, etc., can also enhance the effects of platinum treatment [21, 37]. Using a particular strategy, such as the active flavone saponin made from Erigeron Breviscapus, the lethal effects of cisplatin-induced apoptosis and autophagy may be lessened via the ERK/p53 and c-met/AKT signalling pathways [35]. Additionally, TCM maintenance treatment had such a greater 1-year rate of survival than maintenance cancer treatment for those suffering from advanced NSCLC who have not progressed after first-line chemotherapy [37]. It could have to do with controlling the blood's level of dissolved cytotoxic T lymphocyte-associated antigen 4. (sCTLA-4) [37].

3 Nanoemulsions in the Treatment of Lung Cancer

The bioavailability of chemotherapy drugs in the tumour microenvironment depends on whether the targeting is active when NE is delivered intravenously (TME) [6, 27]. The complexation of 2-hydroxypropyl-cyclodextrin (HP-CD), lipoid E 80 (a compound that improves permeability), and subsequent transient opening of cellular tight junctions boost the bioavailability of injectable chemotherapeutics [20, 36]. The particular NE design is provided with hydrophobic or lipophilic chemotherapeutics to create hydrophilic and lipophilic conditions. The systemic availability of chemotherapy is eventually increased by this sort of NE design, which lowers hepatic bypass, inhibits P-glycoprotein leakage, avoids drug breakdown in unfavourable circumstances, and enhances mucosal penetration [6].

The anti-cancer drug doxorubicin is employed to treat several malignancies. However, due to its wider dispersing special property and shorter half-life, it poses a risk to healthy tissues. An ionic strength lapidated NE with a doxorubicin prodrug

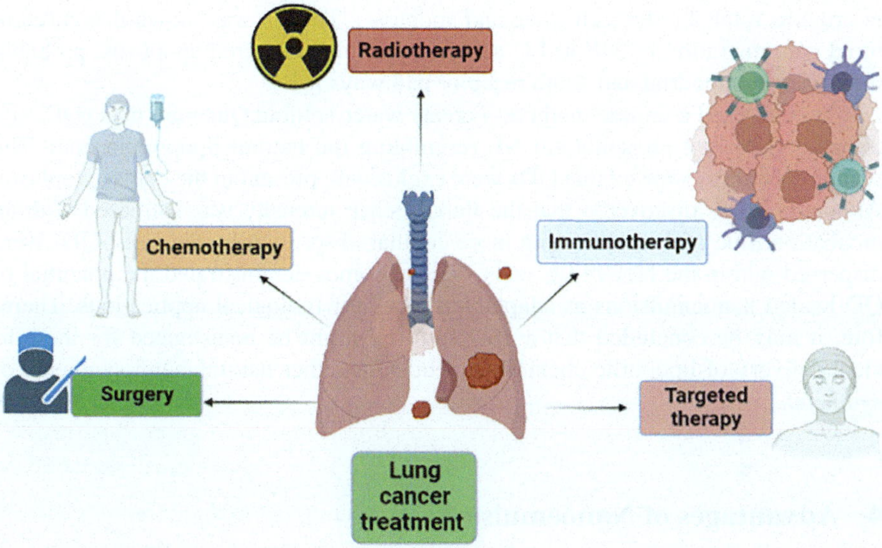

Fig. 1 Traditional treatment of lung cancer

(DNE) with high biocompatibility was created. The mouse model experiment's bioluminescent results demonstrated a significant reduction of DNE in remote lung metastases [9].

Paclitaxel, a popular semi-synthetic taxane chemotherapeutic treatment against solid cancer, is only utilized in specific situations due to its poor solubility, inevitable toxicities, and possibility for P-gp efflux by transporter proteins [10]. As a result, formulation scientists are interested in innovative nanotechnological delivery strategies since they increase this agent's solubility to target cancer cells without causing severe toxicities. Our team has disclosed a Novel approach to enhance the administration of paclitaxel to patients with breast cancer [7]. Using this approach, paclitaxel was more permeable from the Caco-2 cell monolayer, the IC50 value dropped dramatically, DNA was fragmented, and the cytotoxicity was improved.

Using the rhizome of Curcuma longa, Chang and Chen produced a NE of curcuminoids while testing the NE's potency against LC cell lines [5]. Two different cell lines, including those from lung adenocarcinoma (A549 cells) and large lung carcinoma (H460 cells), were treated with curcuminoids and NE of curcuminoids. The scientists suggested a unique strategy for preventing tumour formation for several carcinoma cell lines. It was shown that H460 cells were typically more prone to apoptosis than A549 cells. Enhancing P21 suppression caused the G2/M phase of the cancer cell cycle to be halted, reducing CDK1 and cyclin B interpretation. This curcuminoid NE on H460 cells trial outcomes. These apoptotic pathways opened up new possibilities for identifying other apoptotic pathways. Further research revealed a dose-dependent increase in the expression of caspase-3, caspase-8, and caspase-9 in proteins related to the cell cycle and apoptosis. Thus, it was shown that curcuminoid administration by NE to LC cells successfully triggered apoptosis, probably through mitochondrial and death receptor pathways [5].

Liu et al. used a unique method to create water-soluble Quantum dots (QDs) by trapping QDs in a phospholipid NE resembling the natural lipoprotein core. The scanning pictures showed that QDs were exclusively present in the cell's cytoplasm. Although it was discovered that the fluorescence intensity was increased with an incubation time of 4 h, indicating more cellular absorption by the cells, QDs were dispersed within the H1299 LC cells. The outcomes demonstrated the potential of QD-loaded nanoemulsions as adaptable probes for biological applications. Therefore, it may be concluded that a NE platform might be investigated for the efficient delivery of lipophilic chemotherapeutics, whether natural, semi-synthetic, or synthetic.

4 Advantages of Nanoemulsions

Moreover, nanoemulsions are favourable because of their better drug bioavailability and excellent therapeutic absorption due to their larger surface area. Moreover, it shields the medication against oxidative and UV (ultraviolet) degradation, improving the drug's persistence in the composition [15]. The resilience of nanoemulsions to fungi, viruses, and bacteria is another advantage [15]. While the small-sized globule was easy to infiltrate into the skin through pores and hair follicles, less skin irritation was observed whenever the mean droplet diameter was smaller. This is due to the adjustable droplet size's capacity to reduce the tendency for infiltration (where the surrounding healthy tissues are undisturbed)[15].

5 Nanoemulsions for Lung Cancer: Clinical Aspects and Commercialisation

Clinically, LCNE-based anticancer treatment would be advantageous because tailored drug delivery increases medication efficacy while minimizing side effects. In other words, safety and effectiveness both increases. Despite these benefits, the "bench to bedside" road for LCNE-based treatment has not yet been completed. The NE's stability, sterility, and safety are key problems with cancer cell-targeted NE that may have clinically significant effects [12]. Unstable nanoemulsions can lose their functionality and cause undesirable side effects in some individuals. Maintaining an aseptic environment for manufacturing nanoemulsions is difficult due to the complicated manner of preparation. The inability to apply a dry heat or wet heat approach also makes sterilizing the formulation difficult. Several targeting ligands are now added to the standard emulsion ingredients as inactive additives to create nanoemulsions targeted explicitly to a cancer cell. Critical consideration must be given to evaluating toxicity for those new inactive excipients. A poisonous substance in a NE or an unsterile NE can cause several physiological reactions that are clinically unfavourable and dangerous for the health of the patient. Manufacturing nanoemulsions incorporating anticancer agents is usually difficult in the development stage because it requires reducing the globule size with the proper drug load and directing the medication to the targeted tumour location. A number of challenges must be overcome before it can go on to the preclinical or clinical stages, including immune response, rate of system clearance, targeted effectiveness, and permeability to the natural barrier [2]. A delivery method that targets LC using nanoemulsions has yet to be commercially available. The research product may not always or in some cases be able to address issues with diagnostic effectiveness and safety outcomes, the desired configuration of the delivery mechanism and character traits, commercially relevant production methods, adequate testing procedures for complex drug formulations, a favourable pharmacological and toxicological profile, etc. [8].

6 Conclusion

Recent advancements in nanotechnology have broadened the field's applicability to the treatment, diagnosis, and prevention of cancer, despite the reality that LC is among the primary reasons people die from cancer. Many studies have been conducted on the newly developed field of NE in the treatment and diagnostics of cancer. Via various administration methods, NE has been demonstrated to be an effective drug-delivery technique that helps increase the bioavailability of chemotherapeutic drugs used to treat this fatal illness. Even though many Chinese herbal medicines have anti-cancer properties, we still need to learn more about the precise active substances and distinct processes. In the area of research and development for cancer treatment, NEs are gaining more and more attention., but several obstacles still need to be solved to guarantee their safety and effectiveness as they go from the laboratory bench to the patient's bedside. Prior to considering their commercialization, it is critical to address the following issues: the formulation's stability properties, which guarantee the chemotherapeutics' biological fate; the excipients' security profile, which ensures patients safety against adverse side effects; their efficacy by precision and focused carcinoma cell-specific ligands; and their cheap production-scale expansion costs for NE.

References

1. Abdelaziz HM, Gaber M, Abd-Elwakil MM, Mabrouk MT, Elgohary MM, Kamel NM, Kabary DM, Freag MS, Samaha MW, Mortada SM, Elkhodairy KA, Fang JY, Elzoghby AO (2018)

Inhalable particulate drug delivery systems for lung cancer therapy: Nanoparticles, microparticles, nanocomposites and nanoaggregates. J Control Release 269:374–392. https://doi.org/10.1016/J.JCONREL.2017.11.036
2. Babu A, Templeton AK, Munshi A, Ramesh R (2013) Nanoparticle-based drug delivery for therapy of lung cancer: Progress and challenges. J Nanomater. https://doi.org/10.1155/2013/863951
3. Bray F, Ferlay J, Soerjomataram I, Siegel RL, Torre LA, Jemal A (2018) Global cancer statistics 2018: GLOBOCAN estimates of incidence and mortality worldwide for 36 cancers in 185 countries. CA: Cancer J Clin, 68(6), 394–424. https://doi.org/10.3322/CAAC.21492
4. *Cancer*. (n.d.). Retrieved March 28, 2023, from https://www.who.int/news-room/fact-sheets/detail/cancer
5. Chang HB, Chen BH (2015) Inhibition of lung cancer cells A549 and H460 by curcuminoid extracts and nanoemulsions prepared from—Curcuma longa—Linnaeus. Int J Nanomed 10(1):5059–5080. https://doi.org/10.2147/IJN.S87225
6. Choudhury H, Gorain B, Chatterjee B, Mandal UK, Sengupta P, Tekade RK (2017) Pharmacokinetic and pharmacodynamic features of nanoemulsion following oral, intravenous, topical and nasal route. Curr Pharm Des, 23(17). https://doi.org/10.2174/13816128226661612011 43600
7. Choudhury H, Gorain B, Karmakar S, Biswas E, Dey G, Barik R, Mandal M, Pal TK (2014) Improvement of cellular uptake, in vitro antitumor activity and sustained release profile with increased bioavailability from a nanoemulsion platform. Int J Pharm 460(1–2):131–143. https://doi.org/10.1016/J.IJPHARM.2013.10.055
8. Choudhury H, Pandey M, Gorain B, Chatterjee B, Madheswaran T, Md, S, Mak K-K, Tambuwala M, Chourasia MK, Kesharwani P (2019). Nanoemulsions as effective carriers for the treatment of lung cancer. Nanotechnol-Based Target Drug Deliv Syst Lung Cancer, 217–247. https://doi.org/10.1016/B978-0-12-815720-6.00009-5
9. Dos Santos Câmara AL, Nagel G, Tschiche HR, Cardador CM, Muehlmann LA, De Oliveira DM, Alvim PQ, Azevedo RB, Calderón M, Longo JPF (2017) Acid-sensitive lipidated doxorubicin prodrug entrapped in nanoemulsion impairs lung tumor metastasis in a breast cancer model. https://Doi.Org/https://doi.org/10.2217/Nnm-2017-0091, 12(15), 1751–1765. https://doi.org/10.2217/NNM-2017-0091
10. Du X, Khan AR, Fu M, Ji J, Yu A, Zhai G (2018) Current development in the formulations of non-injection administration of paclitaxel. Int J Pharm 542(1–2):242–252. https://doi.org/10.1016/J.IJPHARM.2018.03.030
11. Fitzmaurice C, Dicker D, Pain A, Hamavid H, Moradi-Lakeh M, MacIntyre MF, Allen C, Hansen G, Woodbrook R, Wolfe C, Hamadeh RR, Moore A, Werdecker A, Gessner BD, Te Ao B, McMahon B, Karimkhani C, Yu C, Cooke GS, Naghavi M (2015) The Global Burden of Cancer 2013. JAMA Oncol 1(4):505–527. https://doi.org/10.1001/JAMAONCOL.2015.0735
12. Ganta S, Talekar M, Singh A, Coleman TP, Amiji MM (2014) Nanoemulsions in translational research-opportunities and challenges in targeted cancer therapy. AAPS Pharm Sci Tech 15(3):694–708. https://doi.org/10.1208/S12249-014-0088-9
13. Hirsch FR, Suda K, Wiens J, Bunn PA (2016) New and emerging targeted treatments in advanced non-small-cell lung cancer. Lancet (London, England) 388(10048):1012–1024. https://doi.org/10.1016/S0140-6736(16)31473-8
14. Huang CY, Ju DT, Chang CF, Muralidhar Reddy P, Velmurugan BK (2017) A review on the effects of current chemotherapy drugs and natural agents in treating non-small cell lung cancer. Biomedicine 7(4):12–23. https://doi.org/10.1051/BMDCN/2017070423
15. Hwang YY, Ramalingam K, Bienek DR, Lee V, You T, Alvarez R (2013) Antimicrobial activity of nanoemulsion in combination with cetylpyridinium chloride in multidrug-resistant acinetobacter baumannii. Antimicrob Agents Chemother 57(8):3568–3575. https://doi.org/10.1128/AAC.02109-12/ASSET/6EB53E77-993B-4C43-B8ED-68AED12DC8DE/ASSETS/GRAPHIC/ZAC9991019910006.JPEG
16. Jiang XQ, Mei XD, Feng D (2016) Air pollution and chronic airway diseases: what should people know and do? J Thorac Dis 8(1):E31–E40. https://doi.org/10.3978/J.ISSN.2072-1439.2015.11.50

17. Jiang Y, Liu LS, Shen LP, Han ZF, Jian H, Liu JX, Xu L, Li HG, Tian JH, Mao ZJ (2016) Traditional Chinese Medicine treatment as maintenance therapy in advanced non-small-cell lung cancer: A randomized controlled trial. Complement Ther Med 24:55–62. https://doi.org/10.1016/J.CTIM.2015.12.006
18. Jiang Y, Wu LB, Shen LP, Zhang P, Jiang WJ, Tian JH, Liu LS (2016) Effect of traditional Chinese medicine treatment as maintenance therapy on regulating the serum concentration of sCTLA-4 in patients with advanced non-small-cell lung cancer and its relationship with prognosis. Zhonghua Zhong Liu Za Zhi [Chinese Journal of Oncology] 38(10):757–762. https://doi.org/10.3760/CMA.J.ISSN.0253-3766.2016.10.009
19. Khan I, Bahuguna A, Kumar P, Bajpai VK, Kang SC (2018) In vitro and in vivo antitumor potential of carvacrol nanoemulsion against human lung adenocarcinoma A549 cells via mitochondrial mediated apoptosis. Sci Rep 2017 8:1, 8(1), 1–15. https://doi.org/10.1038/s41598-017-18644-9
20. Khani S, Keyhanfar F, Amani A (2015) Design and evaluation of oral nanoemulsion drug delivery system of mebudipine. https://Doi.Org/https://doi.org/10.3109/10717544.2015.1088597, 23(6), 2035–2043. https://doi.org/10.3109/10717544.2015.1088597
21. Kim KH, Han CW, Yoon SH, Kim YS, Kim JI, Joo M, Choi JY (2016) The fruit hull of gleditsia sinensis enhances the Anti-Tumor effect of cis-diammine dichloridoplatinum II (Cisplatin). Evid-Based Complement Altern Medicine : ECAM. https://doi.org/10.1155/2016/7480971
22. Kumar S, Sharma AK, Lalhlenmawia H, Kumar D (2021) Natural compounds targeting major signaling pathways in lung cancer. Target Cell Signal PathwS Lung Dis, 821–846. https://doi.org/10.1007/978-981-33-6827-9_37
23. Lemjabbar-Alaoui H, Hassan OUI, Yang YW, Buchanan P (2015) Lung cancer: biology and treatment options. Biochem Biophys Acta 1856(2):189. https://doi.org/10.1016/J.BBCAN.2015.08.002
24. Li H, Ji Y, Zhang S, Gao Z, Hu C, Jiang R, Chen M, Li G, Zhang X (2019) Kangai Injection Combined with Platinum-based Chemotherapy for the Treatment of Stage III/IV Non-Small Cell Lung Cancer: A Meta-analysis and Systematic Review of 35 Randomized Controlled Trials. J Cancer 10(21):5283. https://doi.org/10.7150/JCA.31928
25. Li S, Fang C, Zhang J, Liu B, Wei Z, Fan X, Sui Z, Tan Q (2016) Catanionic lipid nanosystems improve pharmacokinetics and anti-lung cancer activity of curcumin. Nanomedicine: Nanotechnol, Biol Med, 12(6), 1567–1579. https://doi.org/10.1016/J.NANO.2016.02.007
26. Liu G, Yang L, Chen G, Xu F, Yang F, Yu H, Li L, Dong X, Han J, Cao C, Qi J, Su J, Xu X, Li X, Li B (2021) A review on drug delivery system for tumor therapy. Front Pharmacol, 12. https://doi.org/10.3389/FPHAR.2021.735446
27. Mahato R (2017) Nanoemulsion as targeted drug delivery system for cancer therapeutics. J Pharm Sci Pharmacol 3(2):83–97. https://doi.org/10.1166/JPSP.2017.1082
28. Meghani N, Patel P, Kansara K, Ranjan S, Dasgupta N, Ramalingam C, Kumar A (2018) Formulation of vitamin D encapsulated cinnamon oil nanoemulsion: Its potential anti-cancerous activity in human alveolar carcinoma cells. Colloids Surf, B 166:349–357. https://doi.org/10.1016/J.COLSURFB.2018.03.041
29. Panagopoulos N, Leivaditis V, Koletsis E, Prokakis C, Alexopoulos P, Baltayiannis N, Hatzimichalis A, Tsakiridis K, Zarogoulidis P, Zarogoulidis K, Katsikogiannis N, Kougioumtzi I, Machairiotis N, Tsiouda T, Kesisis G, Siminelakis S, Madesis A, Dougenis D (2014). Pancoast tumors: characteristics and preoperative assessment. J Thorac Dis, 6 Suppl 1(Suppl 1). https://doi.org/10.3978/J.ISSN.2072-1439.2013.12.29
30. Pass HI, Carbone DP, Minna JD, Johnson DH, Scagliotti GV, Turrisi AT (2010) Principles and practice of lung cancer: the official reference text of the IASLC, 4e
31. Pérez-Herrero E, Fernández-Medarde A (2015) Advanced targeted therapies in cancer: Drug nano-carriers, the future of chemotherapy. Eur J Pharm Biopharmaceutics : Off J ArbSgemeinschaft Fur Pharm VerfahrStechnik e.V, 93, 52–79. https://doi.org/10.1016/J.EJPB.2015.03.018
32. Pikor LA, Ramnarine VR, Lam S, Lam WL (2013) Genetic alterations defining NSCLC subtypes and their therapeutic implications. Lung Cancer (Amsterdam, Netherlands) 82(2):179–189. https://doi.org/10.1016/J.LUNGCAN.2013.07.025

33. Praveen Kumar G, Divya A, Kumar GP (2015) 272 Citation: Praveen Kumar G, Divya A (2015) Nanoemulsion based targeting in cancer therapeutics. Med Chem, an Open Access Journal 5(6):272–284. https://doi.org/10.4172/2161-0444.1000275
34. Regassa H, Sourirajan A, Kumar V, Pandey S, Kumar D, Dev K (2022) A review of medicinal plants of the himalayas with Anti-Proliferative activity for the treatment of various cancers. Cancers 2022, Vol. 14, Page 3898, 14(16), 3898. https://doi.org/10.3390/CANCERS14163898
35. Sun CY, Zhu Y, Li XF, Wang XQ, Tang LP, Su ZQ, Li CY, Zheng GJ, Feng B (2018). Scutellarin increases cisplatin-induced apoptosis and autophagy to overcome cisplatin resistance in non-small cell lung cancer via ERK/p53 and c-met/AKT signaling pathways. Front Pharmacol, 9 (FEB). https://doi.org/10.3389/FPHAR.2018.00092/FULL
36. Sun D, Wei X, Xue X, Fang Z, Ren M, Lou H, Zhang X (2014) Enhanced oral absorption and therapeutic effect of acetylpuerarin based on D-α-tocopheryl polyethylene glycol 1000 succinate nanoemulsions. Int J Nanomed 9(1):3413–3423. https://doi.org/10.2147/IJN.S63777
37. Tseng CY, Lin CH, Wu LY, Wang JS, Chung MC, Chang JF, Chao MW (2016) Potential combinational anti-cancer therapy in non-small cell lung cancer with traditional chinese medicine Sun-Bai-Pi extract and cisplatin. PloS One, 11(5). https://doi.org/10.1371/JOURNAL.PONE.0155469
38. Wang Q, Jiao L, Wang S, Chen P, Bi L, Zhou D, Yao J, Li J, Wang L, Chen Z, Jia Y, Zhang Z, Shen W, Zhu W, Xu J, Gao Y, Xu L, Gong Y (2020) Adjuvant chemotherapy with chinese herbal medicine formulas versus placebo in patients with lung adenocarcinoma after radical surgery: a multicenter, randomized, Double-Blind, Placebo-Controlled Trial. Biol Proced Online, 22 (1). https://doi.org/10.1186/S12575-020-00117-5
39. Wang SF, Wang Q, Jiao LJ, Huang YL, Garfield D, Zhang J, Xu L (2016) Astragalus-containing traditional Chinese Medicine, with and without prescription based on syndrome differentiation, combined with chemotherapy for advanced non-small-cell lung cancer: a systemic review and meta-analysis. Current Oncology (Toronto, Ont.), 23(3), e188–e195. https://doi.org/10.3747/CO.23.2920
40. Zhang L, Li H, Zhang F, Wang S, Li G (2020) [CAR-T Immunotherapy and Non-small Cell Lung Cancer: Bottleneck and Dawn]. Zhongguo Fei Ai Za Zhi = Chinese Journal of Lung Cancer, 23(10), 916–920. https://doi.org/10.3779/J.ISSN.1009-3419.2020.103.10
41. Zorzetto M, Ferrari S, Saracino L, Inghilleri S, Stella GM (2012) MET genetic lesions in non-small-cell lung cancer: pharmacological and clinical implications. Transl Lung Cancer Res 1(3):194–207. https://doi.org/10.3978/J.ISSN.2218-6751.2012.09.03

Nanoemulsions-Based Systems for Breast Cancer Treatment

Aashna Sinha, Chetan Shrivastava, Atreyi Pramanik, Pranchal Rajput, Kaninika Vyas, Anis Kumar Pal, Kundan Kumar Chaubey, Sujata Jayaraman, S. D. Pandey, Rakesh Kumar Bachheti, and Archana Bachheti

Abstract Breast cancer is the most prevalent malignancy in this group of patients and is the second most deadly disease in women after lung cancer. Nowadays, surgery, chemotherapy, or radiation are recommended to treat breast cancer. Many approaches have been devised to address this clinical problem; however, they all have drawbacks. The unforeseen side effects of the chemotherapy treatment and negligent harm to both healthy and cancerous cells may be regarded as the reason. These challenges fuel the urge to deploy cutting-edge therapy platforms to target cancerous cells. The fundamental issues with traditional drug administration techniques have been resolved by developing a novel medicine delivery technology. Emulsions created in nanoscale diameters are known as nanoemulsions (NE) to improve the transport of pharmacologically active compounds. Targeted delivery has a significant potential to fundamentally revolutionize the way cancer is treated, according to the key targeting techniques employed to deliver medications and are now being explored for many of the reported NEs. The targeted delivery of pharmaceuticals via certain cellular

A. Sinha · C. Shrivastava · A. Pramanik · P. Rajput · A. K. Pal · K. K. Chaubey (✉) · S. Jayaraman
Division of Research and Innovation, Uttaranchal University, Arcadia Grant, P.O. Chandanwari, Prem Nagar, Dehradun Uttarakhand-248007, India
e-mail: kundan2006chaubey@gmail.com

K. Vyas
Department of Biotechnology, Mahila P.G. Mahavidhylay, Affiliated By Jai Narayan Vyas University, Pratap Nagar, Jodhpur Rajasthan-342001, India

K. K. Chaubey
School of Basic and Applied Sciences, Sanskriti University, Mathura, Uttar Pradesh-281401, India

S. D. Pandey
Uttaranchal Institute of Technology, Uttaranchal University, Arcadia Grant, P.O. Chandanwari, Prem Nagar, Dehradun Uttarakhand-248007, India

R. K. Bachheti
Centre of Excellence in Nanotechnology, Department of Industrial Chemistry, Addis Ababa Sciences and Technology University, P.O. Box-16417, Addis Ababa, Ethiopia

A. Bachheti
Department of Environment Science, Graphic Era University, Dehradun-248002, Uttarakhand, India

A. Husen et al. (eds.), *Current Trends in Green Nano-emulsions*, Smart Nanomaterials Technology, https://doi.org/10.1007/978-981-99-5398-1_10

markers may boost their potency and toxicity. A cancer therapy device's efficacy is determined by its capacity to shrink and remove tumors without causing harm to healthy tissue. The extension and quality of the patient's life are the ultimate goals of cancer therapy. Because of their site selectivity, ability to overcome multidrug resistance, and effective agent delivery, NEs significantly improve therapy. This chapter aims to discuss the role of nanoemulsions in breast cancer treatment.

Keywords Breast cancer · Multidrug resistance · Nanoemulsions · Drug delivery · Treatment

1 Introduction

Cancer is a diverse form of the disease that causes the body's abnormal cells to grow and divide out of control, causing it to spread to other bodily tissues [43]. The most prevalent and second-most fatal kind of cancer in women globally is breast cancer (BC). Each breast has lymphatic veins, small bean-shaped lymph nodes, and blood vessels that link to them. The bulk of the lymphatic veins leaves the body through the axillary and internal breast lymph nodes, which are situated in the under the armpit, above the vertebral column, and in the chest near the borders of the breasts. The predominant location of spread regionally in the metastases of the initial breast cancer is often the axillary lymph nodes, which are found beneath the arm [30]. The tumor mass turns malignant when tumor cells invade nearby healthy cells or metastasize (spread to other body parts). Men still have a very tiny chance of having BC, even though women are more frequently diagnosed with the disease [27]. A complete or partial mastectomy, radiotherapy after a partial mastectomy, or breast-conserving surgery are all options for treating breast cancer [27]. With millions of new cases each year, BC has become one of the most harmful malignancies in recent years [40]. According to data from worldwide cancer (GLOBOCAN), with a predicted 2.3 million new cases, or 11.7% of all cancer cases and 6.9% of all cancer deaths globally, female BC will now overtake lung cancer as the top cause of global cancer incidence in 2020 [41]. The greatest rates of breast cancer are found in Australia/New Zealand, Western Europe (Belgium has the highest incidence), Northern America, and Northern Europe, while the lowest rates are found in Central America, Eastern and Middle Africa, and Southern Central Asia (40 per 100,000) [41]. The expected mortality rate in developed nations went from 42,280 in 2019 to 43,780 in 2022 as new BC cases increased from 2.6 lac to 3.3 lac in 2022 [7].

BC is treated using a variety of methods, like surgery, chemotherapy, radiation, and hormone therapy. However, these methods are losing their efficacy as a result of the rise of multi drug resistance (MDR) and its serious adverse effects [29, 31]. Pharmaceutical formulations known as nanoemulsions (NEs) are made up of nanoscale particles. Due to their hydrophobic core, they can encapsulate only weakly soluble medications in water. They are a stable and secure choice for drug delivery because they also contain safe gradient excipients. Using ligands of various types, NEscan be altered to target substances found on the surface of tumor cells or to go over MDR defenses. Many researchers are studying multifunctional NEs in various scientific fields, mainly for treating various cancers, including BC [35]. Nano sized oil-in-water emulsions, the heterogeneous system made up of two immiscible liquids stabilized by emulsifiers or surfactants, offer much potential in medicine because of their attractive features for drug administration. It improves the skin permeability of oils, leading to quicker, more complete absorption of cannabinoids into the bloodstream. Due to its distinct physical characteristics, such as its exceptionally tiny droplet diameter, remarkable optical transparency, and elastic properties compared to other traditional emulsions [38], this drug delivery method has recently attracted interest. Because of their tremendous potential to increase the bioavailability and effectiveness of loaded medications while minimizing their adverse effects, NEs are employed in drug delivery studies. Using *Nigella sativa* L essential oil, Periasamy and coworkers developed a natural-based NE [15], might be a successful breast chemotherapeutic agent. By promoting the death of MCF-7 BC cells, these NEs have anti-cancer capabilities in vitro. In order to treat BC, NEs may help entrap active medications [15]. This chapter aims to educate readers on how BC can be treated with NE.

2 Pathophysiology of Breast Cancer

Although around 20–25% of BC patients have a favourable family history, only 5–10% of instances of BC have an autosomal dominant inheritance. Examples of high-risk predisposition alleles that carry a 40-85% lifetime chance of getting BCinclude mutations in the BRCA1 and BRCA2 genes, the TP53 gene, the PTEN gene, the Peutz-syndrome-causing STK11 gene, the Neurofibromatosis (NF1) gene, and the CDH-1 E-Cadherin gene. BRCA1 and BRCA2 mutations are linked to 50% of BC risk syndromes. Harmful BRCA1 or BRCA2 mutations significantly increase females' risk of BC. The likelihood of acquiring BCover the course of one's lifetime varies depending on whether one carries the BRCA1 or BRCA2 mutation. A 20–40% lifetime risk of BCis associated with moderate risk genes, which include homozygous ataxia–telangiectasia (ATM) mutations, somatic mutations in the tumour suppressor gene CHEK2, BRIP1 and PALB2 that act as BRCA1 and BRCA2 modifiers, and others [37] (Fig. 1).

The discovery of BCsusceptibility genes has illuminated aspects of the etiology of both hereditary and spontaneous BC. According to molecular research, there are two distinct molecular pathways by which breast cancer develops, with the majority of these pathways including ER expression, tumour grade, and tumour proliferation. Gain of 1q, loss of 16q, atypical amplification of 17q12, and a gene expression profile (GEP) with the majority of genes connected to the ER phenotype, diploid or nearly diploid karyotypes, and low cancer grade are characteristics of the first route, the low-grade-like pathway. The luminal A group and, to a lesser extent, the luminal B group are included in this pathway. The second pathway, also known as the high-grade-like pathway, is distinguished by the loss of 13q, the gain of the chromosomal region

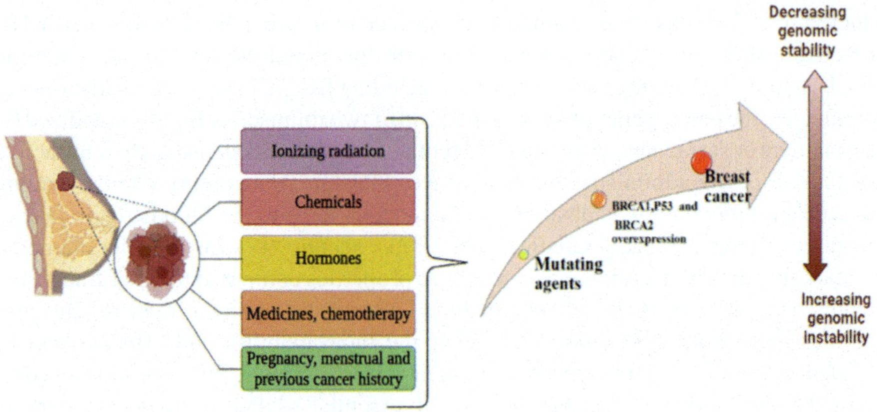

Fig. 1 Pathophysiology of breast cancer

11q13, the amplification of 17q12 (which contains the gene ERBB2, which encodes HER2), and the expression of a signature collection of genes connected to cell cycle and proliferation [11]. This path includes moderate to high grade tumors, such as TNBC and HER2-positive malignancies [22]. Detailed stage-specific symptoms are tabulated in Table 1.

3 Prevalence/Epidemiology

According to the World Health Organisation (WHO), 2.3 million women will be diagnosed with BC globally in 2020, which would result in 685 000 deaths worldwide. With 7.8 million women still alive at the end of 2020 who had gotten a diagnosis during the preceding 5 years, BC is the most prevalent disease in the world. BC is the second most common malignancy in American women after non-melanoma skin cancer. More than 4 million breast cancer survivors live in the US, including both those who are now undergoing treatment and those who have already finished it. In 2023, 300,590 people in the US are expected to receive a BC diagnosis. The estimated number of BC diagnoses in the U.S. in 2023 is 297,790, making it the most common illness in American women. In the United States, a woman is given the diagnosis every two minutes. 2,800 males will likely receive a BC diagnosis in the United States in 2023. According to estimates, 43,700 Americans (43,170 women and 530 men) will pass away from the illness in the U.S. in 2023 [4].

Plans to improve BC outcomes rely on the enhanced fundamental healthcare system to deliver the medicines that are now shown to be successful. They are essential for treating different cancers and other non-communicable, non-malignant conditions (NCDs) and possessing trustworthy referral networks, for instance, between primary care institutions, district hospitals, and cancer centers.

Table 1 Different phases of breast cancer and warning signs

Stage	Symptoms	Key reference
Stage 0	Breast tumors that have not spread, such ductal carcinoma in situ (DCIS) No evidence of malignant or abnormal non-cancerous cells protruding from the breast tissue where they first appeared	[45, 39]
Stage I	Breast cancer that is invasive The cancer has either not gone past the breast or has just slightly reached a lymph node	[24]
Stage IA	The fatty tissue of the breast has been affected by cancer The tumor is about the size of a peanut in its shell	[24]
Stage IB	Just small amounts of cancer cells have been seen in a few lymph nodes	[24]
Stage II	The cancer has spread or grown	[14]
Stage IIA	Demonstrates that any breast tumour is still extremely tiny It's possible that the lymph nodes don't have cancer or that it has spread to three or more	[14]
Stage IIB	The size of the tumor is greater, falling between a walnut and a lime	[14]
Stage III	Cancer is regarded as advanced and more challenging to cure even when it has not spread to the bones or internal organs	[26]
Stage IIIA	Reveals the existence of up to nine lymph nodes from your armpit to your collarbone that have been proven to be cancerous It may have reached the deep breast nodules or enlarged them. Even when the tumor hasn't spread to the lymph nodes, there may be a large tumor in the breast in some patients but not in others	[26]
Stage IIIB	Demonstrates that it has pierced your breast skin or the chest wall	[12]
Stage IV	Breast cancer cells have entered the lymph nodes that surround the breast in addition to the breast itself. The most frequent sites are the lungs, liver, brain, and skeletal bones. The word "metastatic" describes cancer that has spread outside of the area of the body where it was first found	[5, 12]

The WHO Global Breast Cancer Initiative (GBCI) aims to prevent 2.5 million BC deaths between 2020 and 2040 by reducing the annual worldwide BC mortality rate by 2.5%. By 2030 and 2040, respectively, 25% and 40% of BC deaths among women under the age of 70 would be prevented if the worldwide incidence of BC mortality could be reduced by 2.5% each year. Health promotion for early detection, rapid diagnosis, and comprehensive BC management are the three foundations for attaining these goals [3].

4 Traditional Treatment of Breast Cancer and Drawbacks

Despite major advancements in BCtreatment, resistance to traditional chemotherapeutic drugs which cause cancer to return and relapse remains the prime origin of breast cancer-related fatalities. Resistance is innate or may be learnt. Nevertheless, certain cell-originating drug resistance factors in breast cancer that result in drug resistance include P53, microtubules (MT), ATP-binding cassette (ABC), human epidermal growth factor receptor 2 (HER2), permeability glycoprotein (P-gp), topoisomerase, and breast cancer type 1 (BRAC1) [25]. BC survivors frequently experience the long-term side effect of cancer-related fatigue, which is defined as "a severe, persistent, subjective feeling of physical, emotional, and/or cognitive fatigue or exhaustion associated with cancer and/or cancer therapy that is not connected to recent activities and impairs normal functioning." Cancer-related fatigue can differ from ordinary weariness in terms of its physical, cognitive, and emotional components. It also appears to be more acute, enduring, and incapacitating. Depending on the kind and stage of the disease, the treatment for BCoften combines surgery, radiation therapy, chemotherapy, hormone therapy, and targeted therapy [6].

- Surgery: The first course of therapy for BC is frequently surgery, which entails the removal of the tumour and any surrounding tissue. The two surgical treatments that are performed most frequently are mastectomy (removal of the whole breast) and lumpectomy (removal of the tumour and a narrow margin of surrounding tissue [18].
- In radiation therapy, high-energy X-rays are used to kill cancer cells and shrink tumours. It is typically used after surgery to get rid of any cancer cells that could still be around and reduce the risk of a recurrence.
- Chemotherapy: Chemotherapy employs drugs to destroy cancer cells across the body. It is typically used after surgery to get rid of any cancerous cells that could still be around and reduce the risk of a recurrence. Additionally, it can be applied prior to surgery to lessen the tumor's growth and make removal easier.
- As certain BCare hormone-receptor positive, hormone treatment is utilized to prevent the effects of oestrogen on breast cancer cells. To lower the chance of recurrence, this kind of treatment is frequently utilized following surgery.
- Targeted therapy: Frequently used in conjunction with chemotherapy or hormone therapy, targeted therapy employs medications that selectively target certain proteins or genes in cancer cells.

The downsides and side effects of these medicines outweigh their potential effectiveness in treating BC [34]. A few of these are:

- Surgery: Surgery can cause scars and alter the look of the breasts. It can also be physically and emotionally unpleasant. In some circumstances, it could also be necessary to remove lymph nodes, which raises the chance of lymphedema (arm swelling) [44].
- High-energy X-rays: Radiation therapy can make you feel worn out, irritate your skin, and harm your heart and lungs over the long run.

- Chemotherapy: There are a number of adverse effects that chemotherapy patients may have, such as nausea, vomiting, hair loss, exhaustion, and an elevated risk of infection [44].
- Hormone therapy: Hormone therapy increases the risk of osteoporosis and can result in hot flashes, dry vagina, and other side effects.
- Targeted therapy: Targeted therapy raises the risk of blood clots, diarrhea, and skin rashes.

According to several studies, cancer patients and survivors report weariness that is more intense and longer-lasting than people who have never had the disease [17]. With the hurdles of drug resistance and unavoidable side effects of cancer therapy, NE technology comes to the rescue.

5 Overview of Nanoemulsions

Colloidal nanoparticles called NE are produced when water, emulsifying agents, and oil are mixed. The size of nanoemulsions can vary from 10 to 1000 nm. Drug nanocomposites are commonly made of nanoemulsions, which are solid spheres with an amorphous, lipophilic surface and a negative charge. Three common types of nanoemulsions can be produced because they are heterogeneous combinations of oil droplets in the aqueous fluid that are dispersed with microscopic sizes: The oil in the water NEs system, the water in the oil NE system, and the bi-continuous NE are three examples of NEs that spread oil and water in aquatic settings [16]. Over most lipid-based nanomaterials and nanoparticles, NEs have a number of benefits, including optical clarity, thermodynamic stability, a high surface area, simplicity of manufacture, biodegradability, and the optimal pharmacological release profile [13]. The study of membrane-modified nanoemulsions is broad. Co-delivery using nanoemulsions is one way to increase medication effectiveness and bioavailability. The findings of tests on a NE drug carrier system, including paclitaxel PTX and spirulina polysaccharides, revealed that it might enhance the anti-tumor effects of PTX by controlling immunity via TLR4/NF-B signaling pathways [10]. A NE system including temozolomide, rapamycin, and bevacizumab was developed to treat metastatic melanoma. Parenteral dosing enhanced tumor relapse, migration, and angiogenesis inhibition and boosted cytotoxicity against melanoma cells, according to in vitro experiments employing human and animal cell models [9] (Fig. 2).

6 Nanoemulsion for Treatment of Breast Cancer

BC is the primary cause of cancer-related death in women worldwide. The cancer microenvironment is composed of the extracellular matrix (ECM), fibroblasts, cells of the epithelium, immune cell types, pericytes, adipocytes, glial cells (found only

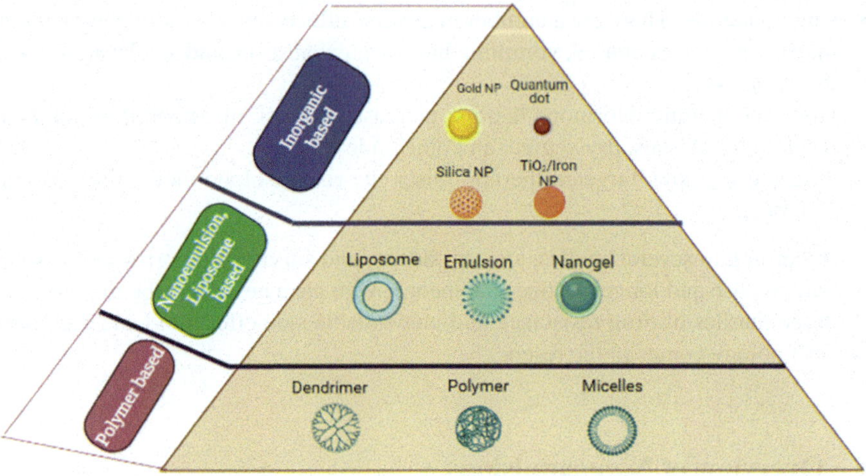

Fig. 2 Overview of nanoemulsions

in the brain system), protein molecules, vascular cells, and lymphatic cells. The tumour cells' proliferation, structure, migration, invasion, and metastasis depend on the ECM. These tumour cells have unique surface markers that can be targeted by medications. When the cancer is less than 2.0 mm^3, oxygen levels are maintained by simple diffusion, but as the tumour grows bigger, the oxygen levels fall, leading to hypoxia and the angiogenic formation of new blood vessels. Therefore, it is also possible to reduce cell development by blocking the angiogenic process. Several anti-angiogenic medications have been created in recent years, including bevacizumab (a VEGF-neutralizing antibody), sorafenib (a VEGF signalling pathway blocker), sunitinib, and pazopanib. These angiogenesis inhibitors, however, are distinguished by pronounced toxicity, increased resistance, and hurdles to drug delivery. Medication toxicity can be decreased and payload delivery improved by using nanoemulsions to encapsulate the medication within its core.

Here are a few ways that BCis being treated with nanoemulsions:

- Delivery of chemotherapy: Chemotherapy medications can be delivered directly to cancer cells using nanoemulsions, boosting the drug's efficacy while lowering adverse effects. For instance, paclitaxel-loaded nanoemulsions have been demonstrated to suppress the development of breast cancer in animal models [23].
- Targeted drug delivery: To deliver medications just to cancer cells, nanoemulsions can also be functionalized with targeting molecules, such as antibodies or peptides. By using this strategy, therapeutic effectiveness can be improved while toxicity to healthy tissues is reduced. To deliver doxorubicin to BC cells, for instance, epidermal growth factor receptor (EGFR)-targeted nanoemulsions have been created [36].
- Radiotherapy sensitization: By delivering radiosensitizing chemicals, such as gold nanoparticles or oxygen to the tumour site, nanoemulsions can be utilised to

sensitise BC cells to radiation treatment. This can improve radiation therapy's efficacy and lower the required dose, minimising negative effects [8].

- Imaging: To make it possible to see breast tumours, imaging agents like fluorescent dyes or magnetic nanoparticles can be added to nanoemulsions. This can help in the detection and tracking of the development of BC [20].

The primary therapeutic method to treatment, chemotherapy has been employed either alone or in conjunction with other forms of therapy. Mitomycin C (MMC), an alkylating drug and potential chemotherapeutic, has been utilized clinically for patients with different forms of cancer [33]. However, the emergence of chemotherapeutic resistance frequently limits the clinical efficacy of mitomycin C (MMC). The therapeutic dosages of MMC also have limited drug accessibility to the target tissue, necessitating a greater clinical dose that is restricted by extremely troublesome systemic toxicities. A novel approach of incorporating the medicine into nanoemulsions (NEs) made with essential oils (ESSO) and manufactured by a high energy process, using co- and surfactants, was presented to address the limitations of the MMC. The therapeutic potential of ESSO of ginger (Gi) and frankincense (Fr) as they function in chemoprevention and cancer suppression has led to their selection. The active ingredients in ginger have been shown to target specific molecular proteins in cancer cells, which has been shown to slow the growth and cause apoptosis in a number of cancer types (Fig. 3) [21, 46]. It has been demonstrated that the bioactive components of Fr oil serve as cytotoxic mediators against a variety of cancer cell types. The genes that cause cell cycle arrest, cell growth inhibition, and apoptosis in cancer cells have been shown to be powerfully activated by Fr oil [1].

Fr oil was selected for the study based on published research studies indicating its exceptional therapeutic potential against BC. The impact of paclitaxel combined with erucin (ER), a natural isothiocyanate isolated from Eruca sativa seeds, loaded in the

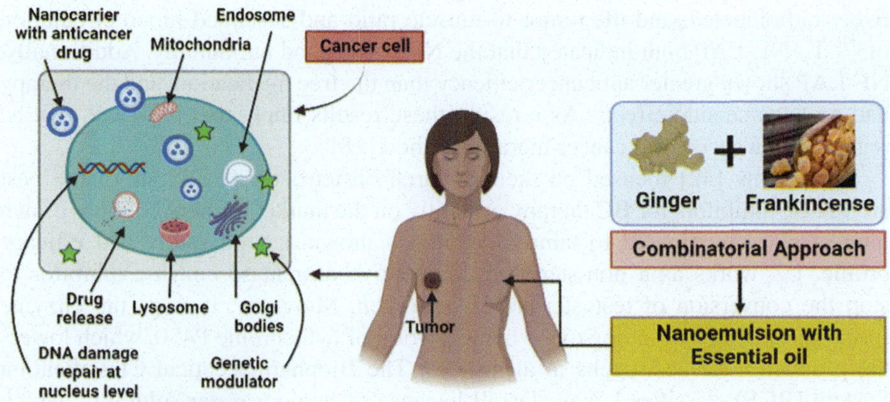

Fig. 3 Nanoemulsion treatment of breast cancer

formulation of frankincense oil-based NE. In comparison to ER and PTX used alone, the combination of paclitaxel and erucin (EPNE) showed increased cytotoxicity toward human epithelial BC cells (T-47D). Further testing of EPNE's anticancer effectiveness was done using a mouse BC model produced by 7, 12-dimethylbenz (a) anthracene (DMBA). EPNE significantly decreased the levels of hepatic and renal indicators in breast cancer animal models in addition to oxidative stress. Also, EPNE found that the inflammatory cytokines TNF- and IL-6 were at reduced levels. In BC mice treated with EPNE, histopathological analyses showed that the tumorous breast had been converted back to normal tissues. As a result, EPNE is a promising lead and therapeutic choice for BC resistant to treatment [19].

A natural substance called lapachol (LAP) has a number of biological characteristics, including anticancer action. However, because of its poor water solubility and probable negative side effects, its clinical applicability is constrained. Drug delivery methods called NE can help provide hydrophobic medications by boosting their bioavailability and shielding them from deterioration. In his work, Sued Eustaquio Mendes Miranda and his team [28], said that the goal was to create an LAP-loaded NE (NE-LAP) and assess its anticancer potential. In order to do this, a hot homogenization technique was used to create the NE. Cryogenic transmission electron microscopy (cryo-TEM) was then used to characterize the NEs morphology. By using DLS, the average diameter, polydispersity index, and zeta potential were assessed. HPLC was used to gauge encapsulation effectiveness. Additionally, it was assessed the drug release, hemolysis in vitro, and short-term storage stability. Additionally, a BC(4T1) tumor model was used to assess the pharmacokinetic, toxicological, and cytotoxic features of ^{99m}Tc-NE-LAP. The NE-LAP's physicochemical characterization revealed a homogenous stable NE with a mean diameter of around 170 nm, a zeta potential of roughly -20 mV, and an encapsulation level of more than 85%, while the cryo-TEM revealed spherical globules. Studies conducted in the lab proved that encapsulation has no effect on LAP's ability to cause cell death. The NE was effectively radiolabeled, and the tumor-to-muscle ratio and sustained blood circulation of ^{99m}Tc-NE-LAP both indicated that the NE was a good tumour ally. Additionally, NE-LAP shown greater anticancer efficacy than the free medication, and the therapy had no adverse side effects. As a result, these results imply that NE-LAP may be regarded as a successful cancer therapy method [28].

Alhamdany [42] focused on their research on letrozole "LZ," one of the best aromatase inhibitors for BC therapy currently on the market. Moreover, it has drawn interest since, compared to tamoxifen, it has shown a good safety and efficacy profile. LZ works as a non-steroidal competitive aromatase enzyme inhibitor to stop the conversion of testosterone to oestrogen. Moreover, it stops the enzyme from functioning by binding to the heme portion of cytochrome P450, which lowers the production of oestrogens in all tissues. The Biopharmaceutical Classification System (BCS) classifies LZ as class II because of its poor water solubility, which will reduce its bioavailability [42]. Oral NE formulations were explicitly developed for chemotherapy drugs like paclitaxel to improve the drugs' poorly water-soluble drug bioavailability.

The use of NEs must improve scientifically for patients with BC who have poor prognoses and elevated recurrence stages, such as those with epidermal growth factor receptor-2 (HER-2 +) positive and triple-negative BC (TNBC). The effectiveness of HER-2 targeting nano therapies based on HER-2 targeting ligands like trastuzumab has dramatically enhanced the treatment of HER-2 + BC. To speed up the destruction of cancer cells, trastuzumab activates immune effector cells that are responsible for antibody-dependent cell-mediated cytotoxicity. It is frequently used either alone or in combination with chemotherapy to improve the patient's outlook and reduce the risk of recurrence [32].

7 Conclusion

In recent years the use of NEs in the treatment of BC has shown encouraging and hopeful results. In addition to better drug solubility and stability, targeted drug administration, higher bioavailability, and efficacy, NEs also provide a variety of benefits. NEs overcome many of the limitations associated with conventional drug delivery systems such as poor water solubility, low bioavailability, and rapid clearance from the body. It is conceivable to increase the effectiveness of anti-cancer medications while lowering their toxicity and adverse effects by utilizing these cutting-edge drug delivery methods. Drug bioavailability may be improved via NEs, enabling more precise and effective treatment. Further research is required to improve the formulation and distribution of NEs being developed for the treatment of BC. The advantages of this technology, however, should not be disregarded. By continuing to explore the use of NEs in BC treatment, we may be able to improve outcomes for patients and reduce the burden of this devastating disease.

References

1. Al-Otaibi WA, Alkhatib MH, Wali AN (2018) Cytotoxicity and apoptosis enhancement in breast and cervical cancer cells upon coadministration of mitomycin c and essential oils in nanoemulsion formulations. Biomed Pharmacother 106:946–955. https://doi.org/10.1016/J.BIOPHA.2018.07.041
2. Alemrayat, Bayan, Abdelbary Elhissi, Husam M Younes (2018) Preparation and characterization of Letrozole-Loaded Poly(d,l-Lactide) nanoparticles for drug delivery in breast cancer therapy. https://Doi.Org/https://doi.org/10.1080/10837450.2018.1455698 24(2):235–42. https://doi.org/10.1080/10837450.2018.1455698
3. Anon. n.d.-a. Breast Cancer. Retrieved March 21, 2023 (https://www.who.int/news-room/fact-sheets/detail/breast-cancer)
4. Anon. n.d.-b. Breast Cancer Statistics And Resources BCRF. Retrieved March 21, 2023 (https://www.bcrf.org/breast-cancer-statistics-and-resources/)
5. Azawi, Shaymaa, Thomas Liehr, Martina Rincic, Mattia Manferrari (2020) Molecular cytogenomic characterization of the murine breast cancer cell lines C-127I, EMT6/P and TA3 Hauschka. Int J Mol Sci, 21, Page 4716 21(13):4716. https://doi.org/10.3390/IJMS21134716

6. Board, PDQ Adult Treatment Editorial. 2023. "Breast Cancer Treatment (PDQ®)." *PDQ Cancer Information Summaries* 1–5.
7. Bray, Freddie, Jacques Ferlay, Isabelle Soerjomataram, Rebecca L Siegel, Lindsey A Torre, Ahmedin Jemal (2018) Global cancer statistics 2018: GLOBOCAN estimates of incidence and mortality worldwide for 36 cancers in 185 countries. CA: Cancer J Clin 68(6):394–424. https://doi.org/10.3322/CAAC.21492.
8. Chen, Yao, Juan Yang, Shaozhi Fu, Jingbo Wu (2020) Gold Nanoparticles as Radiosensitizers in Cancer Radiotherapy. Int J Nanomedicine 15:9407. https://doi.org/10.2147/IJN.S272902
9. Dianzani, Chiara, Chiara Monge, Gianluca Miglio, Loredana Serpe, Katia Martina, Luigi Cangemi, Chiara Ferraris, Silvia Mioletti, Sara Osella, Casimiro Luca Gigliotti, Elena Boggio, Nausicaa Clemente, Umberto Dianzani, Luigi Battaglia (2020) Nanoemulsions as Delivery Systems for Poly-Chemotherapy Aiming at Melanoma Treatment. Cancers 12(5). https://doi.org/10.3390/CANCERS12051198
10. Du, Manling, Zhenjiang Yang, Wenping Lu, Bingyue Wang, Qi Wang, Zhen Chen, Lianyu Chen, Shuyan Han, Tiange Cai, Yu Cai (2020) Design and development of spirulina Polysaccharide-Loaded nanoemulsions with improved the antitumor effects of paclitaxel. J Microencapsul 37(6):403–12. https://doi.org/10.1080/02652048.2020.1767224
11. Ellis, Matthew J, Li Ding, Dong Shen, Jingqin Luo, Vera J Suman, John W Wallis, Brian A Van Tine, Jeremy Hoog, Reece J Goiffon, Theodore C Goldstein, Sam Ng, Li Lin, Robert Crowder, Jacqueline Snider, Karla Ballman, Jason Weber, Ken Chen, Daniel C Koboldt, Cyriac Kandoth, William S Schierding, Joshua F McMichael, Christopher A Miller, Charles Lu, Christopher C Harris, Michael D McLellan, Michael C Wendl, Katherine Deschryver, Craig Allred D, Laura Esserman, Gary Unzeitig, Julie Margenthaler, Babiera GV, Kelly Marcom P, Guenther JM, Marilyn Leitch, Kelly Hunt, John Olson, Yu Tao, Christopher A. Maher, Lucinda Fulton L, Robert S Fulton, Michelle Harrison, Ben Oberkfell, Feiyu Du, Ryan Demeter, Tammi L Vickery, Adnan Elhammali, Helen Piwnica-Worms, Sandra McDonald, Mark Watson, David J Dooling, David Ota, Li Wei Chang, Ron Bose, Timothy J Ley, David Piwnica-Worms, Joshua M Stuart, Richard K Wilson, Elaine R Mardis (2012) Whole-Genome analysis informs breast cancer response to aromatase inhibition. Nature 2012 486:7403 486(7403):353–60. https://doi.org/10.1038/nature11143
12. Fouad TM, Gutierre AM, Barrera JM, Reuben AL, Woodward WA, Stauder MC, Lim B, DeSnyder SM, Arun B, Gildy B, Valero V, Hortobagyi GN, Ueno NT (2017) Inflammatory breast cancer: a proposed conceptual shift in the UICC-AJCC TNM staging system. Lancet Oncol 18(4):e228–e232. https://doi.org/10.1016/S1470-2045(17)30192-4
13. Gorain B, Choudhury H, Nair AB, Dubey SK, Kesharwani P (2020) Theranostic application of nanoemulsions in chemotherapy. Drug Discovery Today 25(7):1174–1188. https://doi.org/10.1016/J.DRUDIS.2020.04.013
14. de Groot, Stefanie, Rieneke T Lugtenberg, Danielle Cohen, Marij JP Welters, Ilina Ehsan, Maaike PG Vreeswijk, Vincent THBM Smit, Hiltje de Graaf, Joan B Heijns, Johanneke EA Portielje, Agnes J van de Wouw, Alex LT Imholz, Lonneke W Kessels, Suzan Vrijaldenhoven, Arnold Baars, Elma Meershoek Klein Kranenbarg, Marjolijn Duijm de Carpentier, Hein Putter, Jacobus JM van der Hoeven, Johan WR Nortier, Valter D Longo, Hanno Pijl, Judith R Kroep, Hiltje de Graaf, Joan B Heijns, Johanneke EA Portielje, Agnes J van de Wouw, Alex LT Imholz, Lonneke W Kessels, Suzan Vrijaldenhoven, Arnold Baars, Emine Göker, Anke JM Pas, Aafke H Honkoop, A Elise van Leeuwen-Stok, Judith R Kroep (2020) Fasting mimicking diet as an adjunct to neoadjuvant chemotherapy for breast cancer in the multicentre randomized phase 2 DIRECT Trial. Nature Communications 11(1). https://doi.org/10.1038/S41467-020-16138-3.
15. JA, Moura, Valduga CJ, Tavares ER, Kretzer IF, Maria DA, and Maranhão RC. (2011) Novel formulation of a methotrexate derivative with a lipid nanoemulsion. Int J Nanomed 6:2285. https://doi.org/10.2147/IJN.S18039
16. Jaiswal, Manjit, Rupesh Dudhe, Sharma PK (2015) Nanoemulsion: An advanced mode of drug delivery system. 3 Biotech 5(2):123. https://doi.org/10.1007/S13205-014-0214-0

17. Joly, Florence, Marie Lange, Melanie Dos Santos, Ines Vaz-Luis, and Antonio Di Meglio (2019) Long-Term fatigue and cognitive disorders in breast cancer survivors. Cancers 11, Page 1896 11(12):1896. https://doi.org/10.3390/CANCERS11121896
18. Jordan, Rebecca M, Rebecca L Vandeverveken, Jacqueline Oxenberg (2022) Breast cancer conservation therapy. 4–5
19. Kaur, Harneetpal, Kirandeep Kaur, Atamjit Singh, Neena Bedi, Balbir Singh, Mansour S Alturki, Mohammed F Aldawsari, Atiah H Almalki, Shafiul Haque, Hae Jeung Lee, Dharmendra K Yadav, Saroj Arora (2022) Frankincense Oil-Loaded nanoemulsion formulation of paclitaxel and erucin: a synergistic combination for ameliorating drug resistance in breast cancer: In vitro and in vivo study. Front Pharmacol 13:4242. https://doi.org/10.3389/FPHAR.2022.1020602/BIBTEX
20. Liu, Christina H, Piotr Grodzinski (2021) Nanotechnology for cancer imaging: advances, challenges, and clinical. Radiology: Imaging Cancer 3(3). https://doi.org/10.1148/RYCAN.2021200052.
21. Longacre, McKenna, Nicole A Snyder, Genevieve Housman, Meghan Leary, Karolina Lapinska, Sarah Heerboth, Amber Willbanks, Sibaji Sarkar (2016) A comparative analysis of genetic and epigenetic events of breast and ovarian cancer related to tumorigenesis. Int J Mol Sci 17(5). https://doi.org/10.3390/IJMS17050759.
22. Lopez-Garcia MA, Geyer FC, Lacroix-Triki M, Marchió C, Reis-Filho JS (2010) Breast cancer precursors revisited: molecular features and progression pathways. Histopathology 57(2):171–192. https://doi.org/10.1111/J.1365-2559.2010.03568.X
23. Ma P, Mumper RJ (2013) Paclitaxel Nano-Delivery systems: a comprehensive review. J Nanomedicine & Nanotechnol 4(2):1000164. https://doi.org/10.4172/2157-7439.1000164
24. Magnuson A, Sedrak MS, Gross CP, Tew WP, Klepin HD, Wildes TM, Muss HB, Dotan E, Freedman RA, O'Connor T, Dale W, Cohen HJ, Katheria V, Arsenyan A, Levi A, Kim H, Mohile S, Hurria A, Sun CL (2021) Development and validation of a risk tool for predicting severe toxicity in older adults receiving chemotherapy for early-stage breast cancer. J Clin Oncol: Off J Am Soc Clin Oncol 39(6):608–618. https://doi.org/10.1200/JCO.20.02063
25. Malik JA, Ahmed S, Jan B, Bender O, Hagbani TA, Alqarni A, Anwar S (2022) Drugs repurposed: an advanced step towards the treatment of breast cancer and associated challenges. Biomed Pharmacother 145:112375. https://doi.org/10.1016/J.BIOPHA.2021.112375
26. Martí C, Yébenes L, Oliver JM, Moreno E, Frías L, Berjón A, Loayza A, Meléndez M, Roca MJ, Córdoba V, Hardisson D, Rodríguez MÁ, Sánchez-Méndez JI (2022) The clinical impact of neoadjuvant endocrine treatment on Luminal-like breast cancers and its prognostic significance: results from a Single-Institution prospective cohort study. Curr Oncol 29(4):2199. https://doi.org/10.3390/CURRONCOL29040179
27. McGhee DE, Steele JR (2020) Biomechanics of Breast Support for Active Women. Exerc Sport Sci Rev 48(3):99–109. https://doi.org/10.1249/JES.0000000000000221
28. Miranda M, Eustáquio S, de Alcântara J, Lemos RS, Fernandes J, de Oliveira Silva, Flaviano M. Ottoni, Danyelle M. Townsend, Domenico Rubello, Ricardo José Alves, Geovanni Dantas Cassali, Lucas Antônio Miranda Ferreira, and Andre Luis Branco de Barros. (2021) Enhanced antitumor efficacy of Lapachol-Loaded nanoemulsion in breast cancer tumor model. Biomed Pharmacother 133:110936. https://doi.org/10.1016/J.BIOPHA.2020.110936
29. Mitra, Sarmistha, Raju Dash (2018) Natural products for the management and prevention of breast cancer. Evid-Based Complement Altern Medicine : ECAM https://doi.org/10.1155/2018/8324696
30. Momenimovahed Z, Salehiniya H (2019) Epidemiological characteristics of and risk factors for breast cancer in the world. Breast Cancer: Targets Ther 11:151. https://doi.org/10.2147/BCTT.S176070
31. Noel B, Singh SK, Lillard JW, Singh R (2020) Role of natural compounds in preventing and treating breast cancer. Front Biosci (Schol Ed) 12(1):137. https://doi.org/10.2741/S544
32. Park, Sae Gwang, Zhujun Jiang, Eric D. Mortenson, Liufu Deng, Olga Radkevich-Brown, Xuanming Yang, Husain Sattar, Yang Wang, Nicholas K. Brown, Mark Greene, Yang Liu, Jie Tang, Shengdian Wang, and Yang Xin Fu (2010) The therapeutic effect of Anti-HER2/Neu

antibody depends on both innate and adaptive immunity. Cancer Cell 18(2):160–70. https://doi.org/10.1016/J.CCR.2010.06.014
33. Patil Y, Amitay Y, Ohana P, Shmeeda H, Gabizon A (2016) Targeting of pegylated liposomal Mitomycin-C prodrug to the folate receptor of cancer cells: intracellular activation and enhanced cytotoxicity. J Control Release 225:87–95. https://doi.org/10.1016/J.JCONREL.2016.01.039
34. Peddie N, Agnew S, Crawford M, Dixon D, MacPherson I, Fleming L (2021) The impact of medication side effects on adherence and persistence to hormone therapy in breast cancer survivors: a qualitative systematic review and thematic synthesis. Breast: Off J Eur Soc Mastology 58:147. https://doi.org/10.1016/J.BREAST.2021.05.005
35. Sánchez-López, Elena, Mariana Guerra, João Dias-Ferreira, Ana Lopez-Machado, Miren Ettcheto, Amanda Cano, Marta Espina, Antoni Camins, Maria Luisa Garcia, Eliana B Souto (2019a) Current applications of nanoemulsions in cancer therapeutics. Nanomater (Basel, Switz) 9(6). https://doi.org/10.3390/NANO9060821
36. Sánchez-López, Elena, Mariana Guerra, João Dias-Ferreira, Ana Lopez-Machado, Miren Ettcheto, Amanda Cano, Marta Espina, Antoni Camins, Maria Luisa Garcia, Eliana B Souto (2019b) Current applications of nanoemulsions in cancer therapeutics. Nanomater (Basel, Switz) 9(6). https://doi.org/10.3390/NANO9060821
37. Shah R, Rosso K, David Nathanson S (2014) Pathogenesis, prevention, diagnosis and treatment of breast cancer. World J Clin Oncol 5(3):283–298. https://doi.org/10.5306/WJCO.V5.I3.283
38. Silva HD, Cerqueira MÂ, Vicente AA (2012) Nanoemulsions for food applications: development and characterization. Food Bioprocess Technol 5(3):854–867. https://doi.org/10.1007/S11947-011-0683-7/METRICS
39. Srinivasan A, Akolkar D, Patil D, Limaye S, Page R, Ranade A, Patil R, Patil S, Mhase V, Datta V, Apurwa S, Pawar S, Datar R (2020) Real-Time Non-invasive chemoresistance profiling of circulating tumor associated cells in breast cancers to determine resistance towards mitotic inhibitors. Eur J Cancer 138:S76-77. https://doi.org/10.1016/S0959-8049(20)30733-4
40. Stevenson JC, Farmer RDT (2020) HRT and breast cancer: a million women ride again. Climacteric: J Int Menopause Soc 23(3):226–228. https://doi.org/10.1080/13697137.2020.1735797
41. Sung, Hyuna, Jacques Ferlay, Rebecca L Siegel, Mathieu Laversanne, Isabelle Soerjomataram, Ahmedin Jemal, and Freddie Bray (2021) Global cancer statistics 2020: GLOBOCAN Estimates of incidence and mortality worldwide for 36 cancers in 185 countries. CA: Cancer J Clin 71(3):209–49. https://doi.org/10.3322/CAAC.21660
42. Alhamdany T, Anas AMH, Saeed, and Maryam Alaayedi. (2021) Nanoemulsion and solid nanoemulsion for improving oral delivery of a breast cancer drug: formulation, evaluation, and a comparison study. Saudi Pharm J 29(11):1278–1288. https://doi.org/10.1016/J.JSPS.2021.09.016
43. Testa U, Castelli G, Pelosi E (2020) Breast cancer: a molecularly heterogenous disease needing subtype-specific treatments. Med Sci (Basel, Switz) 8(1):18. https://doi.org/10.3390/MEDSCI8010018
44. Wanchai A, Armer JM, Stewart BR, Lasinski BB (2016) Breast Cancer-Related lymphedema: a literature review for clinical practice. Int J Nurs Sci 3(2):202–207. https://doi.org/10.1016/J.IJNSS.2016.04.006
45. Weedon-Fekjær, Harald, Xiaoxue Li, Sandra Lee (2021) Estimating the natural progression of non-invasive ductal carcinoma in situ breast cancer lesions using screening data. J Med Screen 28(3). https://doi.org/10.1177/0969141320945736
46. Yallapu MM, Bhusett PK, Nagesh MJ, Chauhan SC (2015) Therapeutic applications of curcumin nanoformulations. AAPS J 17(6):1341. https://doi.org/10.1208/S12248-015-9811-Z

Treatment of Liver and Gastric Cancer Using Nanoemulsion

Sapna Yadav, Manjari Bhatia, Aashna Sinha, Atreyi Pramanik, Pranchal Rajput, Sujata Jayaraman, Anish Kumar Pal, Kundan Kumar Chaubey, Upendra Singh, Rakesh Kumar Bachheti, and Archana Bachheti

Abstract Cancer is one of the leading causes of death worldwide. Current chemotherapy challenges the lack of selectivity, cytotoxicity, the development of multi-drug resistance, and the proliferation of stem-like cells. One of the leading causes of new cancer cases and cancer-related fatalities is a group of deadly malignant tumors known as liver and gastric cancer. Advanced gastric cancer is classified as having liver metastases and is one of the deadliest conditions with a dismal prognosis. As gastric cancer with liver metastases is incurable, systematic chemotherapy is the primary line of treatment, although surgery is typically used to relieve severe gastrointestinal symptoms. The main method for delivering various medications, nucleic acids, and imaging agents is nanoemulsion (NE). Nevertheless, NE is created through many functions, including surface changes and the encapsulation of active cancer-fighting substances. Because of their excellent characteristics to solubilize hydrophobic chemotherapeutics, thermodynamic stability, biocompatibility, and target ability, NEs are an appealing platform for cancer treatment. With their submicron-dispersed colloids in the continuous phase, NEs may be able to perform various tasks, which helps to overcome biological obstacles that prevent traditional chemotherapeutic drug delivery from reaching the target location

S. Yadav · A. Sinha · A. Pramanik · P. Rajput · S. Jayaraman · A. K. Pal · K. K. Chaubey (✉) · U. Singh
Division of Research and Innovation, School of Applied and Life Sciences, Uttaranchal University, Arcadia Grant, P.O. Chandanwari, Premnagar, Dehradun 248007, Uttarakhand, India
e-mail: kundan2006chaubey@gmail.com

M. Bhatia
Department of Biotechnology, GLA University, Uttar Pradesh, Mathura 281406, India

R. K. Bachheti
Department of Industrial Chemistry, Centre of Excellence in Nanotechnology, Addis Ababa Sciences and Technology University, P.O. Box 16417, Addis Ababa, Ethiopia

A. Bachheti
Department of Environment Science, Graphic Era University, Dehradun 248002, Uttarakhand, India

K. K. Chaubey
School of Basic and Applied Sciences, Sanskriti University, Mathura 281401, Uttar Pradesh, India

A. Husen et al. (eds.), *Current Trends in Green Nano-emulsions*, Smart Nanomaterials Technology, https://doi.org/10.1007/978-981-99-5398-1_11

of cancers. This chapter provides a brief overview of the use of NEs in treating gastric and liver cancer.

Keywords NE · Liver cancer · Gastric cancer · Hepatocellular carcinoma · Therapeutic agent · Cancer treatment

1 Introduction

Cancer is the most significant cause of death, accounting for 19.3 million new cases and 10 million fatalities in 2019, according to the IARC GLOBOCAN 2020 report. Compared to normal cells, cancer is the loss of normal growth control. The equilibrium between new cell growth and cellular aging has been upset. A genetic code mutation is the fundamental cause of cancer. Liver cancer claimed 7.1% of all cancer deaths, and stomach cancer claimed 1.01%. The public health impact of gastric cancer is greatest in poorer nations.

Numerous medicines have been developed to treat cancer but have little effectiveness. Several types of therapies are listed (chemotherapy, immunotherapy, radiotherapy). Chemotherapy is the most popular therapy and the main treatment for malignant sickness out of all these remedies. In comparison to other causes of death, it ranks second. The Nano system has a big chance to replace cancer treatment by focusing on the barriers. Due to their effectiveness in delivering drugs, several nanocarriers are being investigated, including NE, nanoparticles (NPs), nanoshells, nanocapsules, core–shell nanoparticles, nanotubes, lipid nanoparticles, dendrimers [32].

With particle sizes between 20 and 200 nm, NEs (NE) are transparent/translucent heterogeneous systems comprising continuous and dispersed phases in droplets [33]. NE differs from emulsions primarily in gravitational stability, i.e., the smaller the size, the greater the stability. Additionally, research indicates that lipophilic compounds become more bioavailable when globule sizes are below 200 nm. NE is also used extensively in the food and pharmaceutical industries despite being thermodynamically unstable, which means that eventually, the phases will separate due to free energy difference [15, 39]. NE differ from microemulsions due to their smaller size and thermodynamic stability; however, they are adversely affected by changes in temperature and composition. Structured emulsifiers, such as monoacylglycerides (MAG) and diacylglycerides (DAG) enriched with medium-chain fatty acids (MCFAs), have been proposed to improve the stability of NE; these emulsifiers are non-ionic small surfactant molecules frequently used in food emulsions as well as cosmetic and pharmaceutical emulsifiers. The ability to choose the length of the chains that make up these acyl glycerides during chemical processes using substances like fats, oils, and glycerol significantly impacts their functionality. One such material is CAPMUL®, which can be used as an emulsifier or primary solubilizer. A combination of MAG, DAG, TAG, glycerol, and free fatty acids is created by the interesterification of oils with glycerol, allowing for improved substrate interaction without needing a solvent [14]. The use of NEs in treating liver and gastric cancer is briefly discussed in this chapter.

2 Statistics of Hepatocellular Carcinoma or Liver Cancer

Hepatocellular carcinoma (HCC) is one of the most common lethal malignancies. Chronic hepatitis B and C virus infection, dietary aflatoxin exposure, fatty liver disease, alcohol-induced cirrhosis, obesity, smoking, diabetes, and iron overload are all known risk factors for liver cancer. For individuals with an early diagnosis, the recommended course of treatment includes chemotherapy, surgery, immunotherapy, radiation therapy, and tumor ablation. In 2020, liver cancer claimed the lives of 830,200 individuals worldwide and was diagnosed in 905,700 people [40]. The release and targeting of medications in HCC are made possible by nanotechnology. NPs, micelles, and liposomes have all been the subject of various receptor-mediated dynamic targeting and delivery studies in HCC. With its large surface area, external charge, the long half-life of circulation, and capacity to be imaged, NE possesses the necessary qualities to produce effective therapeutic effects. Since vascularized tissues around cancer cells, NEs can easily aggregate in these tissues and potentially cross barriers. They may also be used to explain their function, summarize various medications, and select precise targets [47].

3 Statistics of Gastric Cancer

Gastric cancer (GC) is another lethal type of cancer and counts fourth in the list of different types of cancers. It is often called the latest worldwide burden with the most significant percentage [19]. *Helicobacter pylori (H.pylori),* which causes chronic inflammation, carcinogenesis, and several other variables, including bacterial host, is to blame for this [18]. Inflammation associated with oncogenesis may be caused by cell growth and death. The *Epstein-Bar* virus is a pathogen associated with gastric cancer that is more frequently present in cancerous cells than healthy cells [37]. This condition still has a poor prognosis for therapy, and it is presently the fourth most frequent cause of cancer mortality globally. It was the 16th most common cancer to be diagnosed and the 17th leading cause of cancer mortality in the United States in 2021, with an anticipated 26 560 new cases and 11 180 deaths. There are a few distinct causes for it, the first of which is inherited, followed by gastric adenocarcinoma and gastric intestine cancer. Around 45% of families with hereditary conditions have *CDH1* mutations.

Additionally, it has been determined that the genetic cause of stomach cancer is a mutation in the gene *CTNNA1*. Various environmental elements, influence this and daily living habits have a significant role in it [10]. In addition, those with obesity and gastroesophageal reflux illness are blamed for the increased incidence of stomach cancer [41].

4 Gastric Cancer with Liver Metastasis: New Classified Evolution

Despite being extensively utilized in the pathological grading of GCs, the World Health Organization (WHO) and Lauren classifications are insufficient to guide individualized therapy, particularly in Gastric Cancer with Liver Metastasis (GCLM). Therefore, a GCLM's new categorized evaluation is needed. Advanced gastric cancer is classified as having liver metastases and is one of the deadliest conditions with a dismal prognosis. At the time of first diagnosis, 4–14% of patients with gastric cancer had liver metastases [29]. Notably, recent developments in retrospective and prospective research have made it much easier to identify possible candidates.

Referring to the classification of stage IV GC and the clinical investigation of GCLM [49], three categories were created for GCLM patients. Based on the evaluation of clinical judgment in interdisciplinary treatment, GCLM may be categorized into three groups: potentially respectable cancers (category I), tumors with a slim chance of being removed (category II), and tumors that cannot be removed at all (category III).Due to the significantly worse prognosis of patients with peritoneal spread or positive peritoneal cytology, macroscopic peritoneal dissemination, for instance, was regarded as a crucial feature during the classification procedure [17]. The common methods of treatment for liver and GC are systematic chemotherapy followed by medications. These cancer-relieving medications have unavoidable side effects like nausea and vomiting, loss of appetite, hair loss, diarrhea or constipation, mouth sores, deficiency of platelets and white blood cells, and unfavorable fatigue.

5 Use of NE as Therapeutic Agent

Better treatment ways for administering chemotherapeutics were made possible by several innovative nanotechnology-based technologies [43]. NE is the preferred drug delivery method for chemotherapeutics due to its higher stability and bioavailability than traditional drug administration [44]. Other medications are accessible using NEs besides chemotherapeutic ones, particularly for the liver (Table 1) [48]. Because of its bigger surface area compared to other nano systems, NE is preferred and utilized to aid in the higher adsorption of drugs and to boost their bioavailability [24]. In order to increase the stability of the substance, it is protected against UV light and oxidation, and this is the second significant benefit of NE [10]. As a result of the globule NE size's capability to increase the ability of droplets to penetrate, the tiny size of NE droplet, it is highlighted as having low skin irritation because droplets can pass through hair follicles and skin without harming healthy tissues [38].

Compared to traditional emulsions, NEs are highly stable in creaming, breaking, and phase separation and have a larger surface area and free energy [44]. Previously, they ingested many cellular antigen presenters. NE is presently being used swiftly in the pharmaceutical business due to all these important qualities, particularly for drug delivery nanocarrier, cancer-targeting therapy, etc. (Fig. 1) [23]. Due to its prolonged shelf life, it also significantly impacts vaccination distribution [12].

6 Using NE as a Distinct Strategy in Both Cancer Treatments

Researchers are now exploring the potential of NE to deliver various cancer-fighting strategies for treating liver and stomach cancer. The application of multifunctional NEs for the treatment of cancer is shown schematically (Fig. 2). NEs are significant and alluring approaches to hepatocellular carcinoma and gastric cancer due to their stability under thermodynamic conditions, huge surface area, droplet size, stability, and desired drug eliminate study [23]. Research on NE formulations for cancer treatment found that they are a safe grade of excipients (GRAS) with a great potential for physiochemical stability, superior efficacy, and higher bioavailability [11].

One of the most often used medications to treat hepatocellular cancer is sorafenib. To produce anticancer effects, it can block the actions of several tyrosine and serine kinases as well as tumour angiogenesis, cell proliferation, and apoptosis. Sorafenib was delivered using a NE, which can potentially be a safer and more efficient parenteral drug delivery method than the more common tablet form[47]. Medium-chain triglycerides and lecithin made up the dispersed phase of the sorafenib NE, which was created utilising a high-energy emulsification technique. Glycerol and polysorbate 80 made up the continuous phase. The produced emulsion's MTT experiment revealed that while the ideal formulation did not damage normal cells, it could nonetheless eliminate cancer cells.

Combining Gemcitabine (Gem) with doxorubicin (Dox) in a NE delivery method reduced the toxicity of both drugs—Gem for the liver and Dox for the heart—while increasing anticancer activity. Compared to mice treated with Dox solution, which increased the levels of creatine kinase-myocardial band (CK-MB), CK, Hyperlipidemia (HLD), triglycerides (TG), and Lactate dehydrogenase (LDH), tumor-bearing mice received Gem-DoxNE, which decreased the levels of CK-MB, LDH, CK, and CHO while increasing the levels of HDL. In addition, the NE of Vitamin D led to a reduced level of TNFα. The two-step process of water-in-oil-in-water NE of 5-Fluorouracil (5-FU) and Oxaliplatin has shown enhanced membrane permeability compared to 5-FUand oxaliplatin which had poor membrane absorptivity [47]. Abdu et al. [1], reported that NE encapsulated Gem (Gem) and Dox (Dox).GEM increases DOX's ability to prevent tumor growth, which improves its anticancer effects (Abdu,

Table 1 Delivery of several therapeutic agents using NE that have activity against hepatocellular carcinoma

Sl. no	Name	Work	Role of NE	Key reference
1	*Natural drugs*			
(a)	Curcumin	Curcumin has been utilized as a folk medicine and a widely used herb in the family since the dawn of civilization. Curcumin is also known as turmeric. In the in-vivo investigation, it was shown that following therapy, rats' liver and stomachs had 60% more curcumin than their blood did. It demonstrates that oral administration of curcumin results in very low absorption	Curcumin was delivered via a NE-based technology, and it was discovered that the NE made with MCT had the highest bioavailability of curcumin (58.6%). They also noticed that the enhanced lipid content in the NE resulted in higher bioavailability	[3]
(b)	Silymarin	Silibin is Silymarin's most potent and significant component. It influences glutathione oxidation to increase its levels in the colon and liver. It prevents the hepatotoxin from attaching to receptor sites on the hepatocyte membrane. Moreover, it increases hepatocyte regeneration through ribosomal RNA polymerase stimulation	As compared to Silymarin regular suspension, the NE of Silymarin made with sefsol 218, Kolliphor RH40, and PEG-400 had a higher AUC and Cmax and a shorter tmax. Decreased cell viability and enhanced ROS condensation were the results of the cell line investigation on NE	[3, 25, 28]
(c)	Eugenol	Eugenol and related polyphenols are poorly soluble in water. This is the cause of their after-oral administration's comparatively poor bioavailability	In liver cell lines, the apoptotic potential of eugenol loaded NEwas compared to bulk eugenol, and it was shown that both preparations promoted apoptosis via ROS (Reactive Oxygen Species) formation. With eugenol loaded NE, HB8065 cells displayed a greater rate of apoptosis (69%) than with bulk eugenol (53%). In the case of the eugenol-loaded NE, the ROS levels were likewise higher	[22, 30]

(continued)

Table 1 (continued)

Sl. no	Name	Work	Role of NE	Key reference
(d)	Crocin	Crocin, alone in HepG2 cells showed a 59% reduction in telomerase activity after exposure to crocin	Polyglycerol polyricinoleate has a high capacity for producing stable crocin NEs	[4, 31, 34, 42]
(e)	*Ginkgo Biloba (G.biloba)*	Polyprenols found in *G.biloba*have a good track record of treating liver conditions, including cancer	In HepG2 cells, the NE of *G.biloba* demonstrated lower genotoxicity and cytotoxicity than the chitosan nanoparticles with folic acid coupling	[45, 46]
(f)	Resveratrol	Resveratrol, in HepG2 cells and rat hepatoma cells in the S and G2/M phases, downregulates hepatic growth factor to slow the formation of tumors	Resveratrol is shielded against chemical deterioration and isomerization to inactive Z-resveratrol via a NE-based formulation	[7, 20, 36]
2	*Synthetic drug*			
(a)	Sorafenib	It is one of the medications used most frequently to treat hepatocellular carcinoma. To produce the anticancer effects, it can block the actions of several tyrosine and serine kinases as well as tumor angiogenesis, cell proliferation, and apoptosis. Its limited water solubility and hepatic metabolism result in a low bioavailability of 38–49%	Because it contained less medicine and had a greater level of effectiveness than traditional tablets, the sorafenib NE delivery method demonstrated the promise for safe and efficient par-enteral drug administration	[9, 21]

(continued)

Table 1 (continued)

Sl. no	Name	Work	Role of NE	Key reference
(b)	Cisplatin	Together with 5-FU and Dox, cisplatin is a common medication used as a systemic chemotherapeutic agent in the treatment of hepatocellular cancer. Systemic chemotherapy is used as an adjuvant therapy for individuals with unresectable hepatocellular carcinoma and those recovering from surgery	The release of a cisplatin-loaded NE, which was designed for intravesical administration in cases of bladder cancer, was slower and extended (3–4 h), and the amount of medication that reached the tissues increased significantly (2.4–3.3 times). It is necessary to ascribe an increase in tissue concentration of cisplatin to permeation enhancers such terpineol and oleic acid. An analogous strategy could result in much improved cisplatin administration in hepatocellular cancer	[2, 50]
(c)	Gem	Fan et al. investigated the anticancer efficacy of Gem against hepatoma. Gem chlorambucil coupled nanoparticles in the research significantly reduced hepatocellular carcinoma in tumor-bearing mice	To explore the in vitro anticancer activity, Ahmed and Sami combined the vitamin E isomer to-cotrienol with Gem and added it to the NE system. Conjugation was used to boost Gem's anticancer effectiveness by preventing the drug's extensive metabolism through an enzymatic deamination process. Compared to the free medication, the coupled NE demonstrated more anticancer efficacy	[16]
(d)	Dox	Dox is a chemotherapy medication that is frequently used to treat different malignancies. Moreover, it has demonstrated its effectiveness against hepatocellular carcinoma	Alkhatib et al. examined the anticancer effects of Dox-incorporated NEs in Swiss albino mice with Ehrlich ascites carcinoma. In comparison to the animal group treated with Dox solution, it was shown that the serum enzyme levels were decreased and the cardiac tissues suffered less damage in the DoxNE-treated group	[5, 50]

(continued)

Table 1 (continued)

Sl. no	Name	Work	Role of NE	Key reference
(e)	Vitamin D	To compare vitamin D NE to conventional vitamin D and assess its hepatoprotective effects, a comparison research with conventional vitamin D was conducted utilizing substances such as pea protein isolate and canola oil that had been pH-shifted and sonicated	In comparison to vitamin D treated groups, the GPT, GOT, and GGT levels decreased in the vitamin D NE treated group by 10.16, 16.8, and 35.8%, respectively. Also, the levels of inflammatory mediators like TNF increased by 0.53 times. These levels considerably decreased following vitamin D treatment, and the vitamin D NE exhibited a greater drop in TNF levels (20% more than the vitamin D treated group)	[13]
3	*Combination drugs*			
(a)	Dox and gem		Gem and Dox were integrated in the NE delivery method by Abdu et al. Due to the focused drug delivery capabilities of NE, the concentration of the medication in the body is reduced, preventing any negative side effects	[6]

(continued)

Table 1 (continued)

Sl. no	Name	Work	Role of NE	Key reference
(b)	Curcumin and sorafenib	The only molecular targeting drug that has been licensed for advanced hepatocellular carcinoma is sorafenib, which is frequently used as the standard therapy medication for the disease. Nevertheless, only around 30% of those with hepatocellular carcinoma benefit from treatment, and the majority of those who do frequently have resistant tumors and progression. Curcumin may function as a substance that makes cancer cells more susceptible to chemotherapeutic drugs and lowers treatment resistance to increase medication effectiveness	The nanoparticles have demonstrated excellent outcomes, increasing cell apoptosis and cytotoxicity in HepG2 cells. It may provide encouraging results to distribute these together using a delivery mechanism based on NEs	[9]
(c)	5-FU and oxaliplatin	Patients with advanced, unresectable HCC bigger than 10 cm in diameter had demonstrated encouraging outcomes when treated with oxaliplatin, a third-generation cisplatin, in combination with 5-FU. Due to weak membrane absorptivity, 5-FU and oxaliplatin have reduced bioavailability	To increase their membrane permeability, 5-FU and the oxaliplatin combination with the deoxycholic acid derivative (used as a penetration enhancer) were exposed to the preparation of a water-in-oil-in-water NE. A two-step spontaneous emulsification process was used to create the NE. First, a w/o NE containing an Oxaliplatin complex was created, and it was then dispersed in an aqueous phase containing 5-Flurouracil to create a w/o/w NE	[27, 35]

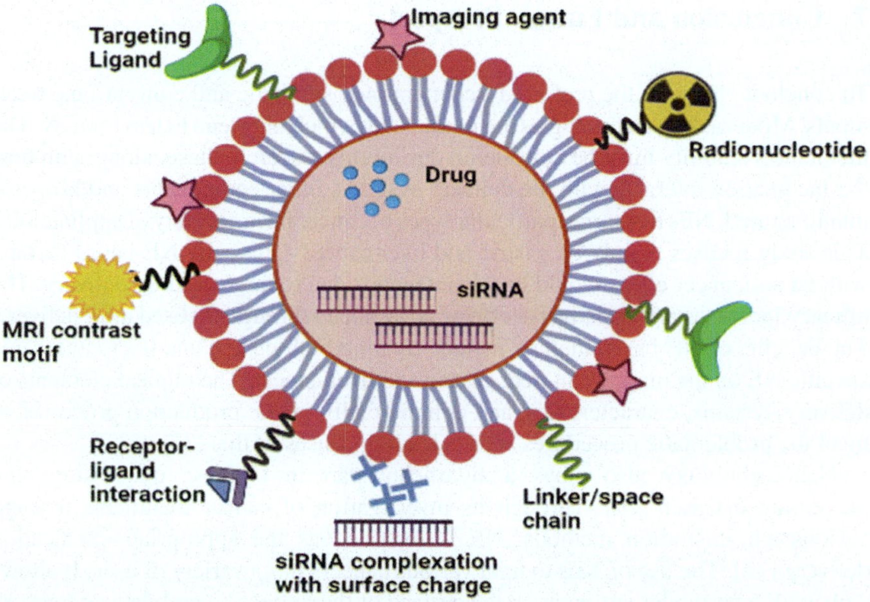

Fig. 1 Structural representation of NE droplet size

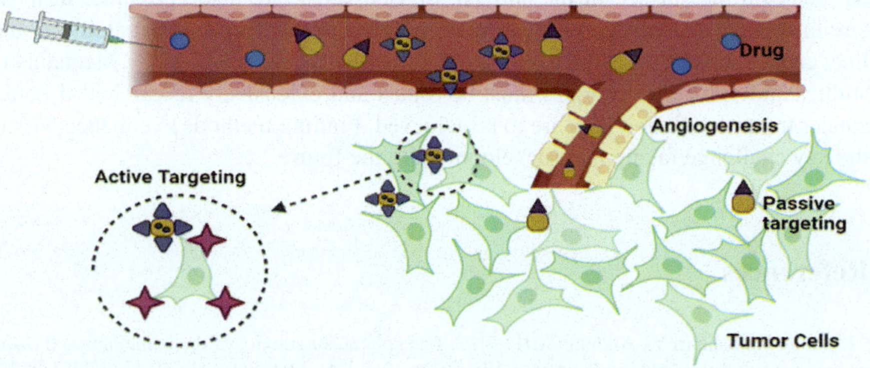

Fig. 2 Structural representation of Multifunctional NE application against cancer

Alshehri, and Alkhatib 2017). Detailed information on therapeutic agents and NE technology for liver and GC treatment are shown in Table 1. Choudhury et al. [10] investigated anti-HIF-1 antibody-linked triblock copolymer nano micelles encapsulated with paclitaxel for targeting the MGC-803 cells. They concluded that antibody-targeted NEs have excellent potential for imaging and therapy targeting clinical tumors [10].

7 Conclusion and Future Prospects

To conclude, NEs are the preferred treatment over synthetic and conventional treatments. Modern drug technologies are modulated, which may lead to less toxicity. The medicine's stability must be considered during this research phase, along with how the medication interacts with the delivery system's other components and how it is manufactured. NEs that target particular types of cancer have a variety of applications. This study focuses mainly on gastric and liver cancer. Clinically NE-based therapy with an anticancer effect would be advantageous due to targeted drug delivery. The efficacy increases and the adverse effects are reduced through focused drug delivery. The benefits of NE-based therapy include treating both gastric and liver cancer. No specific NE on the market can treat gastric and liver cancer. The clinical elements of delivery systems, characteristics, and commercially viable production processes to meet the problematic procedures are the leading causes of this [32].

Nanotechnology also plays a significant part in treating, diagnosing, and preventing diseases [26]. Through its investigation of cancer treatments through various administration methods, NE is regarded as the appropriate medication delivery [19]. The use of NEs to treat various cancers has a variety of uses. It allows cells with a particular antibody on the surface of the antigen to circulate and absorb for an extended period, or it allows the opposite to occur. This study focuses more on liver and gastric cancer. Long circulation times and uptake by cells with a specific antibody on the surface of the antigen, or vice versa, are made possible. In these circumstances, the treatment appears to involve using NE to combat stomach and liver cancer cells. A potential medication carrier that works well is a NE. Mechanistic studies and clinical techniques must be robust and extensively researched if better cancer treatments using NEs are to be achieved. Finding methods to enhance NEs is the key challenge facing their development in the future.

References

1. Abdu F, Alshehri W, Alkhatib MH (2017) The anticancer activity of the combination therapy of gem and dox encapsulated in a NE. Polym Sci 3(4). https://doi.org/10.4172/2471-9935-C1-006
2. Abu-Fayyad A, Nazzal S (2017) Gem-Vitamin E conjugates: synthesis, characterization, entrapment into NEs, and in-vitro deamination and antitumor activity. Int J Pharm 528(1–2):463. https://doi.org/10.1016/J.IJPHARM.2017.06.031
3. Ahmad U, Akhtar J, Singh SP, Badruddeen B, Ahmad FJ, Siddiqui S, Wahajuddin W (2018) Silymarin NE against human hepatocellular carcinoma: development and optimization. Artif Cells, Nanomed Biotechnol 46(2):231–241. https://doi.org/10.1080/21691401.2017.1324465
4. Alavizadeh SH, Hosseinzadeh H (2014) Bioactivity assessment and toxicity of crocin: a comprehensive review. Food Chem Toxicol: Int J Publ Brit Indus Biol Res Assoc 64:65–80. https://doi.org/10.1016/J.FCT.2013.11.016
5. Alkhatib MH, Alkreathy HM, Balamash KS, Abdu F (2016) Antitumor activity of doxe-loaded NE against ehrlich ascites carcinoma-bearing mice. Trop J Pharm Res 15(5):937–943. https://doi.org/10.4314/TJPR.V15I5.6

42. Shamseddine A, Salel
Cancer trends in Leba
tions until 2018. Pop
URES/4
43. Sugumar S, Nirmala J
NE formulation, cha
Basic Microbiol 53(8
44. Tao R, Wang CZ, Ye
genotoxic activity of r
homologs. Lipids Hea
45. Tao R, Wang C, Zhan
and genotoxicity of T
NE. Biol Trace Elem
46. Upadhyay T, Ansari V
therapy. Curr Cancer
0302123336
47. Xu T, Gao CC, Feng
properties of chitosan
Biol Macromol 134:2
48. Yoshida K, Yamaguc
possible in stage iv ga
Gastric Cancer: Off J
https://doi.org/10.100
49. Zhang Y, Ding J, Su
mediated sustained d
Mater Biol Appl 49:2
50. Zhang Y, Guan DX,
Zhou FG, Wu MC, W
chemotherapeutic eff
cancer. J Hepatol 59(

6. Alkhatib MH, Alshehri WS, Abdu FB (2018) In vivo evaluation of the anticancer activity of the gem and dox combined in a NE. J Pharm Bioallied Sci 10(1):35–42. https://doi.org/10.4103/JPBS.JPBS_225_17
7. Alnajjar AsmaM, Elsiesy HussienA (2015) Natural products and hepatocellular carcinoma: a review. Hepatoma Res 1(3):119. https://doi.org/10.4103/2394-5079.167379
8. Cao H, Wang Y, He X, Zhang Z, Yin Q, Chen Y, Yu H, Huang Y, Chen L, Xu M, Gu W, Li Y (2015) Codelivery of sorafenib and curcumin by directed self-assembled nanoparticles enhances therapeutic effect on hepatocellular carcinoma. Mol Pharm 12(3):922–931. https://doi.org/10.1021/MP500755J
9. Choudhury H, Pandey M, Gorain B, Chatterjee B, Madheswaran T, Mak SMKK, Tambuwala M, Chourasia MK, Kesharwani P (2019) NEs as effective carriers for the treatment of lung cancer. Nanotechnol-Based Target Drug Deliv Syst Lung Cancer 217–247. https://doi.org/10.1016/B978-0-12-815720-6.00009-5
10. Efferth T (2017) From ancient herb to modern drug: artemisia annua and artemisinin for cancer therapy. Semin Cancer Biol 46:65–83. https://doi.org/10.1016/J.SEMCANCER.2017.02.009
11. El-Readi MZ, Althubiti MA (2019) Cancer nanomedicine: a new era of successful targeted therapy. J Nanomater. https://doi.org/10.1155/2019/4927312
12. El-Sherbiny M, Eldosoky M, El-Shafey M, Othman G, Elkattawy HA, Bedir T, Elsherbiny NM (2018) Vitamin D NE enhances hepatoprotective effect of conventional Vitamin D in rats fed with a high-fat diet. Chem Biol Interact 288:65–75. https://doi.org/10.1016/J.CBI.2018.04.010
13. Esperón-Rojas AA, Baeza-Jiménez R, Cano-Sarmiento C, García HS (2017) Structured mono- and diacylglycerols with a high content of medium chain fatty acids. J Oleo Sci 66(9):991–996. https://doi.org/10.5650/JOS.ESS17010
14. Esperón-Rojas, Alessa A, Baeza-Jiménez R, García HS (2019) Preparation of curcumin-carrying NEs using mono- and diacylglycerides enzymatically structured with bioactive fatty acids. 38(2):93–103. https://doi.org/10.1080/10242422.2019.1646254
15. Fan M, Liang X, Li Z, Wang H, Yang D, Shi B (2015) Chlorambucil gem conjugate nanomedicine for cancer therapy. Eur J Pharm Sci: Off J Eur Federat Pharm Sci 79:20–26. https://doi.org/10.1016/J.EJPS.2015.08.013
16. Ferlay J, Soerjomataram I, Dikshit R, Eser S, Mathers C, Rebelo M, Parkin DM, Forman D, Bray F (2015) Cancer incidence and mortality worldwide: sources, methods and major patterns in GLOBOCAN 2012. Int J Cancer 136(5):E359–E386. https://doi.org/10.1002/IJC.29210
17. Ganta S, Talekar M, Singh A, Coleman TP, Amiji MM (2014) NEs in translational research-opportunities and challenges in targeted cancer therapy. AAPS PharmSciTech 15(3):694–708. https://doi.org/10.1208/S12249-014-0088-9
18. Gorain B, Choudhury H, Nair AB, Dubey SK, Kesharwani P (2020) Theranostic application of NEs in chemotherapy. Drug Discov Today 25(7):1174–1188. https://doi.org/10.1016/J.DRUDIS.2020.04.013
19. Herneisey M, Mejia G, Pradhan G, Dussor G, Price T, Janjic J (2018) Resveratrol NEs target inflammatory macrophages to prevent. J Pain 19(3):S75-76. https://doi.org/10.1016/j.jpain.2017.12.173
20. Izadiyan Z, Basri M, Masoumi HRF, Karjiban RA, Salim N, Shameli K (2017) Modeling and optimization of NE containing sorafenib for cancer treatment by response surface methodology. Chem Central J 11(1). https://doi.org/10.1186/S13065-017-0248-6
21. Jaganathan SK, Mazumdar A, Mondhe D, Mandal M (2011) Apoptotic effect of eugenol in human colon cancer cell lines. Cell Biol Int 35(6):607–615. https://doi.org/10.1042/CBI20100118
22. Jaiswal M, Dudhe R, Sharma PK (2015) NE: an advanced mode of drug delivery system. 3 Biotech 5(2):123. https://doi.org/10.1007/S13205-014-0214-0
23. Kentish S, Wooster TJ, Ashokkumar M, Balachandran S, Mawson R, Simons L (2008) The use of ultrasonics for NE preparation. Innov Food Sci Emerg Technol 9(2):170–175. https://doi.org/10.1016/J.IFSET.2007.07.005

24. Kim S, Choi
SJ, Yang JH,
9 expression
14(11):4300–
25. Kirtane AR,
approaches fo
1038/s41565-
26. Li JH, Xie XY
Xia JL, Zhang
with lipiodoliz
WJG 21(13):3
27. Ligeret H, B
chondrial pro
J Ethnopharm
28. Luo Z, Rong
Front Oncol 9
29. Majeed H, An
and liver canc
org/10.7314/A
30. Mehrnia MA,
emulsions: fa
Macromol 84:
31. Mohapatra P,
treatment (Ma
32. Naseema A,
of synthesis p
287:102318. h
33. Noureini SK,
inhibition and
https://doi.org
34. Pangeni R, Ch
national deliv
11:6379. https
35. Park SJ, Ahn
Brown AL, K
ameliorates ag
148(3):421–4
36. Plaza-Oliver
nanocarriers f
1007/S13346-
37. Praveen Kum
Open Access
38. Pucek A, Tok
design of phot
12(6):587. htt
39. Rumgay H, A
Soerjomatara
J Hepatol 77(
40. Sánchez-Lópe
M, Camins A,
Nanomaterial
41. Sarfarazi M, J
through respo
10.1007/S121

1 Introduction

In the field of science and technology, nanotechnology is a rapidly growing field. To date, nanotechnology has demonstrated its benefits in several domains [13, 14, 25, 26]. Nanoemulsions are among the best-known applications of nanotechnology. Nanoemulsions are biphasic colloidal dispersions in which the oil droplets are disseminated in an aqueous phase [24, 36, 46]. They are also sometimes known as microemulsions, ultrafine emulsions, and submicron emulsions by certain investigators. These heterogeneous particle systems are usually stabilized by an emulsifier, and their internal phase droplet size is generally below 1000 nm. Water-in-oil (w/o), oil-in-water (o/w), oil-in-water-in-oil (O/W/O), and water-in-oil-in-water (W/O/W) emulsions are the four kinds of emulsions that may be distinguished based on the dispersed and continuous phases (Fig. 1). Due to the existence of a hydrophobic core area that facilitates the administration of medications and bioactive substances that are insoluble in water, O/W nanoemulsions have greater uses in the pharmaceutical sector.

Nanoemulsions have emerged as effective carriers for the delivery of hydrophobic bioactive compounds. In addition to better solubility, nanoemulsions have other advantageous properties. They have higher and longer-term stability, high loading capacity, high biocompatibility, and increased bioavailability. In addition, they may also be used for the regulated release of bioactive substances. Nanoemulsions with beneficial physicochemical properties have been extensively explored as vehicles for the delivery of pesticides [28]. By encapsulating pharmaceuticals in nano formulations, it is possible to obtain sustained drug release and a reduction in the amount of drug that is required. Nanomaterials (NMs) may also be utilized to control and ameliorate infections brought on by resistant strains [39]. Due to their small pore size, nanoemulsions enable the even collection and storage of efficacious substances over the superficial skin [47]. Using just 3–10% surfactants during production, nanoemulsions are more efficient at diffusing substances due to their higher surface area and low overall viscosity [16]. In microemulsions, surfactant absorptions of at least 20% are required [21]. As a consequence, nanoemulsions become more fluid (at lower oil concentrations), particularly when thickeners are not utilized, through appealing physical qualities and skin. The strong steric stability at submicron droplet sizes also prevents emulsion-specific destabilization processes, including creaming or sedimentation, coagulation, and coalescence. Producing nanoemulsions requires careful consideration of the effects of the sequence in which the various ingredients are mixed. It is important to emphasize that before the formulation of nanoemulsions, surfactants must be coupled with the oily process. This allows for the creation of exceptionally favourable circumstances for the growth of nanoemulsions.

On the other hand, the early stages of manufacturing would encourage the production of "macroscopic" emulsions when surfactants and water are mixed [11].

Currently, a variety of approaches are used to produce nanoemulsions, which may be considered either lower- or higher-energy emulsification operations or a combination of both [38]. High-energy techniques are characterized by the use of

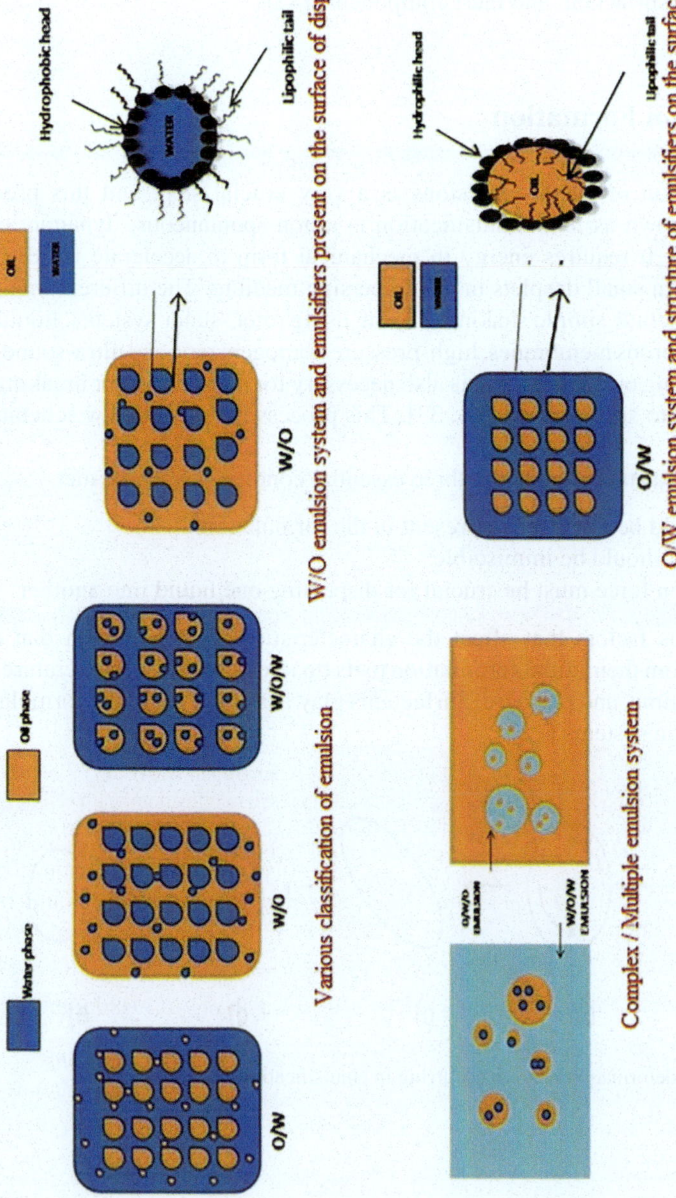

Fig. 1 Types of emulsion

mechanical devices to generate strong, extreme elements that split the oil and water processes to form oil-based droplets. This method makes use of high-pressure homogenizers, microfluidizers, and sonication procedures [9]. Conversely, low-energy techniques emulsify by using the chemical energy of the system. This is accomplished by rerouting the formulation's inherent physicochemical properties through the surfactant, co-surfactant, and inert components [42].

2 Emulsion Formation

The formulation of stable emulsions is a very crucial step, and this process is commonly known as *ion*. Emulsification is a non-spontaneous, dynamic, energy-requiring step. It requires energy in mechanical form to accelerate the dispersion of the liquids in small droplets in the dispersing medium. The different methods of formulation include simple shaking, mixing using rotor–stator systems, liquid injection through porous membranes, high-pressure homogenizers, and ultra-sound generators. A shearing or stirring force is also necessary for deformation or breaking down the droplets into smaller sizes [23, 33]. This process is shown below schematically (Fig. 2).

To develop a stable emulsion, three essential conditions must be met.

- There should be a surfactant present in the formulation.
- The phases should be immiscible.
- An agitation force must be crucial for dispersing one liquid into another.

The various factors that affect the characteristics of an emulsion that can be changeable from their initial formulation to its final resolution are temperature, agitation velocity, time, and pressure. Surfactants play a crucial role in the formulation of stable emulsion systems [12].

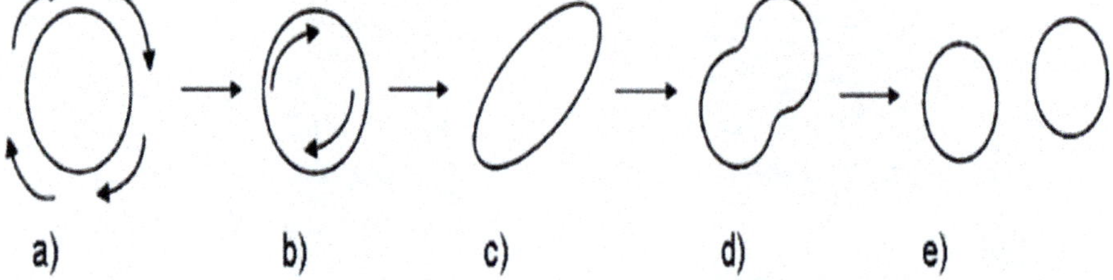

Fig. 2 Droplet deformation process occurring in emulsification

3 Methods of Preparation

The formulation of stable primary (1°) emulsions is the main objective, as it finds application both in the pharmaceutical and food industries. It involves methods that are briefly discussed as following [49].

3.1 Dry Gum Method

- For the primary emulsion, the amounts of oil, water, and emulsifier are calculated.
- A dry porcelain mortar and pestle is used to combine the emulsifier and oil.
- Water is added all at once after the emulsifier has been evenly distributed throughout the oil.
- The primary emulsion is triturated for at least 5 min, stirring constantly but gently in one direction until the mixture thickens under the pestle.
- The emulsion is then poured into a graduated cylinder and diluted with water to reach the desired volume.
- The label for the emulsion should read "Shake well before use."

3.2 Wet Gum Method

- Water is mixed with an emulsifier to create mucilage, and then oil is gradually added.
- For the primary emulsion, the amounts of oil, water, and gum are calculated.
- In a porcelain mortar, emulsifiers and water are triturated to create mucilage.
- The oil is added in small amounts, and the trituration is constant, quick, and light.
- After adding all the oil, the mixture is vigorously stirred for a few minutes. The emulsions are then moved to a graduated cylinder and diluted with water.

3.2.1 Emulsions Containing More Than One Oily Liquid

- The amount of emulsifier needed for each of the two or more oily liquids present is determined, and the emulsion is made by subsequently adding these predetermined amounts.
- Alternately, the oils could be blended first and then emulsified individually.

3.2.2 Emulsions Containing Water-Soluble Substances

- The majority of the ingredients in emulsions, such as salts, syrups, and glycerin, are water-soluble.
- Water-soluble ingredients are used in a diluted state whenever possible because some substances have demulsifying properties and may destabilize the emulsion if added to a concentrated solution.

3.2.3 Emulsion Containing Oil-Soluble Substances

- The amount of emulsifier needed to make an emulsion with oil-soluble ingredients should be increased by 50%.
- In order to prepare the main emulsion, the oil-soluble materials are first dissolved in the oil.

3.2.4 Emulsions Containing Substances Insoluble in Either Oil or Water

- These ingredients must be ground into a fine powder in a mortar before being combined with the emulsifier needed for the main emulsion.
- The main emulsion is then made as normal, with the addition of the oil.
- Examples include phenophthalein in a liquid paraffin emulsion and bismuth carbonate in a castor-oil emulsion.

3.2.5 Emulsions Containing a Small Proportion of Oily Substances

- If the oil content is too low, adjustments must be made. Emulsions with <10% oil have a tendency to cream easily. To raise the amount of oil to 10–20% and avoid this, an inert oil, such as arachis oil, can be added.

4 Wound and the Process of Rebuilding

"*Wound*" refers to the rapid onset of injury that involves:

- a lacerated type of skin or punctured skin (open) or
- a contusion type (closed),

caused by any sort of blunt force, trauma, or compression force. It might be acute or persistent injury to the skin's epidermis. To recover from this damage to the cells and tissues of the epidermal layer, the human body undertakes a series of actions, commonly known as "*wound healing,*" that include replacement of destroyed or damaged tissue with newly produced tissue. A skin wound results from the breakdown of the epidermal layer's integrity [29].

On the basis of severity, wounds can be classified as:

- Acute wound
- Chronic wound

In a normal skin, the epidermis and the dermis layer form a protective barrier against the external environment. When this protective barrier is broken or trespassed, a regulated sequence of biological reactions is set into motion to repair the damage caused to the tissues. The wound healing process or rebuilding the tissues is a highly organized cellular, humoral, molecular and complex but fragile phenomenon.

Wound healing can be classified as:

- Normal type wound healing
- Persistent non-healing wound

It is susceptible to various interruptions or failures, leading to the formation of non-healing chronic wounds. The following factors could be possible reasons that contribute to the chronic wounds: Diabetes, venous or arterial disease infection, and metabolic deficiencies of old age. But, with the help of appropriate wound care depending on the patient's need,

The complex process of healing wounds has four underlying overlapping phases:

- Hemostasis phase
- Inflammatory phase
- Proliferative phase
- Remodelling/maturation phase.

4.1 Hemostasis Phase

- Takes place very quickly.
- Begins with blood leaking from the body, and as a result, blood vessels constrict to reduce blood flow.
- Platelets clump together and attach to the sub-endothelium surface within seconds after an epithelial blood vessel wall rupturing.
- The initial fibrin strands begin to bind within sixty seconds.
- Procoagulants and the release of prothrombin induce the blood to coagulate when the mesh of fibrin forms, altering its liquid condition.
- The creation of a thrombus or clot confines platelets and blood cells to the wound area.
- A thrombus is usually important through various phases of wound healing, however it gets a concern if it splits from the vessel wall and enters the circulatory system, potentially resulting in a heart attack, stroke, or pulmonary embolism [31].

4.2 Inflammatory Phase

- Initiates as soon as the injury occurs when transudation, which is a fluid composed of protein, salt and water, leaks from the injured blood vessels, causing localized swelling.
- Inflammation prevents infection while also controlling bleeding.
- The fluid engorgement enables the migration of repair and healing cells to the wound site.
- Damaged cells, pathogens, and bacteria are cleared from the wound area during the inflammatory phase.
- The swelling, heat, pain, and redness that are frequently present during this stage of wound healing are produced by the white blood cells, growth factors, nutrients, and enzymes.
- Inflammation is a normal part of the healing process for wounds; it only becomes a problem if it lats too long or is excessive [17].

4.3 Proliferative Phase

When the lesion is repaired with fresh collagen-and extracellular matrix-containing tissue,

- The wound shrinks as fresh tissues form.
- To ensure that the granulation tissue is healthy and receives enough nutrients and oxygen, a new network of blood vessels must be built.
- Myo-fibroblasts, which function similarly to smooth muscle cells, grab the wound's margins and pull them together to cause the wound to close.
- Granulation tissue is pink or red in color and has an irregular texture during healthy wound healing stages. Granulation tissue that is healthy does not bleed readily. Dark granulation tissue may indicate ischemia, infection, or inadequate perfusion. Finally, the damaged epithelial cells reappear.
- When wounds are maintained moist, epithelialization occurs more quickly.

To maintain correct tissue humidity in order to optimize the process of epithelialization, the occlusive or semi-occlusive dressing is applied within 48 h of injury [40, 52].

4.4 Remodeling/Maturation Phase

- Type III collagen is converted into type I collagen once the wound fully heals.
- Apoptosis, also known as programmed cell death, is the process by which cells that had been used to treat the wound but were no longer necessary are removed.

- The wound is substantial and the proliferative phase's collagen buildup is disorganised.
- Collagen undergoes remodelling along stress lines, becoming more organised and improving the tensile strength of the healing tissues.
- Fibroblasts release matrix metalloproteinases.
- The enzymes facilitate the conversion of type III collagen to type I collagen.
- Remodeling frequently commences about 21 days following an injury and may extend for a year or longer [34, 53].

The rate and efficiency of wound healing are influenced by a variety of variables, but microbial contamination and subsequent infection have been identified as key factors in slower healing. A localized inflammatory state brought on by a wound site infection promotes the overproduction of cytokines, which suppresses keratinocyte proliferation, interferes with the development of new matrix, delays wound closure, and perhaps starts secondary damage and dermal scarring. These unfavorable circumstances prolong healing timeframes and raise healthcare expenditures. Antibiotics and silver ion-based regimens are two current topical treatments used to treat infected wounds. The rise of multidrug-resistant bacteria and the decreasing effectiveness of conventional antibiotics, however, are of great concern [32].

As an alternative, silver has well-established biocidal characteristics and is regularly used as an antibacterial treatment for burns and severe wounds. Significant data demonstrates that different silver formulations are efficient antibacterial therapies for wound healing. Recent research, however, raises the possibility that silver may be cytotoxic to both fibroblasts and keratinocytes and hinder the healing process. As a result, researchers are exploring a novel therapeutic platform termed hyperosmotic nanoemulsions in order to produce new classes of antimicrobial medicines with low unfavorable host reactions. To hasten the healing of a wound and prevent infection linked to a delayed recovery process, nanoparticle-based distribution has recently been recognized as a prominent strategy for curing dermatological illnesses, including wounds caused by different causes [27, 41].

It has been shown that nanoparticles may penetrate the stratum corneum and reach the deeper layers of the epidermis and dermis of the skin by one or more combination mechanisms, such as their small size, deformable nature, lipid behavior, or lipid-fluid nature [18, 27]. The use of nanoparticles for wound healing is one of the most actively researched areas, where nanoparticles have been endowed with multiple bioactive agents. The purpose of this chapter was to illustrate the new advancements in the diversity of nanoemulsions that are used in the wound healing process (Figs. 3 and 4).

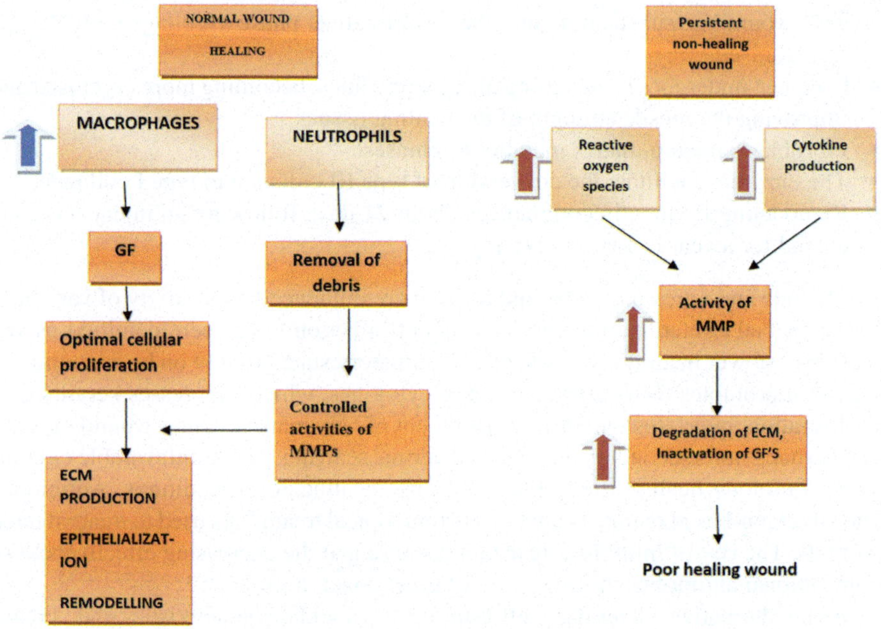

Fig. 3 Comparison of cellular mechanisms in normal and poor wound healing

5 Nanoemulsions Application in Wound Healing

Nanoemulsions have been widely associated with the encapsulation of active pharmaceutical and nutritional ingredients and acting as a carrier system in novel drug delivery approaches. They are thermodynamically stable, having a defined combination of surfactant/co-surfactant, oils, and water [12]. A nanoemulsion facilitates the entry of drug into the superficial layer called the epidermis (stratum corneum) by enhancing the percutaneous absorption of the drug as well as beneath the layers of the skin towards the site of the wound in the dermis and stratum germinativum layer of the skin, where the process of *angiogenesis* during wound healing takes place.

As the outer layer of the skin is lipophilic by nature, the solubility and absorption of water-soluble drugs are poor. The solubility of water-soluble drugs can be enhanced by using nanoemulsions, where drugs are dispersed in an oil droplet phase. Different compositions of nanoemulsions can increase thermodynamic activity by enhancing the penetration of drugs through the diffusional barrier of the skin, which in turn also aids the partitioning of the skin. Nanoemulsions can also function as micro-reservoirs that release the entrapped drug at a slower rate. A nanoemulsion having a lower concentration of surfactant is considered more appropriate for internally administrable drug use. According to leading researchers, it has been found that the nanoemulsions are kinetically stable; therefore, there is a need for external

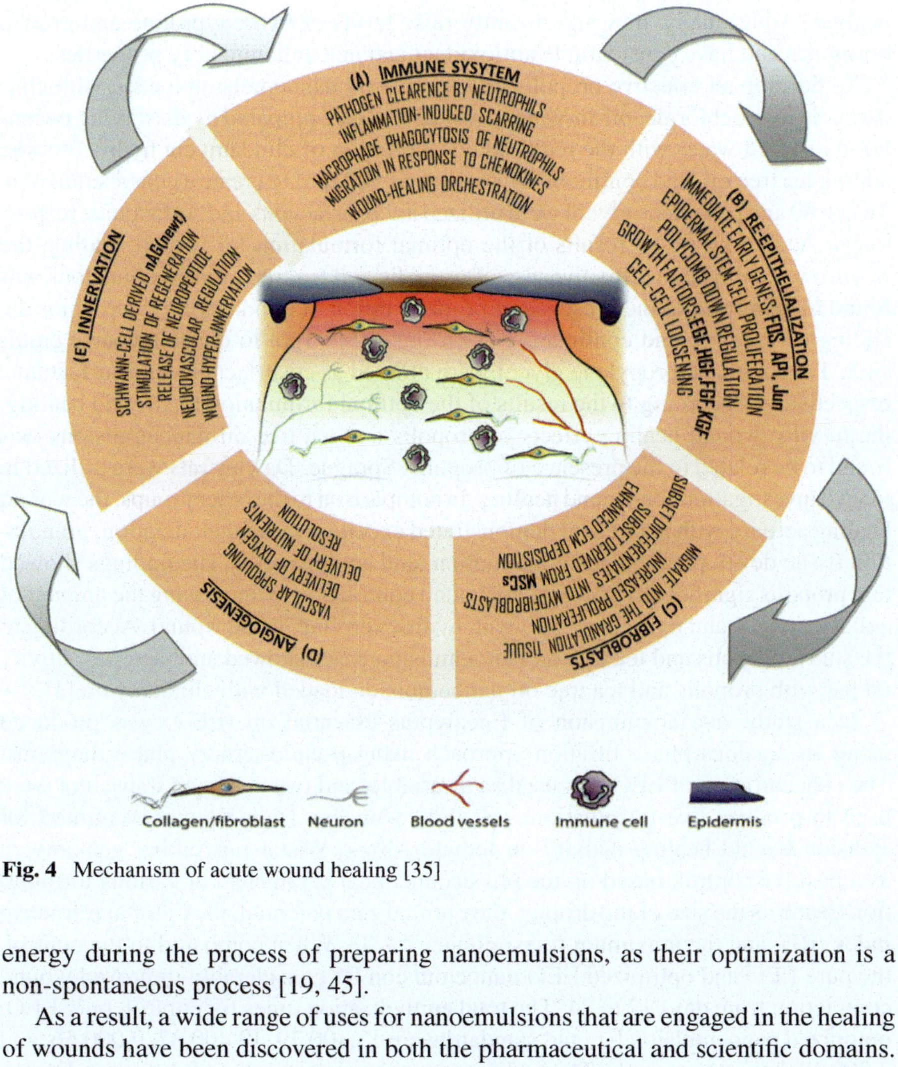

Fig. 4 Mechanism of acute wound healing [35]

energy during the process of preparing nanoemulsions, as their optimization is a non-spontaneous process [19, 45].

As a result, a wide range of uses for nanoemulsions that are engaged in the healing of wounds have been discovered in both the pharmaceutical and scientific domains.

The applications of nanoemulsion in wound healing have been discussed below.

5.1 Nanoemulsions of Essential Oils for Wound Healing

The in vivo wound healing properties of the essential oils (EO) of *Deverra tortuosa* and *Deverra triradiata* by their encapsulation into nanoemulsion were examined in a study by [30]. An aqueous phase titration technique was used to create the EO nanoemulsion. At day 16, topical *D. tortuosa* and *D. triradiate* nanoemulsion administration (1 or 2%) demonstrated approximately 100% wound contraction and full healing. Additionally, they significantly raise levels of hydroxyproline and growth hormones and have considerable antioxidant and anti-inflammatory properties.

To develop an effective propolis and tea tree oil nanoemulsion loaded with clindamycin hydrochloride, oil-in-water (O/W) NEs were prepared by dissolving tween-80 in distilled water with the required concentration of clindamycin hydrochloride, adding tea tree oil, and continuously swirling at 500 rpm to create a coarse emulsion. Tween 80 and propylene glycol were utilized as cosurfactants and surfactants, respectively. According to the results of the optimal formulation for wound healing, the in vitro wound healing effectiveness of propolis and tea tree oil nanoemulsions was found to be related to the presence of propolis ration of clindamycin hydrochloride, adding tea tree oil, and continuously swirling at 500 rpm to create a coarse emulsion. Tween 80 and propylene glycol were utilized as cosurfactants and surfactants, respectively. According to the results of the optimal formulation for wound healing, the in vitro wound healing effects of propolis and tea tree oil nanoemulsions was found to be related to the presence of propolis. Sprague–Dawley rats were utilized in in vivo investigations on wound healing. In comparison to the other groups, the wound healing activity with tea tree oil demonstrated excellent re-epithelialization, granulation tissue development, anti-inflammation, and angiogenesis. The findings showed that propolis significantly accelerated wound contraction by increasing the amount of collagen type 1 and its main component, hydroxyproline, in the wound. According to the study, propolis and tea tree oil nanoemulsion bases showed antibacterial activity on par with propolis and tea tree oil nanoemulsion loaded with clindamycin [1].

In a study, a nanoemulsion of Eucalyptus essential oil (EEO) was produced using an aqueous phase titration approach using pseudo-ternary phase diagrams. The concentration of EEO was used as a variable, and tween 85 and transcutol were used to produce five formulations (E1–E5). Nanoemulsion E1 was examined for excision wound healing activity on female Albino Wistar rats taking gentamycin as a positive control, based on the physicochemical parameters of various formulations, such as the size of the droplet, the optimal zeta potential, the optimal refractive index (RI), and the maximum transmittance (% T). When compared to the control, the pure EEO and optimized EEO nanoemulsion E1 considerably improved wound contraction from days 12 to 24. The total epithelization times for control, pure EEO, optimized nanoemulsion E1, and standard were 13.400.79, 12.400.82, 9.000.56, and 7.800.48 days, respectively. This indicates that the epithelization durations in E1 and standard were much shorter than those in EEO. Drug therapeutic effectiveness and oral absorption are both known to be improved by nanoemulsions [7].

Calophyllum inophyllum oil nanoemulsion (CSONE) formulated using the immense shear homogenizer technique to examine how the independent variables of CSO, Tween 80, high pure water (HPW), and homogenization time affected the responses of CSO nanoemulsions droplet size, PDI, and turbidity. For biological activity, a nanoemulsion with droplet sizes of 45–50 nm and notable stability over 4 weeks was chosen. Cell viability experiments on BSR (BHK-21cells) cells show that CSONE is less toxic compared to CSO at different dosages. Similar results were obtained from the in vitro cell monolayer scratch (wound healing) assay, which showed that CSONE with an amount of CSO that was as small as 0.4% achieved 100%

wound closure after 48 h and was analogous to the fibroblast growth factors. The nanoemulsion size of the droplets, which was stabilized by Tween 80 as a surfactant, allowed CSONE to accelerate wound closure [43].

Skin abrasions and burns are known to contain the Gram-positive opportunistic bacterial strain *Staphylococcus aureus*, which can cause infection and sepsis. Eucalyptus oil (6%), nonionic surfactants (Tween 80 and Tween 20), and water (1:2 v/v) were used to develop an oil-in-water nanoemulsion (NE). Water was added to the oil and surfactant mixture using a magnetic stirrer, and the resulting coarse mixture was then emulsified using a probe sonicator. Chitosan solution and eucalyptus oil nanoemulsion were combined in several ratios (0, 1, 3, and 5% v/v), and the mixture was then homogenized using a sonicator for 10 min. The film solutions were applied to leveled acrylic plates and allowed to dry for 48 h at 50 to 60 °C with a relative humidity of 60 to 2%. By using the conventional plate count method, the antibacterial activity of CH film and various concentrations of NE-CH films against the wound isolate was investigated. The study's findings demonstrate that NE-CH films with increasing essential oil concentrations effectively combated *S. aureus* germs. According to Sugumar et al. [48], NE-CH film demonstrated higher antibacterial activity against the clinical pathogen S. aureus under test than chitosan film alone.

A nanoemulsion of sage oil by spontaneous emulsification method was formulated, and wound healing in rat models showed a significant increase in collagen content (77.52%) and tensile strength (56.20%) in comparison to the control. In comparison to the control, the period of epithelialization was reduced by 42.85% [5].

By using the spontaneous emulsification approach, several nanoemulsions of clove oil (CO) were prepared. The correct proportions of 1% w/w CO and 8% w/w Triacetin were used (oil phase). In a prior combination of CO and triacetin, 15% weight per weight of Tween-80 (a surfactant) and 15% weight per weight of Labrasol (a cosurfactant) were spontaneously added. According to the findings of a study on the effects of pure CO, nanoemulsion F1, and conventional gentamycin on the contraction of the wound area in rats, the time required for full epithelization was 13.21, 12.24, 10.60, and 8.20 days, respectively. When compared to control and pure CO-treated rats, the epithelization duration was much shorter in rats treated with nanoemulsion F1 plus the antibiotic gentamycin. Increased leucine levels in nanoemulsion F1 and antibiotic-treated rats given gentamycin suggested increased collagen levels in these formulations. In contrast to gentamycin, mice treated with CO-nanoemulsion (F1) had significant amounts of granulation tissue, few mononuclear inflammatory cells, restoration of the adnexa, and substantial fibrosis [6].

5.2 Nanoemulsion-Based Gels for Wound Healing

The efficacy of post-burn skin regeneration using a nanoemulsion gel was compared to that of commercially available treatments. Sea buckthorn berry oil (3–10%) served as the oil phase of oil-in-water (O/W) nanoemulsions, while decyl glucoside (4–6%)

served as the emulsifier in various compositions and according to various process parameters, including mixing speed (300–500 rpm), pre-emulsification time (10–15 min.), and ultrasonic homogenization time (60–120 s). By increasing the emulsifier concentration to 6%, decreasing the oil phase concentration to 3%, and lengthening the ultrasonic period to 120 s, a stable basic nanoemulsion was produced. Hyaluronic acid, allantoin, aloe vera gel, and D-panthenol were added to the basic nanoemulsion in order to enhance the moisturizing property of the formulation. Additionally, sodium benzoate and vitamin E were added as preservatives and antioxidants. Finally, the viscosity of the nanoemulsion was enhanced by including carbomer solution to achieve the desired consistency while maintaining pH 6 with the addition of solution of citric acid. As the nanoemulsion contained sea buckthorn oil, which contains palmitoleic acid, a component of skin lipids, it stimulated epidermis regenerative processes and accelerated wound healing processes. The oil also includes a number of other elements that aid in fibroblast growth, collagen formation, tissue repair, and angiogenesis [37].

The hydrogel of *Achyrocline satureioides* extract loaded into nanoemulsions with components of egg yolk lecithin, vitamin E, polysorbate 80, and extract produced by spontaneous technique has a wound-healing effect. Then, gelling agent Carbopol 20 (0.15%) was used to produce hydrogel after NaOH was used to bring the pH level to 7.0. Wistar rats were used to study wound healing activities, and a number of factors, including lesion size, anti-inflammatory markers, oxidative damage, and histology, were examined. According to reports, the nanoemulsion formulations improved the re-epithelialization mechanism in lesions, reduced inflammation, decreased lipid damage, and accelerated angiogenesis to speed up wound healing. Biochemical analysis (TBARS, MPO, IL-1, and TNF-) showed that the nanoemulsion gel that was made decreased tissue inflammation and oxidative damage [15].

Crisaborole-loaded nanoemulsion-based 2% chitosan gel for wound healing was made by a low-energy approach employing Lauroglycol-90 as an oil phase, Tween 80 as a surfactant, Transcutol-HP as a co-surfactant, and water as an aqueous phase. An excision wound healing model was used to analyze the gel. On days 7, 14, and 21, excision wound rats were examined for wound size. Topical use of the standard fusidic acid gel (2%) and Crisaborole-loaded nanoemulsion-loaded chitosan gel on days 7, 14, and 21 post-wounding decreased rat wound diameters compared to the negative control group. Crisaborole-loaded nanoemulsion-loaded chitosan gel performs similarly to fusidic acid gel [10].

A new oil-in-water (o/w) nanoemulsion gel of Levofloxacin in sesame oil cures infected wounds in streptozotocin-induced diabetic rats. Collagen production reduced epithelialization, wound contraction, and inflammatory cells in all groups. Immunohistochemical analysis revealed increased CD31 and TGF- at wound sites in the treatment group [50].

Chakraborty et al. [20] optimized and enhanced the potential to heal wounds of an insulin-loaded nanoemulsion with *Aloe vera* gel in diabetic rat models. The optimum ternary phase diagram nanoemulsions of oleic acid as the oil phase, tween-80 as the surfactant, and polyethylene glycol 400 as the cosurfactant were used to make nine batches (NE1–NE9) of insulin-loaded nanoemulsions. A chosen nanoemulsion

(NE5) was mixed into a topical gel containing *Aloe vera* and a control (no *Aloe vera*). The prepared gels indicated good spread ability, permeation, and stability. Gel micrographs revealed a even distribution of nano-sized spherical particles within the polymeric gels' macromolecular network. The work concluded that *Aloe vera* combined with an insulin-loaded nanoemulsion had synergistic effects on the healing of wounds in diabetic rats and could be an encouraging and efficient method for treating wounds in diabetic patients.

A nano-emulgel was developed by Miastkowska et al. [37] that combined the benefits of an emulsion with a gel by coating the topmost layer of skin with a protective layer that controls its level of hydration. The emulgel quickly absorbed into the skin, showed good spread ability and adherence, and promoted faster wound healing than other medical cosmetics that are freely accessible on the market.

5.3 *Miscellaneous Formulations of Nanoemulsion for Wound Healing*

Ali et al. [8] used a high-speed homogenization process to produce curcumin-encapsulated alpha-tocopherol nanoemulsion systems with polyethylene glycol (PEG) 400, tween 80 with changing compositions, and curcumin and alpha-tocopherol with fixed compositions. Since curcumin is a hydrophobic substance, using a nanoemulsion system can increase its solubility in an aqueous solution. When curcumin is combined with water to form an o/w nanoemulsion, the surfactants and cosurfactants not only help to stabilize the nanoemulsion yet additionally enhance the wettability of hydrophobic substances by depositing on the drug's surface, allowing water to enter the particle core. The optimized formulation accelerated skin tissue regeneration with a significantly higher percent reepithelialization and quick closure of wounds when compared to the untreated animal group, which showed an increased level of collagen fibre deposition and decreased inflammatory response with a high degree of mechanical strength. Curcumin greatly accelerates wound healing and defends against oxidative damage when applied topically. In body tissue with a higher cellular composition, curcumin promotes novo vascularization, speeds up the process of wound repithelialization, and increases granulation generation. Curcumin is only present in the superficial stratum corneum (SC) following topical treatment because of its poor water solubility. Medicines that are weakly water-soluble can have their solubility improved by NEs, which spread the drugs in an oil droplet phase. Different NE compositions increase the drug's ability to pass through the skin's diffusional barrier. Various oils, including clove oil, coconut oil, and olive oil, as well as the cosurfactants PEG 400, PEG, Carbitol, and Labrasol, were combined with the surfactant Tween 80 to create a curcumin-nanoemulsion utilizing a high-energy ultrasonication approach. Cur-NE demonstrated similar wound-treatment effects to fusidic acid due to its nano-sized particles and enhanced penetration, resulting in a more sustained release action. Granulation tissue and mononuclear inflammatory cells were both present in minor amounts in the granulation tissue of Cur-NE-treated rats [2, 3].

Nanoemulsions of betulinen-riched extract and pure spruce balm were made using three oils-jojoba, medium chain triglycerides, and sunflower and 5% lipoid S75 as surfactant by high pressure homogenizer. The impact of NE on wound closure in primary fibroblasts was investigated using an in vitro wound healing test. All three NEs with an active component had considerably lower cell free gaps; jojoba oil-BET NE 83%, jojoba standard drug 77%, and PSB 73% wound closure were noted. In the case of medium-chain triglycerides, NE PSB demonstrated 73% wound closure, while BET demonstrated 84%. With sunflower oil NE, BET demonstrated 91% wound closures and PSB 82%. According to reports, the majority of PSB's wound-healing effects occur during the inflammatory stage by neutralizing bacteria and reducing inflammation [51].

An antimicrobial oil-in-water nanoemulsion containing benzalkonium chloride was produced for use in burn wounds, and the wound healing activity was tested on a porcine burn model. On anesthetized swine, thermal burn wounds were produced, and 3 different formulations of nanoemulsion silver sulfadiazine cream were given to damaged skin on various days after injury. Thermal burn wounds treated with a dermally applied nanoemulsion formulation became full-thickness injuries that could be studied histologically. On day 21, epithelial repair revealed that 77.8% of wounds treated with nanoemulsions had epidermal damage scores of 0, as opposed to 16.7% of burns treated with silver sulfadiazine. Wounds treated with nanoemulsion completely recovered their epithelium, and there was no loss of hair follicles [22].

The current research sought to deliver the medicine in nanoemulsion (NE) form, where the synergistic property of chitosan was realized by the formation of chitosan-coated naringenin NAR-NE (CNNE). The study was inspired by the potential of naringenin (NAR), a natural flavonoid, to be used in the treatment of chronic wounds, as discovered by Akrawi et al. [4]. NAR contains angiogenic, antimicrobial, anti-inflammatory, and antioxidant properties. In light of this, it might be claimed that nanoemulgel formulations based on chitosan and including NAR would give a distinctive viewpoint towards healing chronic wounds with fresh hope and reducing the burden on the globe in the near future.

Shabestani Monfared et al. [44] developed a novel therapeutic strategy for cutaneous wound healing in microfluidic and lab-on-chip systems that detect the mechanisms that impair cell migration and the detrimental molecular and cellular processes that aid in wound healing. In order to produce more accurate in vitro models that more closely mimic the in vivo microenvironment of a wound on a molecular, cellular, and tissue level, these miniaturized cell evaluation platforms enable precise spatial and temporal control over a range of dynamic microenvironmental factors, such as shear stress, biochemical and oxygen gradients. They presented their findings and provided suggestions for further research on cutaneous and vascular wound repair.

6 Conclusion

Nanoemulsions are gaining interest as drug carriers to augment the delivery of ingredients for wound repairing because they have various benefits for the administration of medications in wound healing. This novel technology may be created to circumvent some phytopharmaceuticals' low absorption and poor miscibility with the lipid components of the cell membrane lining, such as curcumin. The nanoemulsions have a huge potential to improve skin regeneration, as this chapter has demonstrated. The desired stability and delivery properties of nanoemulsions are made possible by their small size, which is not possible with ordinary emulsions or microemulsions. However, more work needs to be done to create extremely affordable techniques for creating nanoemulsions from inexpensive, consumer-friendly ingredients and testing them in vivo.

References

1. Abdellatif MM, Elakkad YE, Elwakeel AA, Allam RM, Mousa MR (2021) Formulation and characterization of propolis and tea tree oil nanoemulsion loaded with clindamycin hydrochloride for wound healing: in-vitro and in-vivo wound healing assessment. Saudi Pharm J 29s:1238–1249
2. Ahmad N, Ahmad R, Al-Qudaihi A, Alaseel SE, Fita IZ, Khalid MS, Pottoo FH, Bolla SR (2019) A novel self-nanoemulsifying drug delivery system for curcumin used in the treatment of wound healing and inflammation. Biotech 9:1–20
3. Ahmad N, Ahmad R, Qudaihi AA, Alaseel SE, Fita IZ, Khalid MS, Pottoo FH (2019) Preparation of novel curcumin nanoemulsion by ultrasonication and its comparative effects in wound healing and the treatment of inflammation. RSC Adv 9:20192
4. Akrawi SH, Gorain B, Nair AB, Choudhury H, Pandey M, Shah JN, Venugopala KN (2020) Development and optimization of naringenin-loaded chitosan-coated nanoemulsion for topical therapy in wound healing. Pharmaceutics 12:893–897s
5. Alam MS, Ali MD, Ahmad S, Banu N, Ali MS, Alam N, Mujtaba MA (2021) Evaluation of wound healing potential and biochemical estimation of sage oil nanoemulsion on animal model. Pak J Pharm Sci 34:1385
6. Alam P, Ansari MJ, Anwer MK, Raish M, Kamal YKT, Shakeel F (2017) Wound healing effects of nanoemulsion containing clove essential oil. Nanomed Biotech 45:2017
7. Alam P, Shakeel F, Anwer MK, Foudah AI, Alqarni MH (2018) Wound healing study of Eucalyptus essential oil containing nanoemulsion in rat model. J Oleo Sci 8:957–968
8. Ali M, Khan NR, Subhan Z, Mehmood S, Amin A, Rabbani I, Hafiz MB, Syed HK, Khan IU, Khan MS, Khattak S (2022) Novel curcumin-encapsulated α-tocopherol nanoemulsionsystem and its potential application for wound healing in diabetic animals. Bio Med Res Int 15:7669255
9. Al-Sabagh AM, Emara MM, El-Din MN, Aly WR (2011) Formation of water-in-diesel oil nano-emulsions using high energy method and studying some of their surface active properties. Egypt J Pet 20:17–23
10. Ansari MN, Soliman GA. Rehman NU, Anwer MK (2022) Crisaborole loaded nanoemulsion based chitosan gel: formulation, physicochemical characterization and wound healing studies. Gels 8:318
11. Anton N, Vandamme TF (2011) Nano-emulsions and micro-emulsions: clarifications of the critical differences. Pharm Res 28:978–985

12. Aronson MP (1989) The role of free surfactant in destabilizing oil-in-water emulsions. Langmuir 5:494–501
13. Bachheti A, Bachheti RK, Abate L, Husen A (2022) Current status of Aloe-based nanoparticle fabrication, characterization and their application in some cutting-edge areas. South Afric J Bot 147:1058–1069
14. Bachheti RK, Abate L, Bachheti A, Madhusudhan A, Husen A (2021) Algae-, fungi-, and yeast-mediated biological synthesis of nanoparticles and their various biomedical applications. In: Handbook of greener synthesis of nanomaterials and compounds. Elsevier, Amsterdam, pp 701–734. https://doi.org/10.1016/B978-0-12-821938-6.00022-0
15. Balestrin LA, Back PI, Marques MD, Araújo GD, Carrasco MC, Batista MM, Silveira T, Rodrigues JL, Fachel FN, Koester LS, Bassani VL (2022) Effect of hydrogel containing *Achyroclinesatureioides* (Asteraceae) extract-loaded nanoemulsions on wound healing activity. Pharmaceutics 14:2726
16. Bouchemal K, Briançon S, Perrier E, Fessi H (2004) Nano-emulsion formulation using spontaneous emulsification: solvent, oil and surfactant optimisation. Int J Pharm 280:241–251
17. Broughton G, Janis JE, Attinger CE (2006) The basic science of wound healing. Plast Reconstr Surg 117:12S-34S
18. Bsieso EA, Nasr M, Moftah NH, Sammour OA, Abd El Gawad NA (2015) Could nanovesicles containing a penetration enhancer clinically improve the therapeutic outcome in skin fungal diseases. Nanomed 10:2017–2031
19. Burger Y, Shahzad A, Brummer M, Gerber J (2017) Traversing the skin barrier with nano-emulsions. Curr Drug Deliv 14:458–472
20. Chakraborty T, Gupta S, Nair A, Chauhan S, Saini V (2021) Wound healing potential of insulin-loaded nanoemulsion with *Aloe vera* gel in diabetic rats. J Drug Deliv Sci Technol 64:102601
21. Dehghani F, Farhadian N, Golmohammadzadeh S, Biriaee A, Ebrahimi M, Karimi M (2017) Preparation, characterization and in-vivo evaluation of microemulsions containing tamoxifen citrate anti-cancer drug. Eur J Pharm Sci 96:479–489
22. Dolgachev VA, Ciotti S, Liechty E, Levi B, Wang SC, Baker JR, Hemmila MR (2021) Dermal nanoemulsion treatment reduces burn wound conversion and improves skin healing in a porcine model of thermal burn injury. J Burn Care Res 42:1232–1242
23. Esquena J (2016) Water-in-water (w/w) emulsions. Curr Opin Colloid Interface Sci 25:109–119
24. Ganta S, Singh A, Patel NR, Cacaccio J, Rawal YH, Davis BJ, Amiji MM, Coleman TP (2014) Development of EGFR-targeted nanoemulsion for imaging and novel platinum therapy of ovarian cancer. Pharm Res 31:2490–2502
25. Godeto YG, Ayele A, Ahmed IN, Husen A, Bachheti RK (2023) Medicinal plant-based metabolites in nanoparticles synthesis and their cutting-edge applications: an overview. In: Secondary metabolites from medicinal plants, pp 1–34
26. Gonfa YH, Tessema FB, Tadesse MG, Bachheti A, Bachheti RK (2023) Medicinally important plant roots and their role in nanoparticles synthesis and applications. In: Secondary metabolites from medicinal plants. CRC Press, pp 227–242
27. Hatem S, Nasr M, Elkheshen SA, Geneidi AS (2018) Recent advances in antioxidant cosmeceutical topical delivery. Curr Drug Deliv 15:953–964
28. Hazra DK (2017) Nano-formulations: high definition liquid engineering of pesticides for advanced crop protection in agriculture. Adv Plants Agric Res 6:211–213
29. Heng MC (2011) Wound healing in adult skin: aiming for perfect regeneration. Int J Dermatol 50:1058–1066
30. Kamel RA, Teiama MS, El-Hagrassi AM, Elgayed SH, Khattab MA, El-Sayed EK, Ibrahim MT, Mady MS and Moharram FA (2022) Appraisal on the wound healing potential of *Deverratortuosa* DC and *Deverratriradiata* Hochst essential oil nanoemulsion topical preparation. Front Pharmacol 13:940988
31. Ozgok Kangal MK, Regan JP (2022) Wound Healing. In: StatPearls [Internet]. StatPearls Publishing, Treasure Island (FL)

32. Khan AD, Rastogi V, Lavhale PM, Jain J (2022) Novel approaches for herbal drug delivery in wound healing: a review. Indian J Pharm Sci 84:247–260
33. Kronberg B, Holmberg K, Lindman B (2014) Surface chemistry of surfactants and polymers. Emulsions and emulsifiers. Wiley, West Sussex, UK, pp 431–445
34. Magee DJ, Zachazewski JE, Quillen WS (2008) Pathology and intervention in musculoskeletal rehabilitation. Elsevier Health Sciences
35. Martin P, Nunan R (2015) Cellular and molecular mechanisms of repair in acute and chronic wound healing. Br J Dermatol 173:370–378
36. McClements DJ, Rao J (2011) Food-grade nanoemulsions: formulation, fabrication, properties, performance, biological fate, and potential toxicity. Crit Rev Food Sci Nutr 51:285–330
37. Miastkowska M, Kulawik-Pioro A, Szczurek M (2020) Nanoemulsion gel formulation optimization for burn wounds: analysis of rheological and sensory properties. Processes 8:1416
38. Ngan CL, Basri M, Tripathy M, Karjiban RA, Abdul-Malek E (2014) Physicochemical characterization and thermodynamic studies ofnanoemulsion-based transdermal delivery system for fullerence. Sci World J 2014:12
39. Pentak D, Kozik V, Bąk A, Dybał P, Sochanik A, Jampilek J (2006) Methotrexate and cytarabine loaded nanocarriers for multidrug cancer therapy. Spectroscopic study. Molecules 21:1689
40. Reinke JM, Sorg H (2012) Wound repair and regeneration. Eur Surg Res 49:35–43
41. Rigo C, Ferroni L, Tocco I, Roman M, Munivrana I, Gardin C, Cairns WR, Vindigni V, Azzena B, Barbante C, Zavan B (2013) Active silver nanoparticles for wound healing. Int J Mol Sci 14:4817–4840
42. Saberi AH, Fang Y, McClements DJ (2013) Fabrication of vitamin E-enriched nanoemulsions: factors affecting particle size using spontaneous emulsification. Colloid Int Sci 391:95–102
43. Saki E, Murthy V, Khandanlou R (2022) Optimisation of *Calophylluminophyllum* seed oil nanoemulsion as a potential wound healing agent. BMC Compl Med Ther 22:285
44. ShabestaniMonfared G, Ertl P, Rothbauer M (2021) Microfluidic and lab-on-a-chip systems for cutaneous wound healing studies. Pharmaceutics 13:793
45. Sivakumar M, Tang SY, Tan KW (2014) Cavitation technology—a greener processing technique for the generation of pharmaceutical nanoemulsions. Ultrason Sonochem 21:2069–2083
46. Solans C, Sole I (2012) Nano-emulsions: formation by low-energy methods. Curr Opin Colloid Interf Sci 17:246–254
47. Sonneville-Aubrun O, Simonnet JT, L'alloret F (2004) Nanoemulsions: a new vehicle for skincare products. Adv Colloid Interf Sci 108:145–149
48. Sugumar S, Mukherjee A, Chandrasekaran N (2015) Eucalyptus oil nanoemulsion-impregnated chitosan film: antibacterial effects against a clinical pathogen, *Staphylococcus aureus*, in vitro. Int J Nanomed 10:67–75
49. Tadros T, Izquierdo P, Esquena J, Solans C (2004) Formation and stability of nano-emulsions. Adv Colloid Interf Sci 20:303–318
50. Valizadeh A, Shirzad M, Pourmand MR, Farahmandfar M, Sereshti H, Amani A (2021) Levofloxacin nanoemulsion gel has a powerful healing effect on infected wound in streptozotocin-induced diabetic rats. Drug Deliv Transl Res 11:292–304
51. Vater C, Bosch L, Mitter A, Gols T, Seiser S, Heiss E, Burger AE, Wirth M, Valenta C, Klang V (2022) Lecithin based nanoemulsions of traditional herbal wound healing agents and their effect on human skin cells. Eur J Pharm Biopharm 170:1–9
52. Velazquez OC (2007) Angiogenesis and vasculogenesis: inducing the growth of new blood vessels and wound healing by stimulation of bone marrow-derived progenitor cell mobilization and homing. J Vasc Surg 45:A39-47
53. Wang PH, Huang BS, Horng HC, Yeh CC, Chen YJ (2018) Wound healing. J Chin Med Assoc 81:94–101

Nanoemulsions Challenges and Future Prospects as a Drug Delivery System

Farzad Abaszadeh, Muhammad Hossein Ashoub, and Mahnaz Amiri

Abstract Nanoemulsion during the recent years proved that has the potential to overwhelm numerous difficulties in formulation of drugs. By loading drugs with poor water-solubility in the proper nanoemulsions improves their solubility and/or wettability. Therefore, this expands their pharmacodynamics as well as pharmacokinetics by diverse administration methods. Accompanying with the optimal size of nanodroplets, the droplets act as a drugs pool, facilitating nanoemulsion to do as a multifunctional platform to defect different diseases. A number of significant advantages, which include nanoemulsion qualities, such as well-organized drug release with suitable rate, extended efficacy, control of drug uptake, the ability of drug protection from oxidation and low side effects, have been described in last decades. The great characteristic of nanoemulsion contains also a diversity of engineering procedure options as well as a combination of extensively mixed components such as liquid lipids, surfactants or even drug-conjugates. These structures afford alternatives for designing advanced nanoemulsions pointing at high-value and different applications. This review presents the challenges and prospects of diverse nanoemulsion types and its application as drug delivery methods. Types of nanoemulsions as well as components of a nanoemulsions, their manufacturing via different methods like (High pressure homogenization, Ultrasonication, Microfluidization, Phase Inversion, Temperature Spontaneous Emulsification, Membrane Emulsification, Emulsion Inversion Point) is described. Different drug delivery applications (Oral, Parenteral, Transdermal, Topical, Intranasal and Ocular delivery) is presented. Kinetics of drug release as well as stability of nanoemulsions are also presented.

Keywords Drug delivery system · Nanoemulsion · Manufacturing · Applications

F. Abaszadeh · M. Amiri (✉)
Neuroscience Research Center, Institute of Neuropharmacology, Kerman University of Medical Science, Kerman, Iran
e-mail: ma.amiri@kmu.ac.ir

Student Research Committee, Faculty of Allied Medicine, Kerman University of Medical Sciences, Kerman, Iran

M. H. Ashoub
Department of Hematology and Medical Laboratory Sciences, Faculty of Allied Medicine, Kerman University of Medical Sciences, Kerman, Iran

A. Husen et al. (eds.), *Current Trends in Green Nano-emulsions*, Smart Nanomaterials Technology, https://doi.org/10.1007/978-981-99-5398-1_13

List of abbreviations

(BCS)	Biopharmaceutics Classification System
(BBB)	Blood Brain Barrier
(CNS)	Central Nervous System
(DLVO)	Derjaguin, Landau, Verwey, And Overbeek
(DOPE)	Di-Oleoyl Phosphatidylethanolamine
(DSPC)	Di-Stearoyl Phosphatidylcholine
(EIP)	Emulsion Inversion Point
(FDA)	Food and Drug Administration
(GIT)	Gastrointestinal Tract
(GRAS)	Generally Recognized as Safe
(HPH)	High-Pressure Homogenization
(HLB)	Hydrophilic-Lipophilic Balance
(IV)	Intravenous
(IPM)	Isopropyl Myristate
(LC)	Liquid Chromatography
(MQ)	Methyl Quercetin
(O/W)	Oil-In-Water
(O/W/O)	Oil-In-Water-In-Oil
(PIT)	Phase Inversion Temperature
(PEG)	Polyethylene Glycol
(POE)	Polyoxyethylene
(Ph)	Potential of Hydrogen
(W/O)	Water-In-Oil
(W/O/W)	Water-In-Oil-In-Water

1 Introduction

Pharmaceutical dosage forms have evolved from very basic to extremely sophisticated systems, sometimes referred to as new drug delivery systems, due to technological advancements. With mean droplet sizes generally ranging from roughly 50 to 500 nm in pharmaceutical applications, nano-emulsion-based delivery systems typically consist of a colloidal dispersion of oil and water phases [6–9]. Oil-in-water (O/W) or water-in-oil (W/O) nano-emulsions are possible, where the droplets are made of either oil or water. To stabilize nano-emulsions, pharmaceutically approved surfactants that are regarded as safe (generally recognized as safe (GRAS)) are utilized as emulsifiers. Nano-emulsions are especially excellent medication delivery methods because of their ability to dissolve vast amounts of insufficiently water-soluble medicines while also shielding them from deterioration. Increased drug loading, higher drug solubility, enhanced bioavailability, chemical or enzymatic degradation

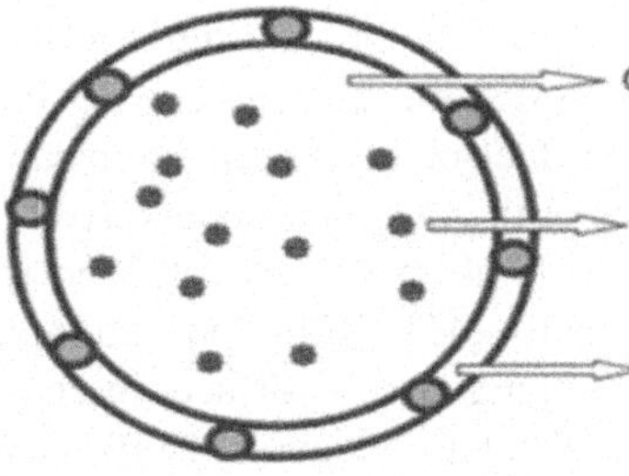

Fig. 1 An illustration of the O/W Nano-emulsion system [55]

resistance, and controlled drug release are among the main benefits of nano-emulsions for this use [31].

Additionally, the wide interfacial area and tiny droplet size may improve the targeted administration of bioactive substances. Nano-emulsions also exhibit other intriguing physicochemical characteristics, including exceptional optical clarity and peculiar viscoelastic behavior, as a result of the comparatively small size of the droplets. Nano-emulsions can have rheological qualities that range from liquid to semisolid and optical properties that range from opaque to almost transparent. It should be noted that a variety of distinct colloidal dispersions with some comparable and some different properties may be created from oil, water, and surfactants (Fig. 1). The dispersed droplets can be divided into three types, depending on their size and thermodynamic stability: (i) Micro emulsions, (thermodynamically stable and diameters between 20 and 100 nm); (ii) Nano-emulsions, (metastable stables and mean particle diameters under 200 nm); and (iii) conventional emulsions (metastable systems and mean particle diameters between 200 nm and several micrometers) [49].

A biphasic liquid system is an emulsion in which one liquid's internal or dispersed phase is scattered as tiny droplets across the exterior or continuous phase of the other liquid. Numerous physicochemical elements have been shown to impact how emulsion nanoparticles behave in living things (Fig. 2). For novel emulsion nanomedicines to treat illness and adhere to strict regulatory standards effectively; these aspects must be regulated. Size is a crucial physical factor that affects how cells react to nanoparticles. The renal and reticuloendothelial systems quickly remove somewhat bigger nanoparticles, >60 nm, and smaller, <20 nm. Nanoparticles with a size of approximately 50 nm have the maximum cellular uptake. As nanoparticle size increases, different cellular uptake pathways are used: small nanoparticles (200 nm or smaller) are taken up via pinocytosis pathways, those between ~250 nm and larger are absorbed by phagocytosis, and those larger than microns are taken up by non-receptor mediated micropinocytosis [27]. For the Food and Drug Administration (FDA) to approve nanomedicine, size distribution, a measurement of the various populations of each size, is becoming more and more crucial. Contrary to polydisperse droplet systems, formulations with monodisperse emulsion droplets exhibit predictable as well as repeatable biological activity. This demonstrates how essential emulsion stability against coalescence is the stability of emulsions against coalescence, as well as their interactions with biological tissues, are governed by their surface characteristics. It is well known that extracellular proteins cover nanoparticle surfaces when injected into the body. The precise surface characteristics of the nanoparticle have a significant role in determining the composition of this layer, known as the "protein corona". The surface charge is crucial in controlling the protein corona and subsequent interactions with cells. Due to electrostatic repulsion, anionic nanoparticle surfaces are not easily absorbed via cells and are often quickly eliminated by liver and macrophage cells [45]. Although cationic nanoparticle surfaces readily attach to the negatively charged cell surface, increasing absorption, there was still an enhanced clearance problem because of opsonization and macrophage uptake. Coating nanoparticles can obtain similar results with hydrophilic uncharged polymers, such as Polyethylene Glycol (PEG), which act as a steric barrier to protein adsorption and lessen electrostatic attraction [23]. A Nano medicine formulation's unique requirement must be carefully considered while choosing a suitable emulsion excipient. To comply with FDA regulations, the excipient must rapidly solubilize the problematic medicine, be non-toxic, biocompatible, and produce monodisperse stable emulsion droplets [73].

A perfect medication delivery system achieves the goal of increasing therapeutic impact while reducing toxicity. With the passage of time and the development of science and technology, dosage forms have changed from straightforward mixes and tablets to incredibly complex systems known as innovative drug delivery systems. The medication delivery technique known as nano-emulsions is one such example.

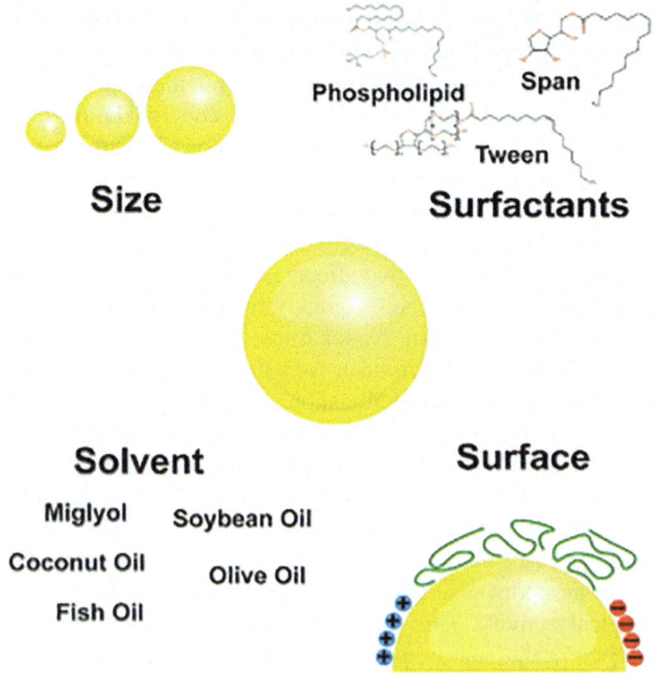

tem.

Fig. 2 Systematic drug distribution using Nano-emulsion [81]

The term "nano-emulsion" refers to a dispersion of two immiscible liquids, such as oil and water, which is thermodynamically stable, isotropically transparent, and stabilized by an interfacial surfactant coating. Nano-emulsions have a uniform, tiny droplet sizes between 20 and 200 nm. Nano-emulsions do not form naturally; external shear is required to separate larger droplets into smaller forms. Compared to microemulsion phases, there isn't much information available about making and managing nano-emulsions. The first nano-emulsions were created in the 1940s and can be classified as bi-continuous, (O/W) or (W/O). Nano-emulsions are also called mini emulsions, submicron emulsions, and ultrafine emulsions. Research shows that nano-emulsion is a far more effective medication delivery than other transdermal drug delivery systems. Although emulsion is thermodynamically unstable but kinetically stable, the primary physical distinction between emulsion and nano-emulsion is that the former is murky while the latter is quite transparent [13–15, 51].

2 Types of Nano-Emulsions

A fine dispersion of pharmaceuticals in nanodroplets makes up a nano-emulsion, an isotropic, transparent, or translucent, heterogeneous system of two immiscible liquids. An interfacial layer of emulsifiers and co-emulsifiers stabilizes it. With small droplet sizes (20 to 400 nm), a uniform size distribution, and various physicochemical and biological properties from other emulsions (>500 nm), they are thermodynamically and kinetically stable systems. The two immiscible phases are often supplemented with substances soluble in either oil or water. In the presence of an emulsifier, mixing oil and water results in a coarse emulsion that can spontaneously transform into a nano-emulsion or be made using high energy [28, 61, 64, 68, 69]. In their simplest form, emulsions consist of two phases, one of which is hydrophilic and the other of which is hydrophobic (Fig. 3). These emulsions are therefore named as (O/W) emulsions, where small oil droplets are scattered through water, or (W/O) emulsions, where small water droplets are disseminated through oil. However, by enclosing an emulsion within an emulsion named as a double emulsion and creating water-in-oil-in-water (W/O/W) or oil-in-water-in-oil (O/W/O) emulsions, more complexity may be added to these straightforward systems. In the past, creating double emulsions required creating an initial internal emulsion, which was encircled via creating a second emulsion on top of the initial emulsion. Double emulsions provide extra difficulties in their formation and stabilization, including the need for lipophilic and hydrophilic surfactants to stabilize each oil–water interface and a higher propensity for degradation and coalescence because of diffusion between the phases. Rekindled research interest has been shown in uniform double emulsions produced by microfluidic devices for use as microreactors or templates for particle production [42, 53].

Oil, water, and a surfactant make up typical nano-emulsions. Choosing the suitable surfactant is essential for creating and maintaining nano-emulsions. Nano-emulsions are kinetically stable but thermodynamically unstable. In other words, nano-emulsion

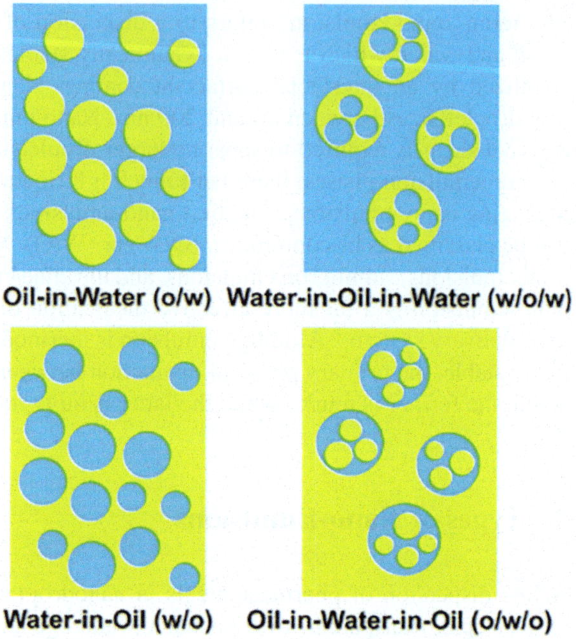

Fig. 3 Emulsion forms are divided into continuous and scattered phases [81]

phase separation happens with enough time. For use in a variety of pharmacological, culinary, and cosmetic applications, nano-emulsions have been produced. They must be toxic-free and biocompatible for all of these purposes. As a result, choosing the right oil and surfactants is crucial. It is preferable to use biocompatible oils and surfactants, such as vegetable or pharmaceutical-grade oils. In addition to surfactants, proteins and lipids are frequently utilized to stabilize nano-emulsions. As it solubilizes lipophilic medications intended to be utilized for a variety of diseases, the oil phase is crucial in the formation of nano-emulsions. Depending on the administration location, the quantity of oil in an o/w type nano-emulsion can range from 2 to 20% w/w. According to Choudhury et al. [18], Biopharmaceutics classification system (BCS) classes II and Intravenous (IV) medications are the best options for creating o/w nano-emulsions since they help increase the solubility of the pharmaceuticals. Isopropyl Myristate (IPM), Triacetin (Glyceryl triacetate), Sefsol 218 (Propylene glycol mono ethyl ether), etc., are among the FDA-approved and GRAS-certified oils that are favored over traditional high-density fixed oils like castor oil, coconut oil, sesame oil, cottonseed oil, fish oil, linseed oil, mineral oil, olive oil, peanut oil, and sunflower oil [18, 19, 76]. Based on their solubility and emulsification power, the top emulsifier systems are chosen from a pool. Due to the less toxicity and irritating than their anionic, and especially their cationic counterparts, non-ionic surfactants are frequently applied. Emulsifiers are chosen based on the solubility in aqueous and oil phases, hydrophilic-lipophilic balance (HLB) value, and lower toxicity, among others. Non-ionic surfactants with an HLB value of 8–16 is ideal for creating an o/w nano-emulsion.

Depending on the composition, three types of nano-emulsions are most likely to form:

- Water with oil where Oil droplets spread in a continuous aqueous phase in nano-emulsions
- Water in a liquid nano-emulsions in which the continuous oil phase is diluted with water droplets
- Bi-continuous Nano-emulsions, in which the system contains interspersed microdomains of water and oil [12, 38, 75].

3 Components of a Nano-Emulsion

Oil: Oil is the second most important carrier after water due to its capacity to solubilize lipophilic medication compounds and improve absorption toward the body's lipid barrier. Due to its exceptional capability to penetrate cell walls, oil is beneficial for administering lipophilic active drugs (Fig. 4).

Surfactant: To aid in the dispersion of all components, the surfactant has to be capable of reducing the interfacial tension as near to zero as possible. When making a W/O nano-emulsion, Surfactants with HLB values between 3–6 are beneficial, however surfactants with increased HLB values between 8 and 18 are useful for making O/W nano-emulsions. Surfactants with an HLB value of 20 or higher operate as co-surfactants to lower surfactant levels to acceptable levels and create microemulsions. The surfactants are non-ionic, anionic, cationic, and Witter ionic surfactants. This impacts ionic surfactants. As a result, they are vulnerable to stability concerns

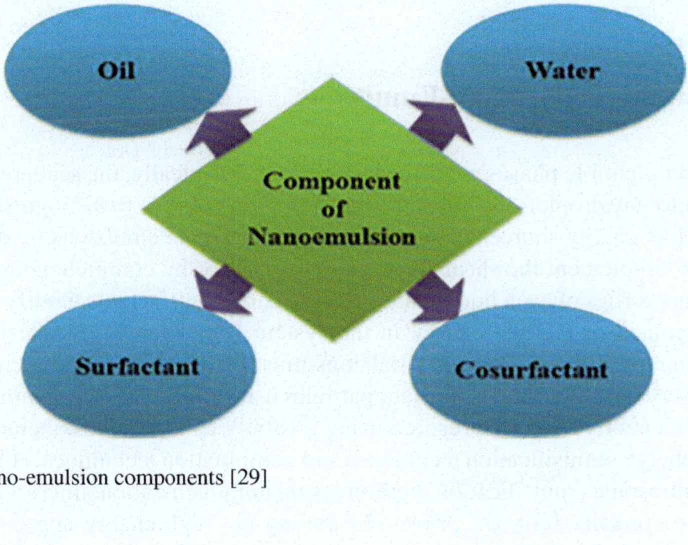

Fig. 4 Nano-emulsion components [29]

and typically are not favored because of toxicity worries. However, non-ionic surfactants are more common than other types because they may be used to make safe medicinal dosage forms.

The following are examples of non-ionic surfactants:

Co-surfactant: High quantities of single-chain surfactants are needed to lower the interfacial tension between oil and water to the point where a nano-emulsion can spontaneously form. Co-surfactants rise the fluidity of the interface because it contains fluidizing groups such as unsaturated bonds, which destroys the liquid crystal or gel structure and changes the HLB value in a way that results in the spontaneous creation of nano-emulsion.

Aqueous Phase: The aqueous phase's characteristics, such as potential of hydrogen (pH), ionic concentration, and electrolytes, have an impact on the stability and droplet size of the nano-emulsion. For the examination of spontaneous nano-emulsification of nano-emulsion, aqueous phases such as plain water, simulated gastric fluid (pH 1.2), Ringer's solution, simulated intestinal fluid (pH 6.8), and phosphate buffered saline can be utilized. Based on the aqueous phase's aforementioned characteristics, when a medication with pH-dependent solubility is added to the system, the phase behavior of nano-emulsions can be drastically affected. Consider the following illustration to comprehend the makeup of a nano-emulsion. Using the medication mebudipine, *Samira Khan* and colleagues developed and assessed an oral nano-emulsion drug delivery system. In this instance, deionized water served as the aqueous phase, Tween 80 and Span 80 served as surfactants, and ethanol served as a co-surfactant. Emulsion formulations are frequently used to dispense medications because they improve the pharmacokinetic profile, increase the solubility of hydrophobic chemicals, and lessen side effects that patients may encounter [36, 62]. Emulsions can be given to patients in many ways, which are shown in Fig. 5.

4 Manufacturing Nano-Emulsions

As two incompatible phases are forced together mechanically, the scattered phase is sheared into tiny droplets, forming an emulsion. It is possible to distinguish between high and low energy sources of shear force used to create emulsions by employing a device to implement the shear force (high energy) or by comprehending how the physical properties of each liquid phase and the chosen surfactants modify as a result of the chemical or thermal energy in the system (low energy). It takes two steps to create nano-emulsions: coarse emulsions must be developed, and then the large droplets must be broken down into nanoparticles using high-pressure homogenization or ultrasonication. They are prepared using a variety of techniques, including high- and low-energy emulsification techniques and combination techniques. High-energy stirring, ultrasonic emulsification, high-pressure homogenization, microfluidics, and membrane emulsification are prioritized among the high-energy approaches. The

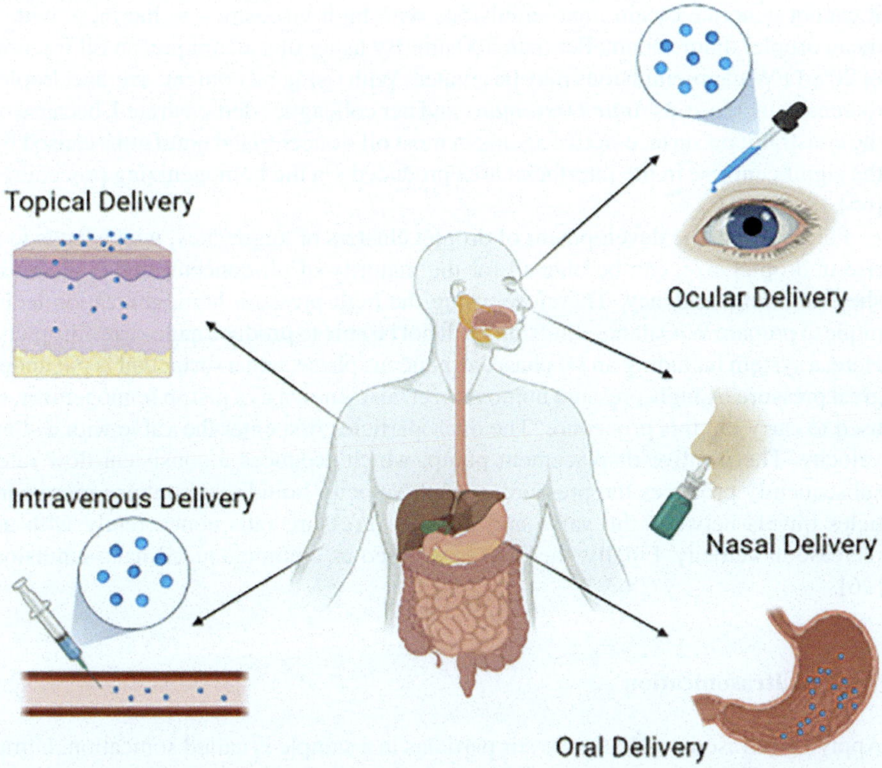

Fig. 5 Various methods of administering drugs delivered via Nano-emulsion [81]

phase inversion temperature method, the emulsion inversion point method, and spontaneous emulsification are the low-energy emulsification techniques that receive the most consideration. Reverse nano-emulsions may be made in dense environments using a combination technique that combines high-energy and low-energy emulsification [44].

4.1 Method of Preparation of Nano-Emulsion

4.1.1 High-Pressure Homogenization (HPH) Method

High pressure is applied to a system consisting of an oil phase, an aqueous phase, and a surfactant or co-surfactant to carry out this operation. With the use of the homogenizer, pressure is exerted. Poor productivity and component degradation owing to excessive heat generation are some issues with homogenizers. This technique can only be used to create liquid O/W nano-emulsions with less than 20% oil phase;

it cannot generate cream nano-emulsions with high viscosities or hardness with a mean droplet diameter smaller than 200 nm. By using this technique, an oil fraction of 20% O/W nano-emulsions may be created. With rising oil content, average droplet diameters increase, as *Anne Desrumaux* and her colleagues demonstrated, because of the constraint on surface-active agents in most oil-concentrated emulsions caused by the significant rise in the interfacial area produced via the homogenizing procedures [65].

Furthermore, the development of droplet clusters or aggregates, which leads to a rise in droplet size, can be blamed for the majority of oil-concentrated emulsions' shear-thinning tendency. Therefore, using the high-pressure homogenization technique to prepare W/O nano-emulsions will not be able to produce nano-sized droplets. Here, a system including an oil phase, an aqueous phase, and a surfactant is put under great pressure. A high-pressure homogenizer, also known as a piston homogenizer, is used to carry out this procedure. The microparticles first enter the valve with a slow velocity. The positive-displacement pump, which produces a consistent flow rate, subsequently produces the pressure. A high-velocity liquid that contains microparticles travels between the valve and the seat. Pressure falls concurrently with an increase in velocity. Finally, the fluid is released as a homogenized nano-emulsion [26].

4.1.2 Ultrasonication

Applying ultrasonic radiation to stir particles in a sample is called sonication. Ultrasonic emulsification mostly happens via two methods. First, applying an acoustic field creates interfacial waves that eventually turn unstable and cause the oil phase to erupt as droplets into the water medium. Second, using low-frequency ultrasound results in acoustic cavitation, which is the production and subsequent collapse of microbubbles caused via a sound wave's pressure variation. High amounts of intensely concentrated turbulence are produced by each bubble collapse (a microscopic-scale implosion) occurrence. The original droplets of dispersed oil are effectively broken up into droplets of sub-micron size using turbulent micro-implosions [34, 80]. There are two primary processes used in ultrasonic emulsification. It is primarily based on the acoustic field, which produces an interfacial wave-forming oil phase that disperses as droplets in the dispersion system. The second process involves ultrasound, which improves acoustic cavitation and causes microbubbles to develop and burst as a consequence of pressure changes brought on by a single sound wave. Similar to this, other types of extremely localized turbulence are produced, eventually leading to micro implosions that ultimately cause the breakup of big droplets into sub-micron-sized ones [35, 79]. Typically, this method uses solid surfaces that vibrate at 29 kHz or higher frequencies to stir a pre-mixed macroemulsion. Devices with focused horns and sharp ends that use ultrasound that create severe shear and cavitation resulting in droplet breaking are used to produce ultrasonic effects. Researchers have discovered that the majority of ultrasonic systems create an inhomogeneous sound field. As a result, recirculating the emulsion through the area of high power is necessary, and all droplets should encounter the highest shear rate. Obtaining emulsions with homogeneous droplet sizes at diluted concentrations is also attainable by repeatedly doing this sort of recirculation [82, 84].

Additionally, there are specific issues with the procedures since it is possible to cause lipid oxidation, polysaccharide depolymerization, and protein denaturation.

The best approach for creating a nano-emulsion is this one. The droplet size of a typical emulsion or microemulsion is lowered with a sonication process. However, one drawback of this approach is that it can only be used to make small batches of nano-emulsions; it is not appropriate for large batches. An effective screening method for choosing excipients was developed by Sneh Priya et al. to produce the best nano-emulsion formulation. The model medication utilized was quetiapine fumarate. The ultrasonication method employs a probe sonicator to produce nano-emulsions. It is possible to produce desired qualities by adjusting the quantity of oil, surfactants, and secondary surfactants [58].

4.1.3 Microfluidization

It is a patented manufacturing technique that uses a tool called a microfluidizer. This device applies a high-pressure displacement pump (500–20,000 psi) to move materials through a chamber with microchannels. The materials go to an impingement area and become microscopic particles as they flow through the microchannels. An inline homogenizer processes the liquid phases to create a coarse emulsion. The fine nano-emulsion is then produced via passing the coarse emulsion across a microfluidizer. Up until the required particle size is achieved, this process is repeated. The bigger particles are subsequently filtered out of the nano-emulsion using nitrogen [63, 66]. Most often, the microfluidic membrane approach is chosen. The pharmaceutical sector uses it the most frequently to create delicate emulsions. This technique makes use of a microfluidizer, which generates high pressures. High pressure throughout the procedure propels the macroemulsion to the interaction chamber, allowing for nano-emulsions with submicron-sized particles. The process can be repeated numerous times while adjusting the operating pressure to produce uniform nano-emulsions with the desired particle size. It's a trademarked mixing technique. An apparatus called a microfluidizer is used in this technique.

The substance is pushed through the interaction chamber using a high-pressure positive displacement pump (500–20,000 psi) in this system. Microchannels, which are tiny channels, make up this chamber. The product flows onto an impingement region, producing submicron-sized small particles. In this instance, an inline homogenizer is used to blend and treat two solutions (the aqueous phase and the oily phase) to create a coarse emulsion. A microfluidizer is used further to convert the coarse emulsion into a stable nano-emulsion. It has been proven to be more effective than ultrasonography. Still, because of the expense of manufacturing, the risk of equipment contamination, and the need for aseptic processing, this technology is less useful [29, 63].

4.1.4 Phase Inversion Temperature

Chemical energy from phase changes resulting from the emulsification process is used to create fine dispersion. The proper phase transitions are produced by varying the composition at a constant temperature or the composition at a constant temperature. Based on the idea that a surfactant of the type polyoxyethylene (POE) changes in solubility with temperature, the phase inversion temperature (PIT) approach was developed. Due to the polymer chain's drying with increasing temperature, this surfactant develops a lipophilic nature. At low temperatures, the surfactant monolayer displays a sizable positive spontaneous curvature, resulting in the formation of an oil-swollen micellar solution phase [65]. Low energy and spontaneous emulsification are characteristics of the emulsion inversion point (EIP) approach. At a steady temperature, it leads to the gradual dilution of thermodynamically stable liquid crystals or microemulsions with W/O, respectively, to make direct or inverse nano-emulsions that are thermodynamically unstable but kinetically stable [11, [25, 40, 60]. The PIT approach involves raising the temperature of the emulsion system to shift the surfactant's solubilizing pattern from hydrophilic to lipophilic, which results in the formation of bicontinuous microemulsions and emulsion inversion. There are four steps in the procedure (Fig. 6). (a) When the temperature is below the PIT, a macro-emulsion and primarily hydrophilic non-ionic surfactants are present. (b) As the temperature rises, the surfactants progressively turn lipophilic and are dissolved via the oil phase. In case (c), bicontinuous microemulsions occur when the temperature reaches the PIT. (d) Oil and lipophilic surfactant are mixed with water when the temperature is raised above the PIT, and the emulsion is inverted. The system is then rapidly cooled via water dilution, which instantly makes the surfactant hydrophilic and triggers spontaneous and quick migration to the aqueous phase. As a result of this turbulent displacement, nano-emulsions are created [3].

4.1.5 Spontaneous Emulsification

There are three primary phases in this procedure, which are as follows:

Creating a homogenous organic solution with hydrophilic and lipophilic surfactants dissolved in a water-soluble solvent. After that, the water-miscible solvent is eliminated by evaporation under decreased pressure after the organic phase is introduced into the aqueous phase while being stirred magnetically. The quantification of quercetin or methyl quercetin (MQ) included in topical nano-emulsions was verified by Daniel Fasolo et al. using an isocratic liquid chromatography (LC) technique. The nano-emulsions were made utilizing the spontaneous emulsification technique. In this procedure, the water phase is stirred magnetically for 15 min while an organic phase containing elements of the oil core is injected into it. The organic solvent was then eliminated using evaporation at 40–45 °C under decreased pressure [10, 11, 24].

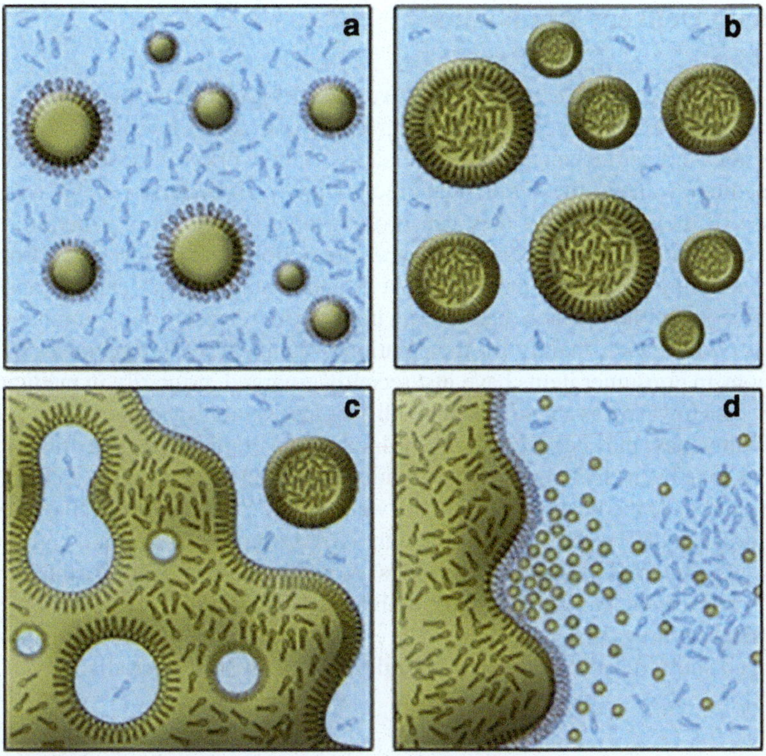

Fig. 6 Making nano-emulsions with the PIT technique. A phase inversion occurs in an oil/water/non-ionic surfactant system [3]

4.1.6 Membrane Emulsification

A low-energy nano-emulsion approach is membrane emulsification. This approach produces an emulsion with a limited size distribution range and requires very little surfactant. This method involves converting a dispersed phase into a continuous phase by passing it through a membrane. This method's disadvantage is that it possesses a negligible dispersed phase flux across the membrane, which can be problematic when scaling up [29].

4.1.7 Emulsion Inversion Point

This method involves varying the system's composition while maintaining a constant temperature. To make the kinetically stable nano-emulsions, dilution with oil or water is done gradually to generate structures [29].

5 Drug Delivery Applications

Nano-emulsions are a new drug delivery technology that enables the regulated or sustained release of genetic material, drugs, and active biological ingredients. Emulsions and surfactants offer a versatile framework for developing various nanomedicines due to the simple manufacture and well-understood characteristics. Because of their small size, they can prolong blood circulation, penetrate deeply into tissues, and engage in special bio-nano interactions. While various hydrophobic cargo, including medicines, photosensitizers, and contrast agents, can be transported in the oil core. Although there has been an increase in interest in nano-emulsions over the past two decades, mainly for the creation of nanoparticles, direct uses of nano-emulsions in consumer items have just recently emerged, primarily in medicine and cosmetics. However, as a result of instability or low solubility in the carrier, medication effectiveness may be severely constrained, so developing effective drug formulations has long been challenging. Nano-emulsion is used to increase the solubility and bioavailability of medications that are not water soluble. The transport characteristics of the drug would be impacted by the nano-sized droplets that would significantly increase the interfacial areas associated with nano-emulsion. This formulation has recently attracted a lot of attention for delivering hydrophilic and hydrophobic pharmaceuticals as drug carriers because of its enhanced ability to solubilize medications, prolonged shelf life, simplicity of production, and increased bioavailability of drugs [41, 72].

5.1 *Oral Delivery*

The oral route is the most practical, straightforward, and economical method for the non-invasive administration of drugs, thus Based on this concept, medication delivery technologies today control the pharmaceutical industry. Additionally, it is the optimal strategy for achieving treatment objectives due to improved patient comfort. Concerning elderly, pediatric, and maybe trauma epileptic patients, this distribution method has several disadvantages. In addition, due to their nonconducive physiochemical characteristics, certain medications are intrinsically challenging to administer via the oral route. In terms of a drug's gastrointestinal tract (GIT) solubility, stability, and absorption, oral delivery of insoluble medications presents some significant challenges. Peptide medications experience hydrolysis and enzymatic degradation, which restricts their bioactivity and intestinal absorption [47]. Additional downsides might result from some medications' inability to penetrate the cell walls of the epithelium. Using aqueous-dispersible particle delivery methods, solid dispersions, complexing with cyclodextrins, amorphization, and micronization/nanonization are just a few of the methods that have been proposed to boost the overall drug bioavailability. In the last ten years, numerous research teams using nano-emulsion technology have reported significant increases in the oral bioavailability of hydrophobic and poorly soluble drugs [30, 56].

5.2 *Parenteral Delivery*

Although drugs with poor solubility are typically thought to be unsuitable for parenteral administration, they are capable of being made into parenteral dosage forms thanks to nano-emulsification techniques. Using biodegradable surfactants guarantees effective pharmacological action without interfering with the body's normal biological processes. According to a recent study, the broadly applied anticonvulsant carbamazepine, which has low solubility, can be made into a nano-emulsion using a spontaneous emulsification process with 2 mg/mL, where 95% of the drug is released in just 11 h. Another study demonstrates that the IV preparation of thalidomide (0.01–0.05% w/w) released 95% of the drug within 4 h of spontaneous emulsification [5, 33]. For this kind of delivery system, choosing the suitable emulsifiers that can create mono- or multilayer structures surrounding the oil droplets to promote the creation of nano-emulsions and increase nano-emulsion stability is crucial. The phospholipids most frequently utilized in the design of parenteral administration systems are semi-synthetic materials, such as di-oleoyl phosphatidylethanolamine (DOPE) and di-stearoyl phosphatidylcholine (DSPC), come from natural sources, such as lecithins. Lecithins can come from either plant or animal sources, making them biocompatible and degradable. The emulsions' long-term stability depends heavily on the proportion of lecithin phospholipids with charged polar head groups. Electrostatic repulsion results from the comparatively high-negative charge that these head groups impart to the droplet surfaces. Although natural surfactants are preferred to synthetic surfactants, adjuvant emulsifying compounds have been utilized to boost emulsion stability since they produce enough emulsification outcomes. Their application is, however, constrained by hemolytic reactions and modifications in the droplet diameter of nano-emulsions stabilized by Tween 80 after autoclaving [67].

5.3 *Transdermal and Topical Delivery*

Although numerous types of nano-emulsion have been investigated as drug delivery vehicles, it is highly desirable to expand research into dermal and transdermal drug delivery using engineered nano-emulsion technology to broaden its exceptional applications. Topical drug delivery based on nano-emulsions can significantly get around this obstacle. Drugs often enter the skin through three different pathways: sweat ducts, stratum corneum, and hair follicles. The small nanoparticles easily penetrate the pores in nano-emulsions (Fig. 7).

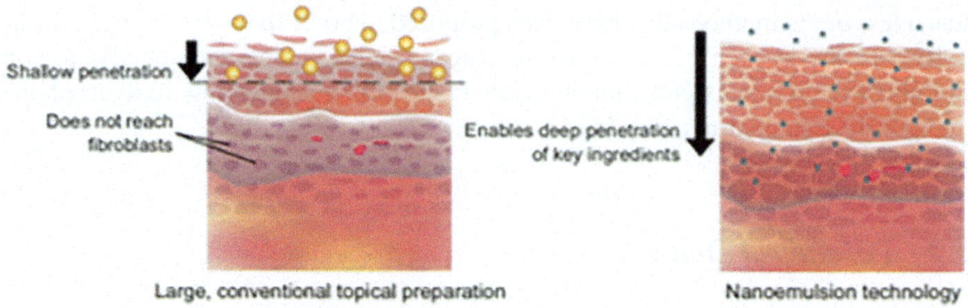

Fig. 7 Comparison of the standard transdermal formulations and nano-emulsions for skin barrier crossing [70]

Although many treatments have been identified, only a few medications have had clinical success. The lower "bioavailability" and poor site specificity are frequently blamed for the lower success rate. The drug administration route, organ physiology, and metabolism primarily influence these measurements [74]. The systemic route shows a rapid fluctuation in drug plasma levels, either below or above, necessitates frequent dosing, and hurts when administered. As a result, systemic medication distribution might occasionally be painful. The skin is the most practical place for medication administration because of its adaptability. It has primarily been used to treat various skin conditions. The transdermal method of medication administration has been regularly used to treat a variety of systemic diseases, including cancer, diabetes, arthritis, and hypertension. It aids in overcoming the limitations of intravenous and oral routes. The molecular size of the drugs and their hydrophilicity or lipophilicity can affect their ability to be delivered topically [22]. Topical medication delivery reduces frequent dosage and helps maintain steady drug plasma levels for longer than oral or parenteral drug administration. The release of medications at the stratum corneum or in deeper layers of the epidermis is ensured by ensuring the interaction of nano-sized globules with skin cells. Because of the smaller droplet size, higher zeta potential, lower polydispersity index, higher elasticity, and a variety of nano-emulsion and emulsifiers, altering the physiochemical properties of drugs enhances their ability to physically interact with cellular membranes and pass through the stratum corneum and epidermal layer [1, 57].

5.4 *Intranasal Delivery*

Another dependable route for administering some medications is intranasal drug delivery. The nasal mucosa has become a therapeutically effective route for administering systemic medications. Also, it appears helpful to get around barriers to direct drug entry into the target site. Since ancient times, The Indian medical system

Ayurveda has employed this technique. More recently, however, it has gained popularity over oral medication delivery since it promotes greater systemic bioavailability by bypassing the drug's gastrointestinal digestion [48].

Additionally, painless, non-invasive, and well-accepted is the intranasal approach. Targeting medications in the brain presents several challenges, particularly for hydrophilic and high-molecular-weight drugs. This is due to the blood–brain barrier's endothelium's impervious nature, which separates systemic circulation from the brain. The nasal mucosa's olfactory area serves as a direct conduit between the brain and the nose. To treat or manage illnesses, nano-emulsions have been loaded with medications. Polar medicines for treating chronic central nervous system (CNS) disorders like Parkinson's or Alzheimer's disease have been delivered to the CNS by nasal administration. Intranasal administration of vaccines, which is covered elsewhere in this chapter, offers several benefits over oral and parenteral administration. Nevertheless, the main limitations of nasal medication administration are their limited capacity, difficulties in achieving dosage precision, and repeatability. Drug penetration, residence duration, and metabolism in the nasal cavity are frequently influenced by the delivery method, formulation, and administration approach [46, 83].

5.5 Ocular Delivery

Due to its excellent solubility and permeability through barriers, it is one of the most often used ocular drug delivery systems. When made into a nano-emulsion, several poorly soluble medications are now soluble. The typical drug delivery strategy has some limitations, including low bioavailability and a diminished pharmacotherapeutic impact because of subsequent lacrimal discharges from the eye. As demonstrated in Fig. 8, most of the medicine supplied is drained away, resulting in a diminished therapeutic impact. The nano-emulsion is prioritized as one of the therapeutic administration methods because it lengthens the time the medication is in touch with the eye, which solves patient issues including needing to administer drugs frequently. Many research scientists have learned how to create nano-emulsions using high- and low-energy techniques [15, 17].

High-energy methods include ultrasonication, high-shear stirring, and other processes involving the input of high energy and the creation of nanodroplets. Low-energy approaches employ internal chemical changes or temperature changes that result in the production of nanoparticles rather than any external pressure. The phase inversion temperature approach and phase inversion composition technique are two examples.

Because of lacrimal secretion and nasolacrimal drainage in the eyes, conventional eye drops used for ophthalmic medication delivery have low bioavailability and pharmacological effects. Cationic nano-emulsions interact with the negatively charged corneal cells and enhance medication absorption, making them superior delivery systems for ophthalmic drugs (Fig. 9). In a study, dorzolamide hydrochloride, a

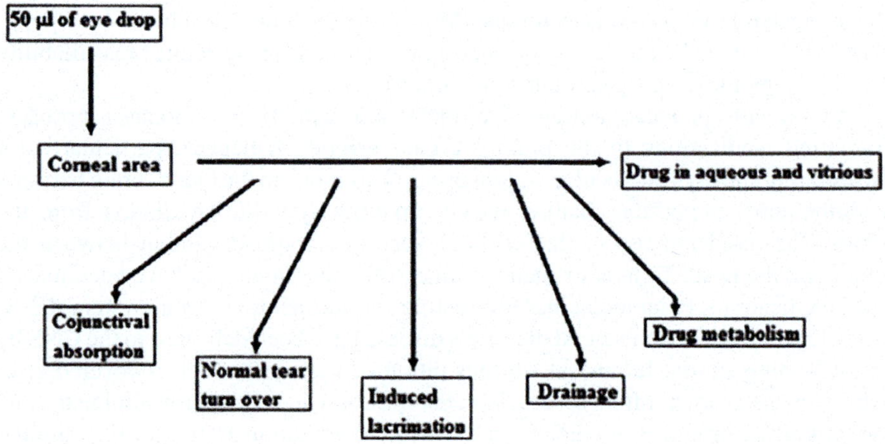

Fig. 8 Challenges to corneal adsorption [17]

possible antiglaucoma medication, was created as an eye nano-emulsion and shown significant therapeutic efficacy and sustained effect [2, 39].

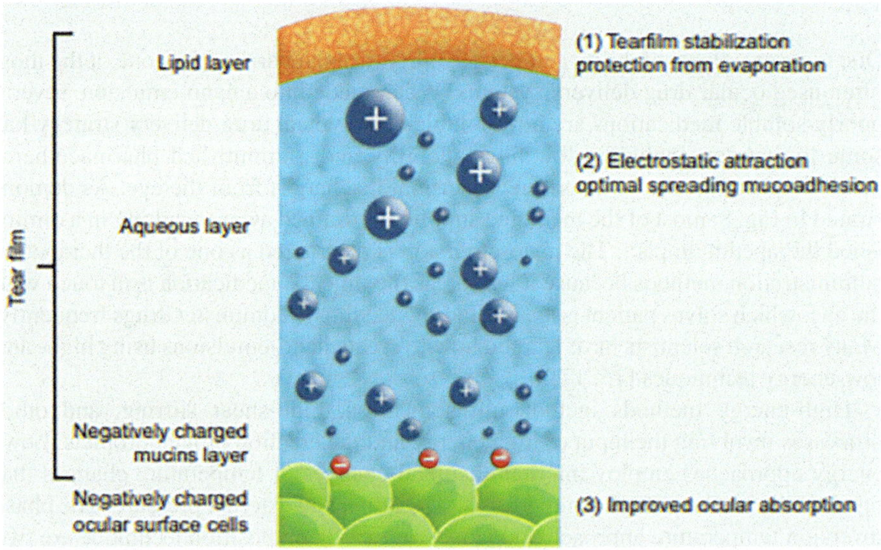

Fig. 9 Cationic nano-emulsions benefits for ophthalmic delivery [39]

6 Kinetics of Drug Release

The foundation of the nanocarrier concept is the notion that extremely tiny nanoparticles can cross biological barriers. The targeted medications are contained by nanocarriers, which can be polymers, amphiphilic lipids, or solid colloidal particles. Gold, ceramic (70), and solid lipid nanoparticles, nanocomposite (71), carbon nanotubes, liposomes, and polymer-drug conjugates are a few types of nanocarriers. The process of encapsulating pharmaceuticals in nanoparticles can be done in several different ways. Nanocarriers are designed to deliver drugs to specific tissues while avoiding the immune system's reaction. One method for making drug-entrapping nanocarriers is to employ a nano-emulsion system. Drugs may either be attached to the surface of nanocarriers or trapped inside them. The nanocarriers are being prepared using a variety of emulsification techniques. The medicine can even be delivered via nanocarriers exceptionally well across the Blood Brain Barrier (BBB) [32, 77].

7 Stability in Nano-Emulsions

Though kinetically stable, there is a chance that the various phases will eventually separate in nano-emulsions. Nano-emulsions are thermodynamically unstable systems that gradually divide into two distinct phases. An effective emulsion can retain its fundamental qualities for weeks or years while having a lengthy shelf life when stabilized by surfactants, making emulsions kinetically stable. This is crucial for translating nano-emulsion into a variety of applications. Still, it's especially essential for emulsion nanomedicine, where gradual formulation changes could negatively impact a patient's health. To ensure that nano-emulsions are prepared and stored correctly, it is crucial to comprehend the processes of emulsion destabilization and stabilization. According to the Derjaguin, Landau, Verwey, and Overbeek (DLVO) hypothesis, attractive van der Waal interactions and repulsive electrostatic double-layer forces combine to provide emulsion stability. Since the two forces are considered separate, adding their sums at certain distances results in the total energy of interaction (FT), which provides a reliable estimate of stability up to around 5 nm. Repulsive forces predominate the contact energy when droplets are far apart, favoring colloid stability. However, when droplets become closer to one another, attractive forces dominate, leading to instability. According to the DLVO hypothesis, the emulsion experiences colloidal instability when the attractive forces take control (Fig. 10) [21, 54].

Larger droplets develop at the expense of smaller droplets during the Ostwald ripening process, which involves the diffusion of the dispersed phase across the continuous phase. The Kelvin effect, which states that particles with a smaller diameter have a higher solubility in solution, causes this process, decreasing the dispersed phase's total surface area and lowering the Gibbs free energy [52]. Ostwald ripening can be prevented by utilizing a dispersed phase with extremely low solubility in

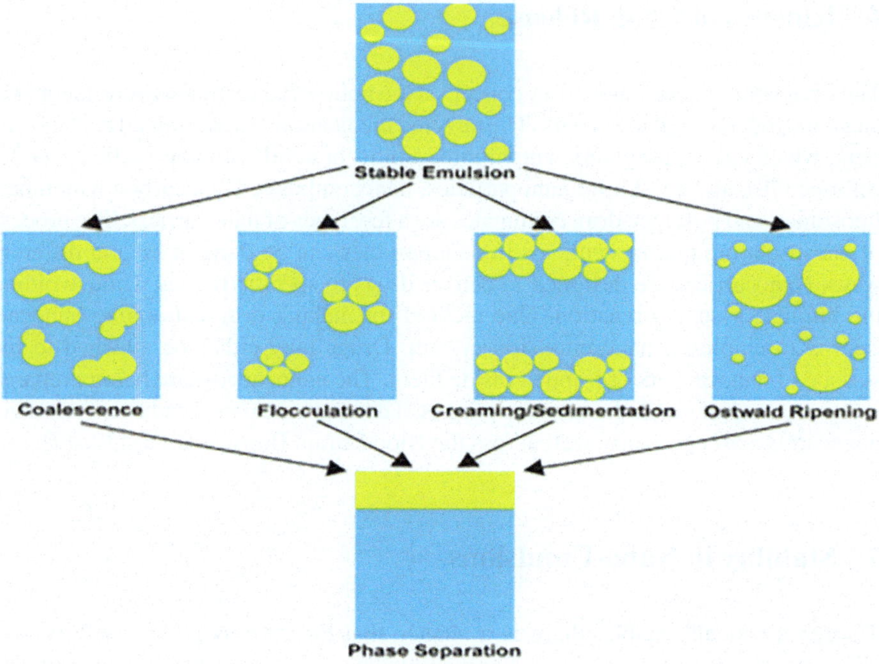

Fig. 10 Emulsion destabilization mechanism illustrated [81]

the continuous phase and making sure that droplet sizes are monodisperse. Ionic surfactant-coated emulsions draw counterions from the solution to create an electrical double layer, which enables electrostatic stability via repulsive forces (Fig. 11) [78]. The Stern layer of tightly bound counter ions, the charged emulsion surface, and a diffuse layer of highly concentrated loosely bound ions and counter ions make up an electrical double layer. The slip plane is where the diffuse layer stops moving together with the emulsion droplets. A crucial indicator of an emulsion's stability is the charge at the slip plane, which is applied to calculate an emulsion's zeta potential and is where particularly positive or negative emulsions are more stable [20].

Droplet size, surface charge, and emulsifier composition directly impact how stable a nano-emulsion is the combined impact of the three factors as mentioned earlier impacts the stability of the nano-emulsion. A good emulsifier mixture creates a flexible interface between two immiscible liquids and aids in suspending the dispersed phase as small droplets in the dispersion medium. Figure 12 provides a typical example. Multi-component biopolymer-based nano-emulsions have been successfully developed using combinations of surfactant, primary electrolyte, and secondary electrolytes [4, 37, 43].

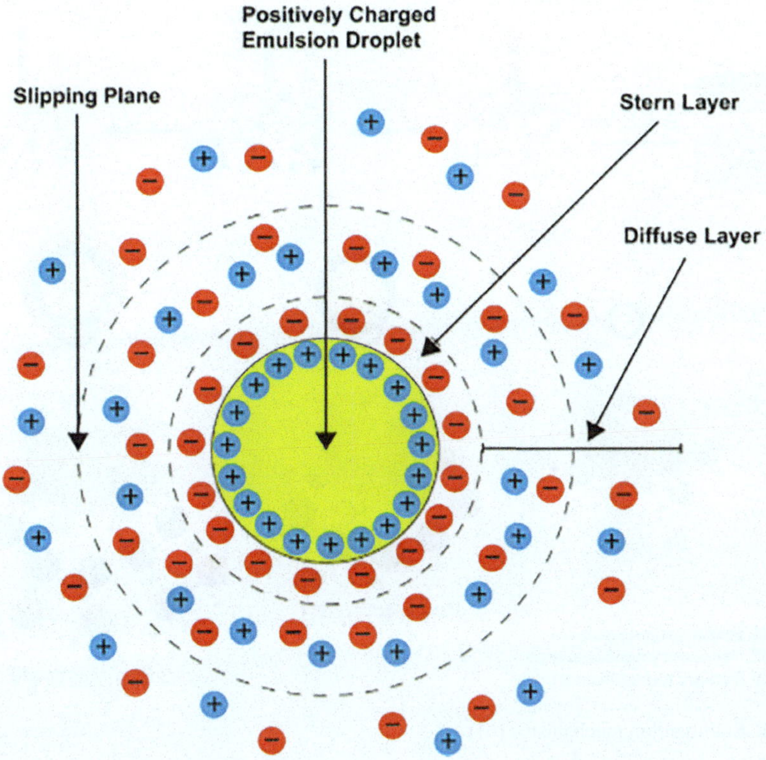

Fig. 11 Illustration of the charged emulsion droplets' electrical double layer [81]

8 Future Prospects

Recently, nano-emulsions have been developed and are currently being tested for various severe diseases and medication therapy-related restrictions. Increases in medication permeability and bioavailability are being made gradually. Recent research by *Mendes* and his colleagues has demonstrated that locoregional injection of lipid nano-emulsion concentrates more in breast cancer while limiting contact with healthy cells. As a result, it could be a potential strategy in neoadjuvant chemotherapy for cancer therapy [50]. *Hyun-Jong Choc* and colleagues discovered that a brand-new lipid nano-emulsion technology might enhance granisetron penetration. Fisetin's nano-emulsion formulation has been shown to increase its bioavailability and anti-tumor efficacy in mice, according to a study team led by Choi et al. [16], Ragelle et al. [59].

Along with nanosuspension and other solubility improvement methods, nano-emulsion has recently emerged as a promising strategy to increase the bioavailability of several medications. Currently, the formulation of several medications as nano-emulsions is being tested.

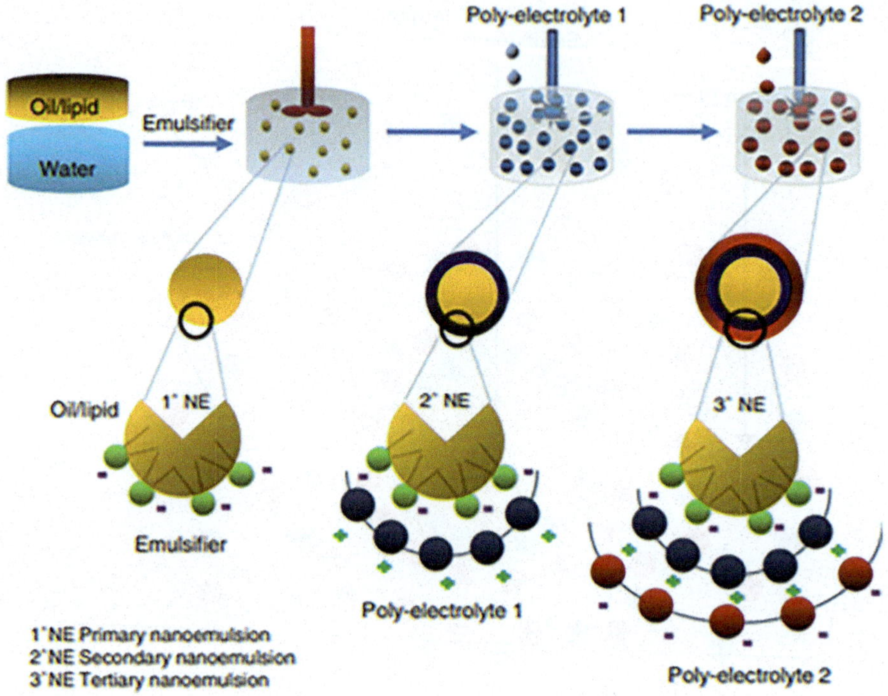

Fig. 12 Stabilization mechanisms [61]

The preparation of nano-emulsions may be done in various ways, giving scientists several options. Nano-emulsification methods are also being used to create nanocarriers, nanoparticles, and nano capsules. The use of nano-emulsions significantly reduces the challenge of piercing physiological membranes. Based on nano emulsification, it is possible to create a variety of medications that would have a considerably lower dosage, be delivered to the intended location, have a higher local concentration, and have fewer adverse effects. Nano-emulsions are erasing barriers to building more specialized and focused medication delivery. As a result, it is emerging as a top formulation choice and has ushered in a new era in the pharmaceutical treatments. In practically all drug administration methods, nano-emulsions are being created and used as pharmaceutical delivery systems. The enhanced solubilization and distribution of poorly soluble actives have received the most outstanding research attention among the many drug delivery applications. To guarantee that this technology is widely used in industry, practical, dependable, safe functional components and cost-efficient, reproducible production techniques are needed [71].

9 Conclusion

Nano-emulsions' ability to solubilize non-polar active chemicals has led to several uses for them as drug delivery methods in the pharmacy field. For the administration of pharmaceuticals, biologicals, or diagnostic agents, nano-emulsion formulations provide some benefits. They can manage drug release, preserve labile pharmaceuticals, improve pharmaceutical solubility, boost bioavailability, and lessen patient variability. Most suggested formulations are self-emulsifying systems due to stability issues, and nano-emulsions are created shortly before use. Although not many applications in other domains have been described, there is a lot of promise for nano-emulsion applications if insoluble oils constrain the Oswald-ripening destabilization process. If breaking and coalescence compete for operations during the process, an optimal shear or temporal shearing may be possible in manufacturing nano-emulsions. Recent work reveals that crossing bicontinuous or continuous aqueous phases during emulsification permits getting O/W nano-emulsions with tiny droplet sizes and low polydispersity. This is relevant to optimization in the creation of nano-emulsions by low-energy techniques. The conclusion is that an ideal surfactant mixture composition or HLB typically exists and that the larger the oil surfactant ratio, the larger the droplet size may be drawn from optimizations by selective adjustment of parameters or experimental designs. If the system is tuned for composition, preparation factors like addition, agitation, or cooling rate often have little to no impact. This conclusion has a crucial derivation: if primary factors have no bearing on the system, it may be scaled up from the lab to the industrial setting with identical outcomes to be anticipated. For over 40 years, clinics have used Nano-emulsions as fluids for whole parenteral feeding. A new path has been made possible by targeting moiety to precisely deliver medications, genes, photosensitizers, and other compounds to the tumor site. As a concluding observation, it appears from the current literature that the preparation and use of nano-emulsions are becoming more popular. It is anticipated that more research will be done in the near future for the clinical implementation of these forms of targeted delivery vehicles.

References

1. Alexander A, Khichariya A, Gupta S, Patel RJ, Giri TK, Tripathi DK (2013) Recent expansions in an emergent novel drug delivery technology: emulgel. J Control Release 171:122–132
2. Ammar HO, Salama HA, Ghorab M, Mahmoud AA (2009) Nanoemulsion as a potential ophthalmic delivery system for dorzolamide hydrochloride. AAPS PharmSciTech 10:808–819
3. Anton N, Vandamme TF (2009) The universality of low-energy nano-emulsification. Int J Pharm 377:142–147
4. Aoki T, Decker EA, Julian McClements D (2005) Influence of environmental stresses on stability of O/W emulsions containing droplets stabilized by multilayered membranes produced by a layer-by-layer electrostatic deposition technique. Food Hydrocoll 19:209–220
5. Araújo FA, Kelmann RG, Araújo BV, Finatto RB, Teixeira HF, Koester LS (2011) Development and characterization of parenteral nanoemulsions containing thalidomide. Eur J Pharm Sci 42:238–245

6. Assadpour E, Jafari S-M (2017) Spray drying of folic acid within nano-emulsions: optimization by Taguchi approach. Drying Technol 35:1152–1160
7. Assadpour E, Jafari S-M, Maghsoudlou Y (2017) Evaluation of folic acid release from spray dried powder particles of pectin-whey protein nano-capsules. Int J Biol Macromol 95:238–247
8. Assadpour E, Maghsoudlou Y, Jafari S-M, Ghorbani M, Aalami M (2016a) Evaluation of folic acid nano-encapsulation by double emulsions. Food Bioprocess Technol 9:2024–2032
9. Assadpour E, Maghsoudlou Y, Jafari S-M, Ghorbani M, Aalami M (2016b) Optimization of folic acid nano-emulsification and encapsulation by maltodextrin-whey protein double emulsions. Int J Biol Macromol 86:197–207
10. Bhosale RR, Osmani RA, Ghodake PP, Shaikh SM, Chavan SR (2014) Nanoemulsion: a review on novel profusion in advanced drug delivery. Indian J Pharm Biol Res 2:122
11. Bouchemal K, Briançon S, Perrier E, Fessi H (2004) Nano-emulsion formulation using spontaneous emulsification: solvent, oil and surfactant optimisation. Int J Pharm 280:241–251
12. Bronaugh RL, Maibach HI (2002) Topical absorption of dermatological products. CRC Press
13. Chavda VP, Shah D (2017) A review on novel emulsification technique: a nanoemulsion. J Pharmacol Toxicol Stud 5:32–33
14. Chellapa P, Mohamed AT, Keleb EI, Elmahgoubi A, Eid AM, Issa YS, Elmarzugi NA (2015) Nanoemulsion and nanoemulgel as a topical formulation. IOSR J Pharm 5:43–47
15. Chime SA, Kenechukwu FC, Attama AA (2014) Nanoemulsions—advances in formulation, characterization and applications in drug delivery. Appl Nanotechnol Drug Deliv 3:77–126
16. Choi AJ, Park CG, Han SH, Park CB (2010) Nanoemulsion formation and characterization of neutraceutical component with the sonication and self-assembly methods. Nanotech 1:928–929
17. Choradiya BR, Patil SB (2021) A comprehensive review on nanoemulsion as an ophthalmic drug delivery system. J Mol Liq 339:116751
18. Choudhury S, Dasgupta S, Patel DK, Ramani YR, Ghosh SK, Mazumder B (2013) Nanoemulsion as a carrier for topical delivery of aceclofenac. In: Advanced nanomaterials and nanotechnology. Springer
19. de Almeida Borges VR, Simon A, Cuello Sena AR, Cabral LM, de Sousa VP (2013) Nanoemulsion containing dapsone for topical administration: a study of in vitro release and epidermal permeation. Int J Nanomed 8:535
20. Delgado ÁV, González-Caballero F, Hunter RJ, Koopal LK, Lyklema J (2007) Measurement and interpretation of electrokinetic phenomena. J Colloid Interface Sci 309:194–224
21. Derjaguin B, Landau L (1993) Theory of the stability of strongly charged lyophobic sols and of the adhesion of strongly charged particles in solutions of electrolytes. Prog Surf Sci 43:30–59
22. Desai P, Patlolla RR, Singh M (2010) Interaction of nanoparticles and cell-penetrating peptides with skin for transdermal drug delivery. Mol Membr Biol 27:247–259
23. Fam SY, Chee CF, Yong CY, Ho KL, Mariatulqabtiah AR, Tan WS (2020) Stealth coating of nanoparticles in drug-delivery systems. Nanomaterials 10:787
24. Fasolo D, Schwingel L, Holzschuh M, Bassani V, Teixeira H (2007) Validation of an isocratic LC method for determination of quercetin and methylquercetin in topical nanoemulsions. J Pharm Biomed Anal 44:1174–1177
25. Fernandez P, André V, Rieger J, Kühnle A (2004) Nano-emulsion formation by emulsion phase inversion. Colloids Surf, A 251:53–58
26. Floury J, Desrumaux A, Axelos MAV, Legrand J (2003) Effect of high pressure homogenisation on methylcellulose as food emulsifier. J Food Eng 58:227–238
27. Foroozandeh P, Aziz AA (2018) Insight into cellular uptake and intracellular trafficking of nanoparticles. Nanoscale Res Lett 13:1–12
28. Gorain B, Choudhury H, Kundu A, Sarkar L, Karmakar S, Jaisankar P, Pal TK (2014) Nanoemulsion strategy for olmesartan medoxomil improves oral absorption and extended antihypertensive activity in hypertensive rats. Colloids Surf B: Biointerfaces 115:286–294
29. Halnor VV, Pande VV, Borawake DD, Nagare HS (2018) Nanoemulsion: a novel platform for drug delivery system. J Mat Sci Nanotechol 6:104
30. Huang Y, Dai W-G (2014) Fundamental aspects of solid dispersion technology for poorly soluble drugs. Acta Pharm Sin B 4:18–25

31. Jafari SM, Fathi M, Mandala I (2021) Emerging product formation. Food Waste Recov 257–75
32. Jain KK (2007) Nanobiotechnology-based drug delivery to the central nervous system. Neurodegener Dis 4:287–291
33. Kelmann RG, Kuminek G, Teixeira HF, Koester LS (2007) Carbamazepine parenteral nanoemulsions prepared by spontaneous emulsification process. Int J Pharm 342:231–239
34. Kentish S, Wooster TJ, Ashokkumar M, Balachandran S, Mawson R, Simons L (2008) The use of ultrasonics for nanoemulsion preparation. Innov Food Sci Emerg Technol 9:170–175
35. Khan W, Ansari VA, Hussain Z, Siddique NF (2018) Nanoemulsion: a droplet nanocarrier system for enhancing bioavailability of poorly water soluble drugs. Res J Pharm Technol 11:5191–5196
36. Khani S, Keyhanfar F, Amani A (2016) Design and evaluation of oral nanoemulsion drug delivery system of mebudipine. Drug Deliv 23:2035–2043
37. Klinkesorn U, McClements DJ (2009) Influence of chitosan on stability and lipase digestibility of lecithin-stabilized tuna oil-in-water emulsions. Food Chem 114:1308–1315
38. Kreilgaard M, Pedersen EJ, Jaroszewski JW (2000) NMR characterisation and transdermal drug delivery potential of microemulsion systems. J Control Release 69:421–433
39. Lallemand F, Daull P, Benita S, Buggage R, Garrigue J-S (2012) Successfully improving ocular drug delivery using the cationic nanoemulsion, novasorb. J Drug Deliv
40. Landreau E, Aguni Y, Hamaide T, Chevalier Y (2009) Preparation of nanoemulsions by spontaneous emulsification and stabilization with poly (caprolactone)-b-poly (ethylene oxide) block copolymers. Emuls Sci Technol 191–207
41. Lawrence MJ, Rees GD (2000) Microemulsion-based media as novel drug delivery systems. Adv Drug Deliv Rev 45:89–121
42. Leister N, Karbstein HP (2020) Evaluating the stability of double emulsions—a review of the measurement techniques for the systematic investigation of instability mechanisms. Colloids Interfaces 4:8
43. Li Y, Hu M, Xiao H, Du Y, Decker EA, McClements DJ (2010) Controlling the functional performance of emulsion-based delivery systems using multi-component biopolymer coatings. Eur J Pharm Biopharm 76:38–47
44. Lifshitz IM, Slyozov VV (1961) The kinetics of precipitation from supersaturated solid solutions. J Phys Chem Solids 19:35–50
45. Longmire M, Choyke PL, Kobayashi H (2008) Clearance properties of nano-sized particles and molecules as imaging agents: considerations and caveats
46. Mahajan HS, Mahajan MS, Nerkar PP, Agrawal A (2014) Nanoemulsion-based intranasal drug delivery system of saquinavir mesylate for brain targeting. Drug Deliv 21:148–154
47. Mallick S, Pattnaik S, Swain K, De PK (2007) Current perspectives of solubilization: potential for improved bioavailability. Drug Dev Ind Pharm 33:865–873
48. Marianecci C, Rinaldi F, Hanieh PN, Paolino D, Di Marzio L, Carafa M (2015) Nose to brain delivery: new trends in amphiphile-based "soft" nanocarriers. Curr Pharm Des 21:5225–5232
49. McClements DJ (2012) Nanoemulsions versus microemulsions: terminology, differences, and similarities. Soft Matter 8:1719–1729
50. Mendes S, Graziani SR, Vitório TS, Padoveze AF, Hegg R, Bydlowski SP, Maranhão RC (2009) Uptake by breast carcinoma of a lipidic nanoemulsion after intralesional injection into the patients: a new strategy for neoadjuvant chemotherapy. Gynecol Oncol 112:400–404
51. Mishra RK, Soni GC, Mishra R (2014) Nanoemulsion: a novel drug delivery tool. Int J Pharma Res Rev 3:32–43
52. Mitra AK, Cholkar K, Mandal A (2017) Emerging nanotechnologies for diagnostics, drug delivery and medical devices. William Andrew
53. Ohshima H, Makino K (2014) Colloid and interface science in pharmaceutical research and development. Elsevier
54. Parker G (2001) Encyclopedia of materials: science and technology
55. Pathak K, Pattnaik S, Swain K (2018) Application of nanoemulsions in drug delivery. Nanoemulsions (Elsevier)

56. Pattnaik S, Pathak K (2017) Mesoporous silica molecular sieve based nanocarriers: transpiring drug dissolution research. Curr Pharm Des 23:467–480
57. Paudel KS, Milewski M, Swadley CL, Brogden NK, Ghosh P, Stinchcomb AL (2010) Challenges and opportunities in dermal/transdermal delivery. Ther Deliv 1:109–131
58. Priya S, Koland M, Kumari S (2015) Nanoemulsion components screening of quetiapine fumarate: effect of surfactant and co surfactant. Asian J Pharm Clin Res 8:136–140
59. Ragelle H, Crauste-Manciet S, Seguin J, Brossard D, Scherman D, Arnaud P, Chabot GG (2012) Nanoemulsion formulation of fisetin improves bioavailability and antitumour activity in mice. Int J Pharm 427:452–459
60. Rahn-Chique K, Puertas AM, Romero-Cano MS, Rojas C, Urbina-Villalba G (2012) Nanoemulsion stability: experimental evaluation of the flocculation rate from turbidity measurements. Adv Coll Interface Sci 178:1–20
61. Rai VK, Mishra N, Yadav KS, Yadav NP (2018) Nanoemulsion as pharmaceutical carrier for dermal and transdermal drug delivery: formulation development, stability issues, basic considerations and applications. J Control Release 270:203–225
62. Sarker A, Shimu IJ, Tuhin RH, Raju AA (2015) Nanoemulsion: an excellent mode for delivery of poorly soluble drug through different routes. J Chem Pharm Res 7:966–976
63. Shah P, Bhalodia D, Shelat P (2010) Nanoemulsion: a pharmaceutical review. Syst Rev Pharm 1(1):24
64. Shakeel F, Shafiq S, Haq N, Alanazi FK, Alsarra IA (2012) Nanoemulsions as potential vehicles for transdermal and dermal delivery of hydrophobic compounds: an overview. Expert Opin Drug Deliv 9:953–974
65. Sharma N, Mishra S, Sharma S, Sharma RK (2013) Preparation and optimization of nanoemulsions for targeting drug delivery. Int J Drug Dev Res 5
66. Sharma N, Bansal M, Visht S, Sharma PK, Kulkarni GT (2010) Nanoemulsion: a new concept of delivery system. Chron Young Sci 1:2–6
67. Söderlind E, Wollbratt M, von Corswant C (2003) The usefulness of sugar surfactants as solubilizing agents in parenteral formulations. Int J Pharm 252:61–71
68. Sood S, Jain K, Gowthamarajan K (2014) Optimization of curcumin nanoemulsion for intranasal delivery using design of experiment and its toxicity assessment. Colloids Surf, B 113:330–337
69. Sugumar S, Clarke SK, Nirmala MJ, Tyagi BK, Mukherjee A, Chandrasekaran N (2014) Nanoemulsion of eucalyptus oil and its larvicidal activity against Culex quinquefasciatus. Bull Entomol Res 104:393–402
70. Sutradhar KB, Amin ML (2014) Nanoemulsions: increasing possibilities in drug delivery. Eur J Nanomedicine 6:53–53
71. Sutradhar KB, Khatun S, Luna IP (2013) Increasing possibilities of nanosuspension. J Nanotechnol
72. Swarbrick J, Boylan J (1988) Encyclopedia of pharmaceutical technology, vol 3. Marcel Dekker, Inc., New York
73. Tadros TF (2005) Surfactants in pharmaceutical formulations. Appl Surfactants: Princ Appl 13
74. Tanaka K-I, Kurotsu S, Asano T, Yamakawa N, Kobayashi D, Yamashita Y, Yamazaki H, Ishihara T, Watanabe H, Maruyama T (2014) Superiority of pulmonary administration of mepenzolate bromide over other routes as treatment for chronic obstructive pulmonary disease. Sci Rep 4:1–11
75. Tenjarla S (1999) Microemulsions: an overview and pharmaceutical applications. Crit Rev™ Ther Drug Carr Syst 16
76. Tiwari SB, Amiji MM (2006) Nanoemulsion formulations for tumor-targeted delivery. In: Nanotechnology for cancer therapy. CRC Press
77. Torchilin V (2009) Multifunctional and stimuli-sensitive pharmaceutical nanocarriers. Eur J Pharm Biopharm 71:431–444
78. Verwey EJW (1940) Electrical double layer and stability of emulsions. Trans Faraday Soc 35:192–203

79. Vickers NJ (2017) Animal communication: when i'm calling you, will you answer too? Curr Biol 27:R713–R715
80. Wadle A, Tesmann H, Leonard M, Förster T (1997) Phase inversion in emulsions: CAPICO–concept and application. surfactants in cosmetics. In: Rieger MM, Rhein LD (eds). Dekker, New York
81. Wilson RJ, Li Y, Yang G, Zhao C-X (2022) Nanoemulsions for drug delivery. Particuology 64:85–97
82. Yilmaz S, Kaya E, Kisacam MA (2017) The effect on oxidative stress of aflatoxin and protective effect of lycopene on aflatoxin damage. Aflatoxin-Control, Anal, Detect Health Risks 30:67–90
83. Zeytunluoglu A, Arslan I (2022) Current perspectives on nanoemulsions in targeted drug delivery: an overview. In: Handbook of research on nanoemulsion applications in agriculture, food, health, and biomedical sciences. pp 118–40
84. Zhang J, Peppard TL, Reineccius GA (2015) Preparation and characterization of nanoemulsions stabilized by food biopolymers using microfluidization. Flavour Fragr J 30:288–294

Role of Nanoemulsions in Cosmetics

Anuj Kandwal, Rakesh Kumar Bachheti, Shama Parveen, Archana Bachheti, and Arun Kumar Khajuria

Abstract Nanoemulsions are colloidal dispersion systems comprised of emulsifying agents and two immiscible liquids and today there is a growing demand for using them in food, pharmaceuticals and cosmetics. Over the last decade, nanotechnology has emerged as the most compelling sector in the cosmetics industry. Among numerous methods, nanoemulsions emerged as a delivery method with enormous potential and tremendous appeal in the cosmetics industry. A great deal of research has been done on nanocosmetics and nano cosmeceuticals for the lips, teeth, nails, hair and skin, and it has been discovered that adding nanoparticles to products increases their effectiveness and customer satisfaction. The skin care business places a high value on nanoemulsions because of their favourable biophysical characteristics (especially their hydrating power) and sensory features (merging textures and rapid penetration). Thus, in present chapter, an attempt has been made to provide in depth information about the use of these nanoemulsions in cosmetic industry.

Keywords Nanoemulsions · Nanocomposite · Pharmaceuticals · Cosmetics · Applications

---

A. Kandwal
Department of Chemistry, Harsh Vidya Mandir (PG) College, Raisi, Haridwar, Uttarakhand, India

R. K. Bachheti
Department of Industrial Chemistry, Addis Ababa Science and Technology University, Addis Ababa, Ethiopia

S. Parveen
Department of Chemistry, Faculty of Science, Motherhood University, District- Haridwar, Uttrakhand 247661, India

A. Bachheti
Department of Environment Science, Graphic Era University, Dehradun, Uttarakhand 248002, India

A. K. Khajuria (✉)
Department of Botany, Cluster University of Jammu, Jammu 180001, India
e-mail: arun.khajuria20@gmail.com

A. Husen et al. (eds.), *Current Trends in Green Nano-emulsions*, Smart Nanomaterials Technology, https://doi.org/10.1007/978-981-99-5398-1_14

1 Introduction

The term nanoemulsion is comprised of the two terms nano and emulsion. The term "emulsion" is a liquid created by the dispersion of either oil in water or water in oil when the presence of a surfactant is present whereas the term "nano" refers to a small size ranging from 1–100 nm. Emulsions are compositions that are both kinetically and thermodynamically stable. An emulsion's composition and structure vary depending on its content, and such emulsions are employed in cosmetic items based on their distinct features. Emulsions can be manly divided into four categories: water in oil, oil in water, oil in water in oil, and water in oil in water. The size of nanoemulsion particles ranges from 10 to 200 nm. The compositions known as nanoemulsions are less viscous, kinetically stable, have a larger surface area at the interface, and have a higher solubilization capacity. They have an appearance that is translucent or transparent. The dispersed phase of nanoemulsions is made up of microscopic particles or extremely small droplets that have very low water or oil interfacial tensions. The core of nanoemulsions is made up of lipophilic parts whereas phospholipids make up the outside layer. Since the lipophilic part makes up the core of nanoemulsions, they are appropriate for the transport of lipophilic substances. Sedimentation, creaming, coalescence, flocculation and are the main issues that are solved with the use of nanoemulsions [3, 10, 21, 41, 56].

Co-solvents that are water soluble and surfactants, oils, lipids, and water are all components of nanoemulsion systems. Free fatty acids, vegetable or mineral oils, tri-, di-, or mono-acylglycerols and other triglycerides may be used in the oil phase of nanoemulsion formulations [22]. Drug Typically, solubility is taken into account when choosing an oil. The development of nanoemulsions typically begins with oil phases with high drug loading [5, 52]. Frequent emulsifiers employed in nanoemulsion processes for food and medicine shipment include polyoxyethylene (POE) derivatives of sorbitan fatty acid ester (spans), lauroyl macrogol glycerides (GelucireR 44/14), carbohydrates (derived from gum and starch), and lipid [34, 59, 60]. Incredibly a little negative surface tension is necessary to regard nanoemulsion production. Co-solvents or co-surfactants are used in conjunction with a surfactant in order to achieve this. Transcutol-P (diethylene glyco monoethyl ether), ethylene glycol, propylene glycol, ethanol, polyethylene glycol, propylene glycol, glycerin, and Often used in the creation of nanoemulsion systems, propanol is a co-surfactant or co-solvent [33, 43, 60, 66].

Because of the fact that nanoemulsions are an extremely effective medium for the regulated release of functional ingredients, there is a substantial market for their application in numerous cosmetic products. Nanoemulsions are used in numerous cosmetic items, including body lotions, face and hair serums, hair conditioners, perfumes, and many more. The substance having nanoemulsions can be absorbed into the skin quickly and easily. It also moisturizes the skin and delivers active substances on target. Employing nanoemulsions has several important advantages, including higher absorption rate, simplicity of formulation into foams, liquids, creams or sprays, lack of toxicity and irritation, and compatibility for both lipophilic and hydrophilic drug delivery [64, 67]. The silicone oil-in-water nanoemulsions can improve the silicone oil's absorption on the surface of the hair. The nanoemulsion system is useful for making shampoos because they are thermodynamically and self-assembled stable systems that are simple to create, transport and maintain [24]. Additionally, because of its tiny particle size, it can disperse quickly into the hair. Furthermore, it may efficiently combine washing and hair care into a single step, saving both time and effort. Thus, this chapter's goal is to give in-depth information on the application of nanoemulsions in cosmetics.

2 Method of Synthesis

The best way to create nanoemulsions, which have extremely small particle sizes, is with high-pressure machinery. High-pressure homogenization and microfluidization, which are applied on both a laboratory and an industrial scale, are the most often utilized techniques for creating nanoemulsions [46]. These processes are also appropriate for ultrasonication and in-situ emulsification [7, 25]

2.1 High Energy Methods

(a) **High Pressure Homogenization**
Higher pressure homogenization is essential to the production of nanoemulsions. With the help of a strong homogenizer or piston homogenizer, this technique produces nanoemulsions with very tiny dimensions of the particles (up to 5 nm). Moving the concoction of aqueous phase and oily phase via a small input aperture at a strong pressure (500–5000 psi) generates high hydraulic shear and turbulence to occur with the product, which results in the production of exceedingly minute emulsion particles. The resultant particles are composed of an aqueous monomolecular coating of phospholipids surrounding a liquid, lipophilic core. The only downsides of this approach, which has a high level of efficiency, are During processing, there is a large energy expenditure and colloid temperature rise [15, 22].

(b) **Micro Fluidization**
In the mixing procedure known as micro-fluidization, the use of a tool known as a micro-fluidizer is made. The object needs to be driven via the interaction chamber, which is comprised of microscopic channels known as "micro-channels," using a rising positively displacement pump in this device (500 to 20,000 psi). Submicron-sized particles are produced as a result of the material flowing via micro-channels and onto an impingement area. The two solutions, the oily and the aqueous phase, are mixed and made into a coarse emulsified using an inline homogenizer. The unstable nanoemulsion is produced by further processing the coarse emulsion in a micro-fluidizer. The interaction chamber's

micro-fluidizer is used to continually push the rough emulsion through until the target particle size is achieved. To remove big droplets and create a homogenous nanoemulsion, the bulk emulsion is run through a filter that is submerged in nitrogen [51].

(c) **Ultrasonication**

Many academic articles that aim to use ultrasonic sound frequency to reduce droplet size report on the creation of nanoemulsions. An amplitude and frequency sonotrode is used in a different technique at system pressures higher than ambient pressure. It is generally known that when external pressure rises, the cavitation threshold in an ultra—sonic field sharp rise, resulting in a reduction in the formation of bubbles. However, raising the pressure also raises the cavitation bubbles' collapse pressure.

As a result, when cavitation takes place, the bubbles collapse more violently and with greater strength than when the pressure is at atmospheric levels. The main process of energy dissipation in a lower frequencies ultrasonic system is cavitation, hence fluctuations in the power density can be directly correlated with variations in the navigational intensity. To maintain an ideal temperature, the system additionally uses a water jacket [47, 58].

2.2 *Low Energy Methods*

(a) **Phase inversion emulsifying**

To create fine dispersion, this method leverages chemical energy from stage changes brought on by the emulsification pathway. In this approach, phase change occurs during the emulsification process due to the surfactant's spontaneous curvature. Surfactant spontaneous curvature varies according to changes in variables like temperature and composition. The phase transition is caused by altering the composition of the emulsion while keeping the temperature constant, or vice versa. Phase inversion temperature (PIT) and phase inversion composition (PIC) are the two categories of transitional phase inversion [62].

The PIT procedure inverts the surfactant's spontaneous curvature by changing the temperature. Dehydration of polyoxyethylene (POE) groups of polyethoxylated nonionic surfactants (i.e., polyethoxylated surfactants) results in increased lipophilicity and subsequently causes variations to the surfactant's curvature. Phase inversion consequently occurs, leading to the formation of nanoemulsion [18].

PIC involves adding one of the constituents, like water, to a combination and then adding oil-surfactant or oil to the water-surfactant mixture. In order to create nanoemulsions, POE type nonionic surfactants are frequently used in the PIC process [36].

(b) **Automatic emulsification**
There are three primary steps.

(a) Using surfactants that are both hydrophilic and lipophilic and dissolving them in a water-soluble solvent to produce a homogeneous organic solution.
(b) The liquid layer was combined with the organic layer and stirred magnetically to produce an o/w emulsion.
(c) The component that was water-miscible evaporated when the pressure was lowered [50, 59].

(c) **Membrane emulsification**
Membrane emulsification is a method for creating low energy nanoemulsions. With relatively little surfactant used, this method creates an emulsion with a restricted dimension distribution range. In this method, a dispersed phase is transferred over a membrane to create a continuous phase. Due to the poor flux of the dispersed phase via the membrane, this approach has a problem scaling up, which is one of its biggest drawbacks [19].

3 Challenges of Nanoemulsions in Cosmetics

The development of products that are dependable, stable, efficient, and appealing while still offering customers the best value is an ongoing challenge for the cosmetics sector. Early on in the process of developing new products, considerations for safety, the environment, productivity, and quality must be made. Nanoemulsions have considerable adaptability, allowing users to choose their preferred manufacturing method and a variety of chemicals, including active substances, lipids and surfactants [28]. The impact of method or formulation component variations on end products should be investigated logically [69]. The choice of an acceptable process for a given system depends on factors including reproducibility, simplicity of scaling up, and manufacturing time optimisation. It is also essential to comprehend how surfactant-oil-water (SOW) phases behave whether they are processed, stored, or in touch with skin or hair. The type of oil phase utilised to create O/W nanoemulsions affects the solubilization and effectiveness of lipophilic active components [70]. Additionally, the use of nanoemulsions as a carrier system must ensure the skin's protection by preventing the infiltration of harmful exterior chemicals [12]. Knowing the main destabilising mechanisms well is necessary to guarantee long-term stability in both standard and stressful situations without changing a product's ideal sensory qualities or performance [63]. The usage of nanoemulsions is limited for several applications due to incompatibilities with specific cosmetic chemicals, including as polymers, multivalent salts, pigments, and oxidants, which still need to be resolved. It is recognised that variables like ethanol, water activity and pH, have an impact on how long the colloidal systems of nanoemulsions preserve [20]. But it's also important to have a better understanding of how other factors, including particle size, impacts on the preservation of colloidal solutions [17].

4 Nanoemulsions as a Potential Carrier in Cosmeceuticals

Cosmeceuticals are herbal active ingredients in cosmetic formulations that when applied to the skin provide therapeutic advantages. Cosmeceuticals have the greatest benefit of being free of adverse effects, which is why they are so well-liked and preferred in the market and their claim of boosting beauty has been demonstrated over the past several decades. The use of cosmeceuticals has grown significantly over the years in virtually every aspect of personal care, including the treatment of psoriasis, bloating, and hair loss as well as wrinkles and hyperpigmentation [6, 13].

Numerous innovative techniques have been developed for their simplicity in compatibility with cosmeceuticals, including liposomes, nanoemulsions, gold nanoparticles, niosomes and solid lipid nanostructured materials. Nanoemulsions have been widely employed as a carrier for the regulated administration of cosmeceuticals such as sunscreens, shampoos, lotions, nail polishes, hair serums and conditioners. Nanoemulsions provide therapeutically active drugs with rapid skin penetration and active transdermal delivery. The thermal stability and moisturising effects of an *Opuntia ficus-indica* (L.) Mill extract-formulated nanoemulsions-based moisturising compounds were assessed by Ribeiro et al. [57]. Results showed improved stratum corneum water content and stability for up to 60 days. Using the nano emulsification technology, [23] developed gold nanoparticles that were loaded with lipoic acid, Calendula officinalis extract, and Nigella sativa oil and they investigated their potential in vitro cell culture studies for cell-based and cell-free antioxidant properties and wound healing. The produced formulation, compared to the standard *Nigella sativa* emulsion, showed improved wound healing and antioxidant property, based on the results. This research created new opportunities for the development of herbal medications that are more enriched and effective using other bioactive ingredients. In their formulation of peppermint oil-loaded nanoemulsions, [38] used medium-chain triglycerides that had been stabilised with food-grade biopolymers and further converted to starch. The proposed formulation was tested for its antimicrobial activities against *Staphylococcus aureus* and *Listeria monocytogenes* Scott A, two Gram-positive bacterial strains, using the MIC and time-kill dynamic methods. As compared to bulk peppermint oil, the results showed that nanoemulsions formulation effectively suppressed both strains over the long term. Tatiya et al. [65] synthesized a nanoemulsions gel for transdermal administration that contains *G. glabra* extract, and they tested the formulation ex vivo and in vitro (antimicrobial investigation). Results indicated that nanoemulsion formulation significantly increased the amount of *G. glabra* extract that permeated it in comparison to aqueous extract solution, and it also significantly increased the amount of antimicrobial activity. For example, [40] successfully utilized a nanoemulsion as the target delivery technique and successfully encapsulated vitamin E (an oil-soluble vitamin). In fact, water-in-oil nanoemulsions are a more dependable approach in dermatology because a lipid film structure on the skin makes the oil-soluble active ingredients more favourable. Nanosized range and structure nanoemulsions have thus far been connected to mechanisms explaining the improved skin penetration by these substances [9]. Surfactants in the formulation may be able to cross the stratum corneum's diffusional boundary, resulting in great penetration of cosmetic products. The water composition of nanoemulsions performs a crucial functional task. Due to the stratum corneum's hydration effect, percutaneous penetration of cosmetic products is improved when the water content of the formulation is sufficiently higher. Thus, the active ingredient can now be distributed throughout the skin barrier via nanoscale droplets disseminated in the continuous phase of the nanoemulsions that can migrate easily through the stratum corneum [8]. Dirschka et al. [16] studied that the research on the treatment of actinic keratosis with BF-200 ALA gel and MAL cream shows a trend toward increased effectiveness in the following three months and significantly decreased recurrence rates after a year of follow-up. Abd et al. [1] investigated how to increase more caffeine transfollicular using nanoemulsion formulations with oleic acid and eucalyptol as chemical penetration enhancers. In this study, caffeine was used as a hydrophilic model drug that was successfully transported into and through hair follicles. It was also revealed that the optimized pharmaceutical formulations may have selectively targeted follicles and surrounding areas for the specific medications. Kong et al. [35] suggested that HA-GMS (hyaluronan-glycerol-α-monostearate) based nanoemulsion produces an appealing colloidal transdermal carrier suitable for use in skin care and cosmetic preparation. Nevertheless, the use of nanotechnology in cosmeceuticals is not a very novel recent application. In the 1980s, it was initially introduced to the cosmeceutical market in the preparation of liposomes. Numerous other nanoproducts that contain stem cells, peptides, proteomics and epigenetic factors have since been developed and are still being offered to customers [42]. Every year, major cosmetic companies publish a number of patents relating to nanotechnology. This is why, these investment and patents are crucial at this time for this reason (Table 1; Fig. 1).

Retinol is nano emulsified in porous polymer particles with a size range of 50–200 nm to create the retinol polymer nanocapsule. The stabilization of retinol is then accomplished using lecithin and mung bean medium-chain triglyceride extract [30]. A Japanese firm, Ands Corporation, is continuing to use a patent that describes the creation of a 100 nm-sized nanoemulsion made up of phospholipid and lysophospholipid nanoparticles distributed in water. A method for creating a dry collagen face lotion that combines gelatin protein or collagen with the formation of nanoparticle powder with particle sizes between 10 and 40 nm was patented by Iwamoto Shigemi and the cosmetic items and nanoemulsions were compounded [45]. A nanoemulsion containing a skin-care formula and ginseng saponin metabolites for anti-aging that also contains the same was converted into liposomes using a dermotropic emulsifier for improved skin penetration to promote the synthesis of collagen fibroblast proliferation [68]. Based on the rising number of patents, nanotechnology in cosmetics has been increasingly successful. As consumers become more aware of the advantages of nano-based cosmetic products compared to conventional cosmeceutical treatments, it is clear that they prefer nano-based cosmetics (Fig. 2).

Nanoemulsions are a perfect delivery system for use in cosmetics because of their simplicity in formulation, controllable particle size, enhanced stability, and relatively low surfactant levels. nanoemulsions give formulators a bigger platform to create a variety of modified formulations with different effective properties and

Table 1 Summary of various nanoemulsions -based cosmetic formulations

Cosmeceutical products	Nanoemulsion based formulation	Method of preparation	Action	Key references
Squalene and coconut oil mediated nanoemulsions	Nanocream	Shock method via a phase inversion temperature	The developed NE formulation helped to shrink the droplet size from 3000 to 400 nm, resulting in increased solubility and penetrating power	[11]
Rapeseed oil-mediated nanoemulsions	Topical creams	Sonicator (20 kHz, 500 W, 13 mm)	Amplify the skin membrane's ability to absorb topical lipophilic active substances	[31]
Centella asiatica extract mediated nanoemulsions	Lotion	High-pressure homogenization	Possesses stable physical characteristics while being stored and has greater penetration than non-nanoemulsion lotion	[26]
Curcumin mediated nanoemulsions	Nanoemulsion as a carrier	An aqueous microtitration method with high-energy ultrasonication	Improve transdermal medication delivery permeability for inflammation and wound healing	[4]
Norcanthridin mediated nanoemulsions	Nanoemulsion as a carrier	Aqueous phase titration method	Improve the insecticidal efficacy	[71]
Fluvastatin mediated nanoemulsions	Hydrogel	Aqueous phase titration method	Effective through-skin-layer penetration	[32]
Hydrocortisone mediated nanoemulsions	Gel	Emulsification Method	Effective treatment for topical atopic dermatitis	[61]
Glycyrrhiza glabra Extract mediated nanoemulsions	Gel		Significant antibacterial activity and faster penetration rate	[65]
Lemon myrtle and anise myrtle-mediated nanoemulsions	Nanoemulsion as a carrier	Ultrasonication	NE showed good stability and improved antimicrobial properties	[48]

(continued)

Table 1 (continued)

Cosmeceutical products	Nanoemulsion based formulation	Method of preparation	Action	Key references
Nanoemulsions incorporating citral essential oil	Nanoemulsion as a carrier	Ultrasonication	NE can be used in antimicrobial activity in the agrochemicals, cosmetics and pharmaceutical industries	[39]
Vellozia squamata mediated nanoemulsions	Cream	Phase inversion method	Showed god anti-oxidant activity	[53]
Opuntia ficus indica mediated nanoemulsions	Nanoemulsion as a carrier	Emulsification	The developed NE increased water content of stratum corneum showing its moisturizing efficacy and thus used as a commercial moisturizer	[57]

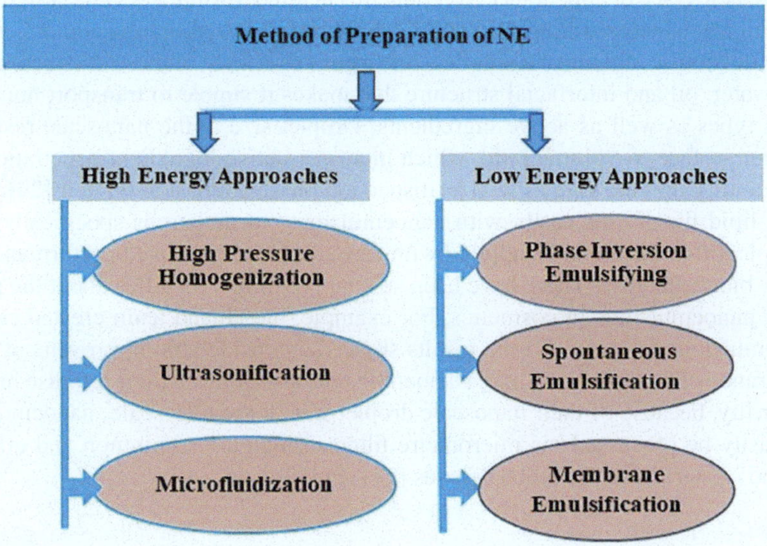

Fig. 1 Method of formation of nanoemulsions

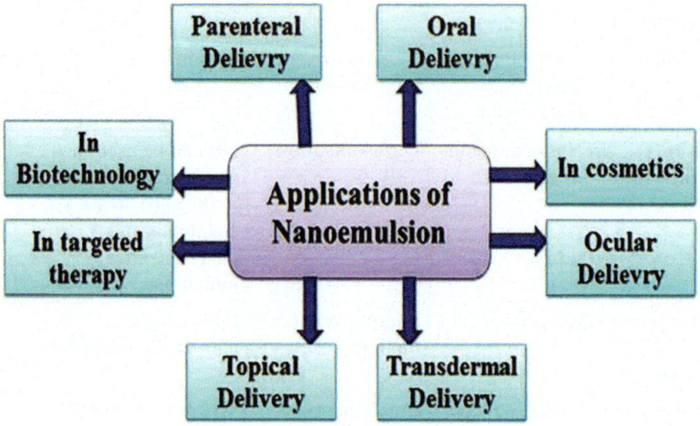

Fig. 2 Application of nanoemulsions (Adopted from [25])

sensory benefits, such as gels, foams, spray, fluids, and creams [14]. A list of several NE-based cosmetic formulations is shown in Table 1. The effectiveness of cosmetic goods is determined by their ability to incorporate functional and active ingredients as well as their grace of appearance. Keeping this in mind, formulators face a significant problem when incorporating amphiphilic, lipophilic and hydrophilic components but nanoemulsions could make it simple to incorporate all three of these. Nanoemulsions have water, oil and interfacial structure that makes it simple to transport numerous media types as well as active ingredients. Droplet size at the nanoscale results in a higher surface-to-volume ratio, which increases active ingredient penetration and permeability into the skin as well as dispersion phase admission on hair [29]. Since a thin lipid film forms easily with nanoemulsions, formulations specifically made for topical treatment can be applied with ease, enabling improved performance and higher bioavailability. There have been several preliminary studies about the potential of nanoemulsions in cosmetics. For example, the Huand team created cationic nanoemulsions for hair, and the results showed a considerable improvement in the appearance of dry hair over time. Nanoemulsions also showed their promise in terms of sterility, because of their nanoscale droplet size at the nanoscale, nanoemulsions can easily be processed via micrometre filters, which are a common and effective method of sterilizing for all businesses [27] (Fig. 3).

4.1 Dermal Application

Recently, nanoemulsions have captured interest as potential solutions for the dispersion of active ingredients in specific skin layers and the controlled release of foundation. It is currently well-established that nanoemulsion can improve the effectiveness of drugs when applied topically and transdermally [54]. Nanoemulsions,

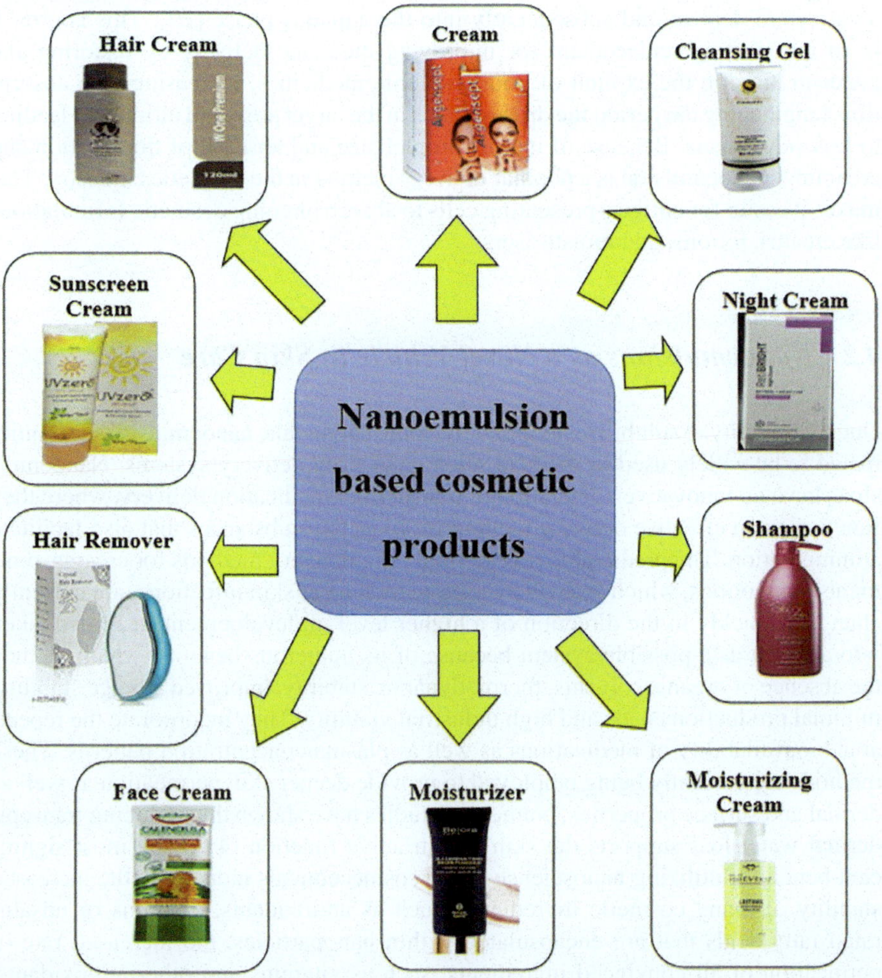

Fig. 3 Applications of nanoemulsions in cosmetics

which are emulsions with narrow distributions and a size range between 20 and 200 nm, has many advantages for the topical and transdermal delivery of cosmeceutical molecules. Controlled droplet size, the capacity to liquefy lipophilic pharmaceuticals successfully, enhanced skin permeation, and prolonged discharge of both lipophilic and hydrophilic drugs are some of these benefits. Additionally, they have outstanding sensory as well as physical qualities including even dispersion on skin and skin nourishment in medications for cosmetics and beauty [2]. When nanoemulsions are applied topically, small particles and solubilized hydrophobic constituents provide a method to greatly speed up the rate of therapeutic dissolution and, as a result, systemic bioavailability via the transcellular pathway. In order to prevent occlusive effects drug discharges from nanoemulsions include partitioning them from oil into a surfactant deposit and subsequently into the aqueous phase [29]. This approach is an intoxicating replacement for improving medicine delivery by restoring and concentrating on the explicit of poorly soluble medicines, increasing skin absorption, lengthening the period the drug remains in the target area, and ultimately leading to less side effects. Because of its tiny droplet size and absence of flocculation and creaming, nanoemulsion is a popular delivery method in the cosmetics industry. This makes it easier for antigen-presenting cells to absorb ongoing cosmetic formulations like creams, lotions, and moisturisers.

4.2 Nanoemulsions as a Novel Vehicle in Skin Care

Due to its ability to solubilize nonpolar dynamic chemicals, nanoemulsions are anticipated to be widely used in medical sciences as drug delivery systems. Nanoemulsions have an innovative application in transdermal medication delivery, where they function as an effective delivery method for bioactive substances that also facilitate administration. The predictable transdermal drug delivery methods for creams, ointments, and lotions, which are widely used to treat local skin infections, are currently changing quickly in the direction of a higher level of development. It is also taken into account as a possible system because of its numerous benefits, which include the absence of organic solvents, thermodynamic stability, improved storage stability, minimal production costs, and high industrial viability. They incorporate the repeatable bioavailability of medications as well as plasma concentration patterns. These methods are currently being employed to provide deeper skin permeation as well as dermal and surface properties. Numerous studies have shown that reducing transepidermal water loss supports the skin's obstructive function [44]. The most significant benefit of utilizing nanoscience in the cosmeceuticals industry is the increased stability of many cosmetic ingredients, such as antioxidants, vitamins or unsaturated fatty acids that are encapsulated within nanoparticles, the increased rate of permeation of all convinced ingredients, such as vitamins and other antioxidants, and the improved aesthetics of the preparations, in addition to improved coordination and approval of UV filters on the stratum corneum [55]. Dispersed nanosized droplets in the continuous phase of the nanoemulsions can successfully penetrate the stratum corneum and spread the active component across the epidermal barrier. O/W nanoemulsions with lipophilic nano-TiO_2 and octyl methoxycinnamate, which contain both physical as well as chemical UV filters, were treated by ultrasonic using conventional ethoxylated alkyl ethers. The effects of structuring the oil phase of nanoemulsions incorporating chemical UV filters were researched by other scientists. Several UV filters were formulated with nanostructured lipid carriers (NLC) and nanoemulsions in the oil phase.

4.3 Nanoemulsions in Hair System Care

Nanoscience has evolved into a very simple tool for creating a solution to this pathology, enabling targeted drug administration with increased local bioavailability, which may reduce the negative effects of the pharmaceuticals. Considering the significance of hair on the value of existence, it is no wonder that people are challenged to change their hair for a number of innovative purposes. Due to their unique and inherent properties, nanoparticles are increasingly being utilized in the creation of superior skin-care medications that particularly target the hair shaft and follicle. The uptake and permeation of topically applied nanostructured materials are considerably facilitated by the hair follicles. According to some evidence, follicular allocation varies depending on the body region, with the forehead supporting a significantly higher density of follicular orifices [49]. They claimed that whereas follicular apertures might give nanoparticle components access to the hair follicle, there may be considerable variations in the percutaneous absorption of appendage-free and abundant parts. Results can reveal a mechanical influence as opposed to a particle size-specific effect when they are compared to earlier findings. Hair growth procedures could be used to push nanoparticles deeper inside hair follicles [37]. Therefore, a variety of aesthetic techniques can be used to specifically target the arrangement surrounding the follicle by varying the particle size. The use of nanomaterials to enhance hair cosmesis (health of hair, smoothness and preserving shine) has been the subject of extensive prior research. Nanostructured materials have been added to shampoos to optimize resident contact time with the hair follicle and scalp, enabling active molecules to coat the cuticles with a protective layer and retain moisture (i.e., averting trans follicular water loss). However, while being fabricated into nanostructure particles, silicone oil may diffuse swiftly into hair fibres despite being manufactured into nanoparticles due to its tiny size. It was noticed that when the size of the nanoemulsion decreases, more nanodroplets were absorbed into the hairs. This is explained by the fact that the interactions between nanoemulsions droplets increase as their droplets size decreases, which raises nanoemulsions viscosity and enhances silicone oil deposition. This improved lubrication, shiny finish, and hair moisture without harming the hair fiber's cuticles, permits absorption into the hydro-lipid emulsion layer [63]. This demonstrates that even when using shampoo frequently, cationic nanoemulsions with droplet size of less than 1–100 nm considerably improve dry hair.

5 Conclusion

In conclusion, it can be concluded that the era of nanoparticles has already begun and that the research around them will continue to advance and be applied to the creation of new testing procedures for a lovely, healthier, and safer future. Over the last decade, nanotechnology has emerged as the most compelling sector in the cosmetics industry.

As new technology and delivery methods continue to advance at a steady rate, researchers are better able to create newer, more adaptable cosmetic goods. Among numerous methods, nanoemulsions emerged as a delivery method with enormous potential and tremendous appeal in the cosmetics industry. This chapter discussed possible nanoemulsion system capabilities in cosmetics and personal care items used for diverse applications with supporting data. nanoemulsions smaller droplet size satisfies all basic criteria including delivery qualities, rheological stability and optical stability, which are necessary for an effective delivery method and which are not satisfied by conventional microemulsions. Because of their greater penetration abilities, nanoemulsions are an ideal option for hair care and skin products, whether they use cosmeceuticals or synthetic cosmetics.

References

1. Abd E, Benson HA, Roberts MS, Grice JE (2018) Follicular penetration of caffeine from topically applied nanoemulsion formulations containing penetration enhancers: in vitro human skin studies. Ski Pharmacol Physiol 31:252–260
2. Abolmaali SS, Tamaddon AM, Farvadi FS, Daneshamuz S, Moghimi H (2011) Pharmaceutical nanoemulsions and their potential topical and transdermal applications. Iran J Pharm Sci 7(3):139–150
3. Acosta E (2009) Bioavailability of nanoparticles in nutrient and nutraceutical delivery. Curr Opin Colloid Interface Sci 14(1):3–15
4. Ahmad N, Ahmad R, Al-Qudaihi A, Alaseel SE, Fita IZ, Khalid MS, Pottoo FH (2019) Preparation of a novel curcumin nanoemulsion by ultrasonication and its comparative effects in wound healing and the treatment of inflammation. RSC Adv 9(35):20192–20206
5. Ashaolu TJ (2021) Nanoemulsions for health, food, and cosmetics: a review. Environ Chem Lett 19(4):3381–3395
6. Azmi NAN, Elgharbawy AA, Motlagh SR, Samsudin N, Salleh HM (2019) Nanoemulsions: factory for food, pharmaceutical and cosmetics. Processes 7(9):617
7. Bissessur R (2020) Nanomaterials applications. In: Polymer science and nanotechnology. Elsevier
8. Boonme P (2007) Applications of microemulsions in cosmetics. J Cosmet Dermatol 6(4):223–228
9. Boonme P, Junyaprasert VB, Suksawad N, Songkro S (2009) Microemulsions and nanoemulsions: novel vehicles for whitening cosmeceuticals. J Biomed Nanotechnol 5(4):373–383
10. Čerpnjak K, Zvonar A, Gašperlin M, Vrečer F (2013) Lipid-based systems as promising approach for enhancing the bioavailability of poorly water-soluble drugs. Acta Pharm 63(4):427–445
11. Charoenkul K, Phromyothin D (2017) Development and characterization of nano-cream preparation containing natural extract using nanoemulsion techniques. Mater Today: Proc 4(5):6105–6110
12. Clares B, Calpena AC, Parra A, Abrego G, Alvarado H, Fangueiro JF, Souto EB (2014) Nanoemulsions (NEs), liposomes (LPs) and solid lipid nanoparticles (SLNs) for retinyl palmitate: effect on skin permeation. Int J Pharm 473(1–2):591–598
13. De Canha MN, Blom van Staden A, Fibrich BD, Lambrechts IA, Denga L, Lall N (2019) Book review: cosmeceuticals and active cosmetics. Front Pharmacol
14. De Souza ML, Oliveira DD, Ribeiro PL, de Paula Pereira N, Druzian JI (2018) Nanoemulsions for cosmetic applications: what innovation status? Recent Pat Nanotechnol 12(2):101–109

15. Dhawan S, Sharma P, Nanda S (2020) Cosmetic nanoformulations and their intended use. In: Nanocosmetics. Elsevier, pp 141–169
16. Dirschka T, Ekanayake-Bohlig S, Dominicus R, Aschoff R, Herrera-Ceballos E, Botella-Estrada R, Hunfeld A, Kremser M, Schmitz B, Lubbert H, Puig S (2019) A randomized, intraindividual, non-inferiority, Phase III study comparing daylight photodynamic therapy with BF-200 ALA gel and MAL cream for the treatment of actinic keratosis. J Eur Acad Dermatol Venereol 33(2):288–297
17. Fang B, Yu M, Zhang W, Wang F (2016) A new alternative to cosmetics preservation and the effect of the particle size of the emulsion droplets on preservation efficacy. Int J Cosmet Sci 38(5):496–503
18. Friberg SE, Corkery RW, Blute IA (2011) Phase inversion temperature (PIT) emulsification process. J Chem Eng Data 56(12):4282–4290
19. Gadhave AD (2014) Nanoemulsions: formation, stability and applications. Int J Res Sci Adv Technol 3(2):38–43
20. Geis PA (ed) (2020) Cosmetic microbiology: a practical approach. CRC Press
21. Gibaud S, Attivi D (2012) Microemulsions for oral administration and their therapeutic applications. Expert Opin Drug Deliv 9(8):937–951
22. Gonçalves A, Nikmaram N, Roohinejad S, Estevinho BN, Rocha F, Greiner R, McClements DJ (2018) Production, properties, and applications of solid self-emulsifying delivery systems (S-SEDS) in the food and pharmaceutical industries. Colloids Surf, A 538:108–126
23. Guler E, Barlas FB, Yavuz M, Demir B, Gumus ZP, Baspinar Y, Coskunol H, Timur S (2014) Bio-active nanoemulsions enriched with gold nanoparticle, marigold extracts and lipoic acid: In vitro investigations. Colloids Surf, B 121:299–306
24. Gupta V, Mohapatra S, Mishra H, Farooq U, Kumar K, Ansari MJ, Aldawsari MF, Alalaiwe AS, Mirza MA, Iqbal Z (2022) Nanotechnology in cosmetics and cosmeceuticals—a review of latest advancements. Gels 8(3):173
25. Halnor VV, Pande VV, Borawake DD, Nagare HS (2018) Nanoemulsion: a novel platform for drug delivery system. J Mat Sci Nanotechol 6(1):104
26. Hanifah M, Jufri M (2018) Formulation and stability testing of nanoemulsion lotion containing centella asiatica extract. J Young Pharm 10(4):404
27. Hu Z, Liao M, Chen Y, Cai Y, Meng L, Liu Y, Lv N, Liu Z, Yuan W (2012) A novel preparation method for silicone oil nanoemulsions and its application for coating hair with silicone. Int J Nanomed 5719–5724
28. Jafari SM, Paximada P, Mandala I, Assadpour E, Mehrnia MA (2017) Encapsulation by nanoemulsions. In: Nanoencapsulation technologies for the food and nutraceutical industries. Academic Press, pp 36–73
29. Jaiswal M, Dudhe R, Sharma PK (2015) Nanoemulsion: an advanced mode of drug delivery system. 3 Biotech 5:123–127
30. Jeong SH, Son JH, Jang SJ, Kim YJ, Cheon JW (2015) U.S. patent no. 8,980,293. U.S. Patent and Trademark Office, Washington, DC
31. Kaci M, Belhaffef A, Meziane S, Dostert G, Menu P, Velot E, Desobry S, Arab-Tehrany E (2018) Nanoemulsions and topical creams for the safe and effective delivery of lipophilic antioxidant coenzyme Q10. Colloids Surf, B 167:165–175
32. Kaur R, Ajitha M (2019) Transdermal delivery of fluvastatin loaded nanoemulsion gel: Preparation, characterization and in vivo anti-osteoporosis activity. Eur J Pharm Sci 136:1–10
33. Khan AW, Kotta S, Ansari SH, Sharma RK, Ali J (2012) Potentials and challenges in self-nanoemulsifying drug delivery systems. Expert Opin Drug Deliv 9(10):1305–1317
34. Komaiko JS, McClements DJ (2016) Formation of food-grade nanoemulsions using low-energy preparation methods: a review of available methods. Compr Rev Food Sci Food Saf 15(2):331–352
35. Kong M, Chen XG, Kweon DK, Park HJ (2011) Investigations on skin permeation of hyaluronic acid based nanoemulsion as transdermal carrier. Carbohyd Polym 86(2):837–843
36. Kumar M, Bishnoi RS, Shukla AK, Jain CP (2019) Techniques for formulation of nanoemulsion drug delivery system: a review. Prev Nutr Food Sci 24(3):225

37. Lademann J, Patzelt A, Richter H, Antoniou C, Sterry W, Knorr F (2009) Determination of the cuticula thickness of human and porcine hairs and their potential influence on the penetration of nanoparticles into the hair follicles. J Biomed Opt 14(2):1–4
38. Liang R, Xu S, Shoemaker CF, Li Y, Zhong F, Huang Q (2012) Physical and antimicrobial properties of peppermint oil nanoemulsions. J Agric Food Chem 60(30):7548–7555
39. Lu WC, Huang DW, Wang CC, Yeh CH, Tsai JC, Huang YT, Li PH (2018) Preparation, characterization, and antimicrobial activity of nanoemulsions incorporating citral essential oil. J Food Drug Anal 26(1):82–89
40. Mayer S, Weiss J, McClements DJ (2013) Vitamin E-enriched nanoemulsions formed by emulsion phase inversion: factors influencing droplet size and stability. J Colloid Interface Sci 402:122–130
41. McClements DJ (2012) Nanoemulsions versus microemulsions: terminology, differences, and similarities. Soft Matter 8(6):1719–1729
42. Mihranyan A, Ferraz N, Strømme M (2012) Current status and future prospects of nanotechnology in cosmetics. Prog Mater Sci 57(5):875–910
43. Mishra PK, Pavelek O, Rasticova M, Mishra H, Ekielski A (2022) Nanocellulose-based biomedical scaffolds in future bioeconomy: a techno-legal assessment of the state-of-the-art. Front Bioeng Biotechnol 9:1–15
44. Müller RH, Benita S, Bohm B (eds) (1998) Emulsions and nanosuspensions for the formulation of poorly soluble drugs. CRC Press
45. Nanbu T (2009) U.S. patent application no. 12/051,987
46. Nanda A, Nanda S, Nguyen TA, Rajendran S, Slimani Y (eds) (2020) Nanocosmetics: fundamentals, applications and toxicity. Elsevier
47. Naseema A, Kovooru L, Behera AK, Kumar KP, Srivastava P (2021) A critical review of synthesis procedures, applications and future potential of nanoemulsions. Adv Coll Interface Sci 287:102318
48. Nirmal NP, Mereddy R, Li L, Sultanbawa Y (2018) Formulation, characterisation and antibacterial activity of lemon myrtle and anise myrtle essential oil in water nanoemulsion. Food Chem 254:1–7
49. Otberg N, Richter H, Schaefer H, Blume-Peytavi U, Sterry W, Lademann J (2004) Variations of hair follicle size and distribution in different body sites. J Investig Dermatol 122(1):14–19
50. Ozogul Y, Karsli GT, Durmuş M, Yazgan H, Oztop HM, McClements DJ, Ozogul F (2022) Recent developments in industrial applications of nanoemulsions. Adv Colloid Interface Sci 102685
51. Pandey V, Shukla R, Garg A, Kori ML, Rai G (2020) Nanoemulsion in cosmetic: from laboratory to market. Elsevier, Nanocosmetics
52. Qadir A, Faiyazuddin MD, Hussain MT, Alshammari TM, Shakeel F (2016) Critical steps and energetics involved in a successful development of a stable nanoemulsion. J Mol Liq 214:7–18
53. Quintão FJ, Tavares RS, Vieira-Filho SA, Souza GH, Santos OD (2013) Hydroalcoholic extracts of Vellozia squamata: study of its nanoemulsions for pharmaceutical or cosmetic applications. Rev Bras 23(1):101–107
54. Rai VK, Mishra N, Yadav KS, Yadav NP (2018) Nanoemulsion as pharmaceutical carrier for dermal and transdermal drug delivery: formulation development, stability issues, basic considerations and applications. J Control Release 270:203–225
55. Raj S, Jose S, Sumod US, Sabitha M (2012) Nanotechnology in cosmetics: opportunities and challenges. J Pharm Bioallied Sci 4(3):186–193
56. Rehman FU, Shah KU, Shah SU, Khan IU, Khan GM, Khan A (2017) From nanoemulsions to self-nanoemulsions, with recent advances in self-nanoemulsifying drug delivery systems (SNEDDS). Expert Opin Drug Deliv 14(11):1325–1340
57. Ribeiro RCDA, Barreto SMDAG, Ostrosky EA, Rocha-Filho PA, Veríssimo LM, Ferrari M (2015) Production and characterization of cosmetic nanoemulsions containing Opuntia ficus-indica (L.) Mill extract as moisturizing agent. Molecules 20(2):2492–2509
58. Romes NB, Abdul Wahab R, Abdul Hamid M (2021) The role of bioactive phytoconstituents-loaded nanoemulsions for skin improvement: a review. Biotechnol Biotechnol Equip 35(1):711–730

59. Singh RD, Kapila S, Ganesan NG, Rangarajan V (2022) A review on green nanoemulsions for cosmetic applications with special emphasis on microbial surfactants as impending emulsifying agents. J Surfactants Deterg 25(3):303–319
60. Singh Y, Meher JG, Raval K, Khan FA, Chaurasia M, Jain NK, Chourasia MK (2017) Nanoemulsion: concepts, development and applications in drug delivery. J Control Release 252:28–49
61. Sinha P, Srivastava S, Mishra N, Singh DK, Luqman S, Chanda D, Yadav NP (2016) Development, optimization, and characterization of a novel tea tree oil nanogel using response surface methodology. Drug Dev Ind Pharm 42(9):1434–1445
62. Solans C, Solé I (2012) Nano-emulsions: formation by low-energy methods. Curr Opin Colloid Interface Sci 17(5):246–254
63. Sonneville-Aubrun O, Simonnet JT, L'alloret F (2004) Nanoemulsions: a new vehicle for skincare products. Adv Coll Interface Sci 108:145–149
64. Souto EB, Cano A, Martins-Gomes C, Coutinho TE, Zielińska A, Silva AM (2022) Microemulsions and nanoemulsions in skin drug delivery. Bioengineering 9(4):158
65. Tatiya A, Bhavsar S, Mahajan H, Surana S (2019) Experimental design and characterization of nanoemulsion based topical herbal gel developed for site-specific activity of glycyrrhiza glabra extract: in vitro and ex vivo studies. Biosc Biotech Res Comm 12(2):1–12
66. Thakur P, Sonawane S, Potoroko I, Sonawane SH (2021) Recent advances in ultrasound-assisted synthesis of nano-emulsions and their industrial applications. Curr Pharm Biotechnol 22(13):1748–1758
67. Tiwari S, Talreja MS (2020) A concept of nanotechnology in cosmetics: a complete overview. Adalya J 9(11):14–23
68. Yoo BH, Kang BY, Yeom MH, Sung DS, Ju HK, Han SH, Kim HK (2003) European Patent No. EP 1327434A1. European Patent Office, Munich, Germany
69. Yukuyama MN, Ghisleni DDM, Pinto TDJA, Bou-Chacra NA (2016) Nanoemulsion: process selection and application in cosmetics–a review. Int J Cosmet Sci 38(1):13–24
70. Yukuyama MN, Kato ETM, Lobenberg R, Bou-Chacra NA (2017) Challenges and future prospects of nanoemulsion as a drug delivery system. Curr Pharm Des 23(3):495–508
71. Zeng L, Liu Y, Pan J, Liu X (2019) Formulation and evaluation of norcanthridin nanoemulsions against the Plutella xylostella (Lepidotera: Plutellidae). BMC Biotechnol 19(1):1–11

Uses of Nanoemulsions in Pharmaceuticals Industries

Abdelsamed I. Elshamy, Walaa A. El-Kashak, Rehab F. Taher, Mai M. Elghonemy, Rania F. Ahmed, Tarik A. Mohamed, Ahmed F. Essa, Tamer I. M. Ragab, Mohamed F. Abdelhameed, Abd El-Nasser G. El-Gendy, Mahmoud I. Nassar, Ahmed M. Abd-ELGawad, Abdel Razik H. Farrag, Howaida I. Abd-Alla, and Mohamed-Elamir F. Hegazy

Abstract Nanoemulsions have attracted great attention in research, dosage form design and pharmacotherapy. This is as a result of a number of attributes peculiar to nanoemulsions such as optical clarity, ease of preparation, thermodynamic stability and increased surface area. Nanoemulsions also known as submicron emulsions serve as vehicles for the delivery of active pharmaceutical ingredients as well as other bioactives. They are designed to address some of the problems associated with conventional drug delivery systems such as low bioavailability and noncompliance. The importance of design and development of emulsion nanocarrier systems aimed at controlling and/or improving required bioavailability levels of therapeutic agents cannot be overemphasized. Reducing droplet sizes to the nanoscale leads to

---

A. I. Elshamy (✉) · W. A. El-Kashak · R. F. Taher · M. M. Elghonemy · R. F. Ahmed · A. F. Essa · M. I. Nassar · H. I. Abd-Alla
Chemistry of Natural Compounds Department, National Research Centre, 33 El Bohouth St., Dokki, Giza 12622, Egypt
e-mail: elshamy@yahoo.com; ai.el-shamy@nrc.sci.eg

T. A. Mohamed · M.-E. F. Hegazy
Chemistry of Medicinal Plants Department, National Research Centre, 33 El-Bohouth St., Dokki, Giza 12622, Egypt

T. I. M. Ragab
Chemistry of Natural and Microbial Products Department, National Research Centre, Dokki, Giza 12622, Egypt

M. F. Abdelhameed
Pharmacology Department, National Research Centre, 33 El Bohouth St., Dokki, Giza 12622, Egypt

A. E.-N. G. El-Gendy
Medicinal and Aromatic Plants Research Department, National Research Centre, 33 El Bohouth St., Dokki, Giza 12622, Egypt

A. M. Abd-ELGawad
Department of Botany, Faculty of Science, Mansoura University, Mansoura 35516, Egypt

A. R. H. Farrag
Pathology Department, National Research Centre, 33 El Bohouth St., Dokki, Giza 12622, Egypt

A. Husen et al. (eds.), *Current Trends in Green Nano-emulsions*, Smart Nanomaterials Technology, https://doi.org/10.1007/978-981-99-5398-1_15

some very interesting physical properties, such as optical transparency and unusual elastic behaviour. This review sheds light on the current state of nanoemulsions in the delivery of drugs and other bioactives. The morphology, formulation, characteristics and characterization of nanoemulsions were also addressed. Plant-derived natural products have been the primary source of medicinal drugs since humanity's inception on the planet. Recently, nanotechnology has become the most important strategy for the development of pharmaceutical drugs around the world. Nowadays, nanoemulsions have significant roles in pharmaceutical applications along with foods and cosmetics. The emulsification techniques are used for the production of nanoscale dispersions of droplets of two immiscible liquids. Due to their stability, NEs can persist for a long time because of the stabilising surfactants that inhibit the droplets' coalescence. The biological potentialities of nanoemulsions might be ascribed to their small particles, high surface area/volume unit, and improved active hydrophobic component dispersion and absorption. Many applications of nanoemulsions are reported in nanomedicines and drug delivery, such as anti-inflammatory, antipyretic, antimicrobial, antiulcer, anti-hepatic and renal diseases, etc. The present chapter aims to summarize the recent contributions of the nanoemulsions in biological and pharmaceutical applications and industrial processes.

Keywords Natural products · Nanoemulsions · Pharmaceutical applications

List of Abbreviations

ABTS	2,2′-Azino-bis(3-ethylbenzothiazoline-6-sulphonic acid)
AChE	Acetylcholinesterase
BAX	BCL2 associated X
CCA	Cellular antioxidant activity
COX	Cyclooxygenase
DOC	Docetaxel
DPPH	Diphenyl picrylhydrazyl
Eos	Essential oils
FASN	Fatty acid synthase
FEGO-NE	*Ferula gummosa* Essential oil nano-emulsion
FEO	Frankincense EO
GC–MS	Gas chromatography–mass spectrometry
HAV	Hepatitis A virus
HMGCR	3-Hydroxy-3-methylglutaryl-CoA reductase
HSV	Herpes simplex virus
HSV-1	Herpes simplex type-1 virus
HTh-7	Thyroid cancer cell line
IFO	Ifosfamide
MBC/MFC	Minimum bactericidal/fungicidal concentration
MCF-7	Michigan Cancer Foundation-7

MIC	Minimum inhibitory concentration
MMC	Mitomycin C
MTT	Methylthiazolyl-diphenyl tetrazolium bromide
NEs	Nanoemulsions
NFs	Nanofibers
PIT	Phase inversion temperature
PVA	Polyvinyl alcohol
QC	Quercetin
QCNE	Quercetin nanoemulsion
SAS	Squamous carcinoma cells
SME	Self-microemulsifying
SREPB1	Sterol regulatory element-binding protein
TA	Thioacetamide
TSB	Tryptic soy broth

1 Introduction

From the first step on the earth, nature's products represented the backbone of the life needs of all humanity, including the feedings and remedies [3, 44]. More than 21,000 medicinal plants have been used to treat diseases all over the world [44]. Essential oils (EOs) are valuable bioactive resources for the treatment of a variety of ailments [1, 3, 17, 21, 51, 109]. The use of nanotechnology has become one of the main developed strategies in pharmaceutical and medicinal applications [45, 50, 52]. The most common application of nanotechnology in pharmaceutical applications is the preparation of nanocarriers as drug delivery systems [128]. Many nanocarrier types are used, like NEs, lipoproteins, liposomes, microcapsules, solid lipid particles, noisome, dendrimers, and micelles [94]. NEs represent the most advanced drug delivery technique and are a promising nanocarrier for several medicinal applications. These nanocarriers exhibited thermo-dynamical stability with the mean range of diameters of 50 and 1000 nm^2 and others [94]. Also, they displayed the advantages of better delivery of poorly soluble drugs [61].

Loading natural products such as extracts, EOs, and bioactive constituents into nanosystems has become a significant strategy for the development of pharmaceuticals [25]. The loading of the EOs in nanosystems is a significant technique for enhancement of the medicinal effects via (i) toxicity decreasing, (ii) volatility reducing, (iii) stability increasing of the active constituents, and (iv) penetration increasing in the cellular uptake and tissues [25, 45] in addition to others as described in Fig. 1.

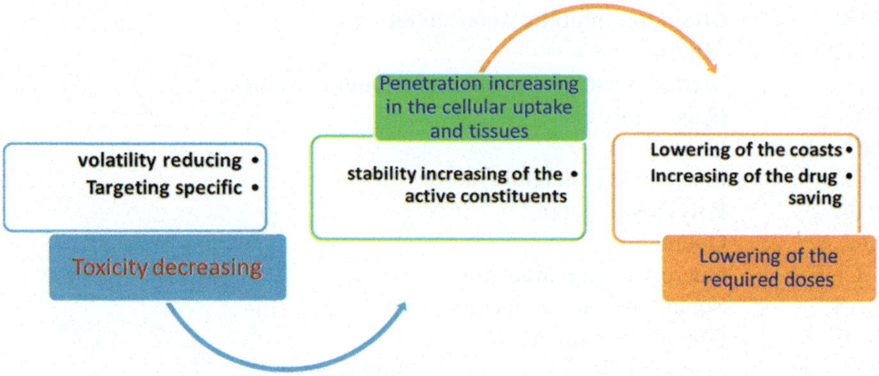

Fig. 1 Some advantages of NEs

2 Nanoemulsified Essential Oils Can Overcome the Limitations of Pure Oils

The potential use of EOs as a natural component in food preservation, with great potential usage in food industries, has been studied [24, 46]. The low aqueous phase solubility, high volatility, and limited long-term stability, on the other hand, are features that restrict its usage as a natural preservative to replacing currently employed preservatives. The creation of EO nanoemulsions has emerged as a possible option thanks to the development of nanotechnology. Due to their nanometric dimension, NEs have special properties like enhanced contact surface area and increased physico-chemical stability. The capacity of the EO nanoemulsified to improve the interaction with particular targets in products and exert antibacterial and antioxidant activity makes it a more environmentally friendly alternative to pure oil. In order to manufacture nanoscale emulsions and reach droplet sizes under 100 nm, both high-energy and low-energy processes can be applied. Bioactivity is strongly correlated with particle size. NEs can therefore be a technical substitute to get around the drawbacks of pure oils and be used in products as a natural preservative.

3 Methods of Nanoemulsion Preparation

Low-energy, high-energy, and combination techniques are used to create NEs. Researchers initially attempted to prepare NEs by using high-energy techniques in their experiments. Because low-energy technologies are "soft," "non-destructive," and don't harm enclosed molecules, interest in them has recently increased significantly. These techniques also use less energy, making them more desirable for large-scale production [68].

3.1 High-Energy Emulsification Methods

High-shear stirring, ultrasonic emulsification, high-pressure homogenization, particularly microfluidics, and membrane emulsification are some of the high-energy techniques [5]. The interior phase droplet size frequently exceeds the nanometer range when NEs are made using high-energy procedures. This is mostly caused by the substantial energy required for dispersion, particularly in viscous systems. Inadequate surfactant concentrations may also contribute to an increase in droplet size by failing to completely adsorb all of the droplets produced during dispersion on the interface. The adsorption velocity of the surfactant has a significant impact on the final size of the droplets [68].

3.1.1 High-Shear Stirring

Initially, NEs were created using rotor–stator systems and high-energy mixers. The internal phase droplet size can be greatly reduced in these devices by increasing the mixing intensity. However, it is challenging to generate emulsions with an average droplet size of less than 200–300 nm [67]. The components of the emulsion are drawn into the rotor–stator unit by the high rarefaction that is formed in the disintegrating head at high rotor speed. Modern equipment is built with relatively little aeration occurring during emulsification. The multi-pass regime is typically used because the single-pass regime frequently fails to achieve the system's maximal degree of dispersion. When using viscous media, the effectiveness of high shear stirring significantly declines, resulting in the formation of emulsions with a high fraction of the internal phase and droplet sizes larger than 1 mm [23, 68].

3.1.2 Ultrasonic Emulsification

When an emulsion is sonicated, cavitation bubbles collapse and a significant amount of energy is released locally, forming emulsion droplets that are only a few nanometers in size [59]. Smaller internal phase droplets are produced when ultrasonic power is increased up to a particular point [137]. The size of the droplet does not significantly alter as the ultrasonic power is increased further. High-energy sonication should not be used because it can cause water to thermally decompose into radicals $H^{\cdot}$ and $^{\cdot}OH$, whose action may cause surfactant molecules gathered on the surface of cavitation bubbles to decompose [112]. Only in the near area of the waveguide radiator does effective emulsification take place. Mechanically churning the mixture is necessary for the emulsification of huge quantities. Because of this, using this procedure to manufacture NEs for research reasons or in small batches is most appropriate.

3.1.3 High-Pressure Homogenization

The emulsification of medium- and low-viscosity systems is the most frequent application of high-pressure homogenization. The systems of high pressure are always in use. The homogenizers are separated into nozzle devices, radial diffusers, and jet dispersers depending on the type of flow directions present in the dispersion cell. The emulsion flow departs from the initial direction by 90° in a standard radial diffuser. With the help of a moveable valve, it is possible to adjust the slit width and, consequently, the flow rate and pressure of the emulsified liquids. The most popular homogenizers for the industrial production of emulsions have flat valves. For laboratory research, homogenizers of various designs are made, in particular, those with tooth-valves of various configurations, inclined seats that divert the flow at varying angles, and cutting-edge valves for still another flow deviation [66]. The capillary-jet homogenizers' basic working theory relies on the collision of two or more highly moving flows. The best-known examples of this kind of equipment are the "Nanojet" gadget and a variety of jet dispersers from Microfluidics Corp [132].

3.1.4 Microfluidic and Membrane Methods

The controlled formation of liquid droplets and gas bubbles, as well as their transportation in microchannels, are made possible by microfluidic methods. The potential to establish a lab-on-a-chip is linked to the recent widespread development of microfluidics techniques [120, 122]. By delivering microscopic amounts of a liquid in the form of drops and bubbles through a network of microchannels to a specific location, these devices enable the miniaturization of chemical and biological processes. This allows for the immediate investigation of the drop composition, for example. The development of such devices for protein isolation, DNA analysis, enzyme analysis, cell encapsulation, and the production of biosensors is currently being studied [66]. Emulsification in microfluidizers happens when two liquids are not capable of combining to form a homogeneous mixture travelling in microchannels collide with a high pressure condition [123]. The stability of the emulsification regime was determined by the wetting of the channel walls exerted by the emulsion components. Typically, devices with hydrophilic and hydrophobic surfaces are used to create oil in water (O/W) and water in oil (W/O) emulsions, respectively [111]. In comparison to other high energy devices, the manufacturing of microfluidizers is pricey. Because channels might be obstructed by solid particles and gas bubbles, they are less useful in actual applications [55]. Additionally, issues with microfluidic sterilization occur when they are utilized to create emulsions for pharmacological and medical applications. Membrane methods of emulsification, like microfluidic methods and in contrast to the emulsification in the turbulent regime with low fluid flow rate, pertain to ways of forming individual droplets [134]. In the membrane methods, fluid is forced through numerous microchannels or pores in a membrane to create droplets of the internal phase [130]. During one-step emulsification and when the internal phase is extruded through the membrane, droplets are created at the membrane/continuous phase interface. Membrane emulsification which entails forcing a coarse emulsion across a membrane, with smaller pore size than the coarse emulsion's droplet size, is used to create emulsions with smaller internal phase droplets. The droplet size is reduced by passing the coarse pre-mixed emulsion across a hydrophilic membrane directly. When the coarse direct emulsion is forced (through a hydrophobic barrier), the inversion phase results in the creation of emulsion in a reversal phase [125]. If the membrane is fixed in a device, stirring or recirculation of the emulsion is created. This is usually carried out to hasten the detachment of droplets from the surface. To hasten the detachment of droplets, devices characterized by vibrating or rotating membranes are used, nevertheless, this greatly complicates the designs of the apparatuses. This approach is currently the subject of active development. However, high-output valve homogenizers are most frequently used to prepare O/W NEs for large-scale applications.

3.2 Low-Energy Emulsification Methods

Techniques of low-energy NE preparation came into being much later than high-energy ones. These techniques rely on the phase inversion that occurs in an emulsion when the composition or temperature changes. The low-energy techniques are based on the ideas put out by Japanese researchers in the 1960s [118]. With temperature, the surfactant adsorbed layer's curvature varies. The system reaches a condition of interfacial tension with low value ($10^{-2} \pm 10^{-5}$ mN m^{-1}) and the curvature surfactant monolayer of zero value at a specific temperature [66].

3.2.1 Method of Phase Inversion Temperature (PIT)

PIT approach is based on a unique characteristic of surfactants, particularly those that are typical of nonionic ethoxylated surfactants, namely the capacity to alter the oil and water affinity with temperature. Hydration of polar groups affects the molecules of ethoxylated surfactants. At low temperatures, they were hydrophilic. The polar groups in the hydrated state occupy a higher surface area than the hydrophobic hydrocarbon chains do. O/W emulsions are created and the surfactant monolayer flexes as a known result. Oxyethylene groups are easily dehydrated as the temperature rises, which makes the molecules more lipophilic. In this scenario, Occupation of the surface area by chains of hydrocarbon increases relative to the surface area occupied by the polar groups, favouring the development of W/O emulsions [117]. The PIT approach has currently received the most attention. The manufacture of NEs stabilised by ethoxylated surfactants uses it successfully. When utilizing ionic and nonionic surfactants, however, this method is not relevant since their hydrophilic lipophilic balance is far less sensitive to temperature variations [119]. Commercially available ethoxylated surfactants are a blend of different oligomers, which might pose issues when the PIT method is used in industry. As a result, the concentration ranges and emulsification temperature regime must be adjusted [108].

3.2.2 Method of Emulsion Inversion Point (EIP)

The phase inversion composition or EIP entails the addition of chemicals that alter the system's hydrophilic-lipophilic balance while maintaining a steady temperature. This cause catastrophic phase inversion in the emulsion as well as changes occurs by itself, without outside assistance, in the radius of curvature of the surfactant monolayers. For instance, gradual addition of aqueous phase to organic phase results in water droplets production in oil. The formation of organic phase droplets in water and phase inversion is caused by a rise in the aqueous phase concentration above a specific threshold value [99]. In systems including different mixes of surfactants, this method was utilized to generate O/W emulsions and measure the minimum amount of the aqueous phase needed for the phase inversion [138].

3.2.3 Spontaneous Nanoemulsification

In 1878, spontaneous emulsification was first documented. A pair of liquid phases that are not equilibrated together should be considered. Sometimes, bringing them together results in a spontaneous emulsification that happens later [72]. As substances diffuse through the interface, the surface tension gradient causes interphase instability, which leads to dispersion and the formation of individual droplets. A brief decrease, to nearly zero, in interfacial tension is results in the dispersion. This is usually combined by self-efficacy to increase the surface area of the interface [139]. The first two mechanisms have to do with dispersion, or the physical rupturing of the current contact. The third process results from the creation of a heterogeneous system from a homogeneous one; toluene emulsification is a well-known example [66]. When it comes to encapsulating unstable substances like peptides, proteins, and nucleic acids-which are frequently used for pharmaceutical and medical purposes-high-energy methods, should not be used because emulsification may result in the destruction, denaturation, or loss of activity of the pharmaceuticals.

3.3 Methods for Preparation of Reverse Nanoemulsions

Methods of combining low- and high-energy are used for the preparation of reverse emulsions in the nano-sized form [85]. Reverse emulsions are extremely viscous systems and are therefore significantly more challenging to create, especially when they include significant amounts of the internal phase. When applying high energy methods, an increase in the internal phase fraction causes a rapid increase in the droplet size because the emulsion's viscosity increases. Reverse NEs can be produced using low-energy techniques with smaller droplet sizes, but there isn't much internal phase present. For instance, stirring and sonication were used to create reverse NEs that contained 20 mass percent of the aqueous phase [85]. Span 80 or poly (ethylene-co-butylene)-β-poly(ethylene oxide), KLE3729 stabilized the emulsions. The average droplet size (80–200 nm) depending on the make-up of the phases and the emulsifier concentration. By using high-pressure homogenization, reverse NEs made of physiological saline dispersed in fluorocarbons were created [38]. Droplet sizes in the NEs ranged from 30 to 100 nm, and less than 5 mass% of them were in the aqueous phase. While the second stage of the process reduced the size of large droplets to 200–230 nm, this method does not allow for the production of exclusively small droplets (during low energy emulsification). Since there were far more nanodroplets than big droplets, the specific surface area of the interface in such emulsions increased greatly.

4 Pharmaceutical and Medicinal Activities and Applications of Nanoemulsions

Table 1 summarizes several plant extracts, EOs, and natural compounds that have been shown to have significant pharmaceutical and biomedical activities such as antioxidant, antimicrobial, antiviral, anti-inflammatory, anticancer, hepatoprotective, antipyretic, and acetylcholinesterase (AChE) activity (Fig. 2).

4.1 Antioxidant Activity

The EO derived from one of *Citrus* species (*C. medica* L. var. sarcodactylis) and its nanoemulsified scavenging potentiality against the hydroxyl radicals were reported to be increased with concentrations increasing. At the two concentrations 0.06 and 0.48 mg/ml, the EO radicals scavenging activity (9.6 and 26.1%) was strongly lower than its nanoemulsified (35.7 and 58.7%), respectively. By the same, EO and its nanoemulsified were reported to have diphenyl picrylhydrazyl (DPPH) scavenging abilities increased with the concentration increasing. At the two concentrations 0.12 and 0.48 mg/mL of the EO and its nanoemulsified, scavenging activity of the EO nanoemulsified (51.6 and 72.4%) was described to be higher than that of EO itself (30.5 and 44.3%), respectively [73].

The nanoemulsion of the etanolic extract of *Brazalian propolis* was documented to have poor antioxidant activity through tests of 2,2′-azino-*bis*(3-ethylbenzothiazoline-6-sulphonic acid) (ABTS) and 2,2-diphenyl-1-picrylhydrazyl (DPPH) assays. The free radical scavenging against ABTS and DPPH were EC_{50} 0.004 and 0.023, respectively [114]. The results of Sedaghat Doost and his team work described that the nanoemulsion of the thymol have strong antioxidant potentiality via DPPH, FRAP,

Table 1 List of plants reported as sources of oil nanoemulsions and their biological activities

Plant name	Family	Biological activity
Anethum graveolens (dill)	Apiaceae	Anticancer
Apium graveolens (celery)	Apiaceae	Anticancer
Araucaria bidiwillii	Araucariaceae	Anti-inflammatory, antipyretic
Araucaria heterophylla	Araucariaceae	Anti-inflammatory
Bergamot orange (Bergamot)	Rutaceae	Anticancer, antimicrobial
Carum carvi	Apiaceae	Anticancer
Casearia sylvestris	Salicaceae	Anticancer
Chamomile	Asteraceae	Hepatoprotective
Cinnamomum zeylanicum (Cinnamon)	Lauraceae	Anti-inflammatory, antimicrobial
Citrus limon (Lemon)	Rutaceae	Anticancer, anti-inflammatory
Citrus medica L. var. sarcodactylis	Rutaceae	Antioxidant, antimicrobial
Cocos nucifera (coconut)	Arecaceae	Anti-inflammatory
Cuminum cyminum (cumin)	Apiaceae	Anticancer, hepatoprotective
Curcuma longa (Turmeric)	Zingiberaceae	Anti-inflammatory
Cymbopogon flexuosus (Lemongrass)	Poaceae	Antimicrobial
Cymbopogon martinii (palmarosa)	Poaceae	Antimicrobial
Deverra tortuosa	Apiaceae	Anti-inflammatory
Deverra triradiate	Apiaceae	Anti-inflammatory
Ferula gummosa	Apiaceae	Anticancer
Foeniculum vulgare (Fennel)	Apiaceae	Hepatoprotective
Heracleum persicum	Apiaceae	Anticancer
Jasminum grandiflorum	Oleaceae	Antiviral
Jasminum humile	Oleaceae	Antiviral
Mentha × piperita (Mentha)	Lamiaceae	Anticancer
Nigella sativa	Ranunculaceae	Anticancer
Origanum vulgare	Lamiaceae	Anticancer, antimicrobial
Pinus koraiensis	Pinaceae	Anticancer
Pinus morrisonicola	Pinaceae	Anticancer
Rosmarinus officinalis	Lamiaceae	Anti-inflammatory
Salvia rosmarinus (rosemary)	Lamiaceae	Hepatoprotective
Syzygium aromaticum (clove)	Myrtaceae	Antimicrobial
Syzygium aroticum	Myrtaceae	Anticancer
Thymus vulgaris (thyme)	Lamiaceae	Antimicrobial
Zataria multifora	Lamiaceae	Anticancer
Zingiber officinale (Ginger)	Zingiberaceae	Anti-inflammatory
Zingiber ottensii	Zingiberaceae	Anticancer

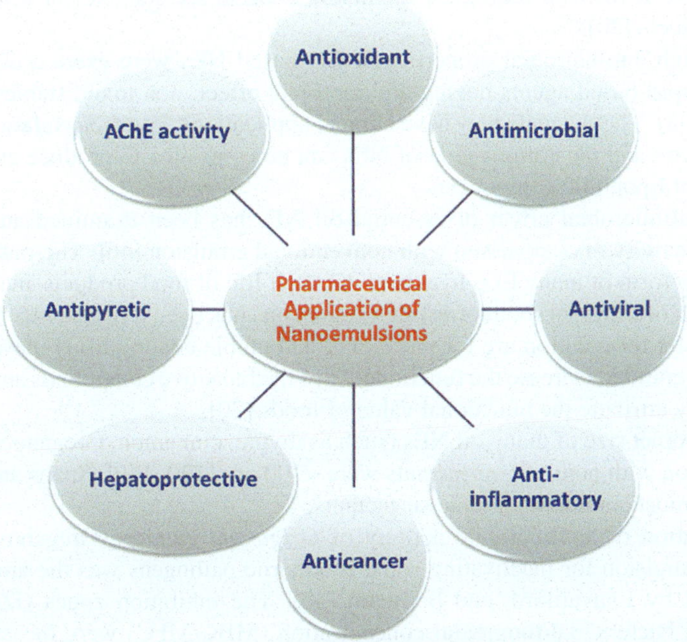

Fig. 2 Brief systematic diagram of the pharmaceutical application of NEs

and CUPRAC assays than the thymol. The *Spirulina* sp. LEB18 develop lipid-based nanoemulsions were described to show significant antioxidant activity using ABTS and DPPH assays with inhibition % of (12.14–21.31%) and (11.01–23.83%) respectively [37].

4.2 Antimicrobial Activity

EOs have been considered as a valid alternative therapy against microbial infections, as natural antimicrobial additives and in preservation applications with high effectiveness for food safety [20, 46].

Nanoemulsions NEs display unique properties that can solve the issue of low solubility of EOs in water as a technological limitation in industry. Nanoemulsions are one of drug delivery nanometric systems that are used for bioactive natural products. These systems are characterized by their potency of ameliorating biological activity of hydrophobic components (such as EOs) and by their stability [46, 57]. With the help of NE technique, the hydrophobic components are distributed in the aqueous phase uniformly and the treat the low dissolvability of water and interactive binding problems. NEs form increases the interactions between the surface areas of active ingredients of oil with biological membranes; these are suitable for utilization in food products [20].

The high antimicrobial properties of EO based NEs were exerted due to non-phospholipid-based agents having antimicrobial effect, non-toxic, stable and inexpensive [98]. These agents may have clinical applications. For encapsulation of EOs, the structure and the composition of NEs can be controlled to produce an effective delivery of lipophilic components.

The antimicrobial effect of essential oil NEs has been examined and showed superior activity in comparison with conventional emulsion [46]. The possibility of using NEs form of many EOs to extend the shelf life of food products and to inhibit the growth of different foodborne microorganisms has been reported [46, 57]. Other uses of NEs formulation are for packaging and edible coating films' development. Their use could to increase the functionality of the bioactive compounds and enhance the quality attribute the nutritional value of foods [20].

The droplet size of many oil NEs (such as thyme, cinnamon, Oregano and clove) formulation with nonionic surfactants were <100 nm [120, 140]. Spans and Tweens were the most common nonionic surfactants.

Evaluation of antimicrobial activity of *Origanum vulgare* (Oregano) essential oil nanoemulsion for inactivating some foodborne pathogens was the aim of work published by Enayatifard, and his team [46]. The inhibition zones (IZ) and the minimum bactericidal/fungicidal concentration (MBC/MFC) were measured. The minimum inhibitory concentration (MIC) against the tested fungi and the tested bacteria was 0.156–0.078 and 0.312–0.156 mg/mL, respectively. A significant antimicrobial activity against *Aspergillus niger*, *Candida albicans* and *Staphylococcus aureus* has been displayed by the nanoemulsion form (IZ 22.3–8.7 mm and MBC/MFC 5.00–0.31 mg/mL) [46]. An appropriate emulsifier in the nanoemulsion form of Organo EO (1%) has been exhibited significant activity against foodborne pathogens.

A lower concentration of antimicrobial component (such as essential oil) in the aqueous phase could be achieved by suitable delivery system [46].

Another interesting study on the antimicrobial activity of EO loaded into NE using the microfluidization method that enhanced the activity of EO loaded into NE (compared with sonication and coarse emulsion techniques). The impact of the production method on the antimicrobial activity of palmarosa, thyme, or clove-loaded nanoemulsions and lemongrass EO NE, were reported. Microfluidization method enhanced the activity of EO of clove or lemongrass loaded into NE respect to the coarse emulsion, sonication compromised its bioactivity. A complete loss of lemongrass EO antimicrobial action was exerted with the amplitude and time of the ultrasound fabrication process [110].

High bactericidal activity of lemongrass oil-loaded nanoemulsion against *E. coli* after 30 min of contact time was reported (4.1 log-reductions) compared with palmarosa- (3.9), thyme- (2.8), or clove- (3.6) loaded nanoemulsions.

Chang et al. reported how large molar volume substances, such as medium-chain triglycerides or vegetable oil (corn oil), defined as "ripening inhibitors", reduce the antimicrobial activity of the thyme EO loaded into NE [28].

Nanoemulsions NEs form could be used as delivery systems and vehicle for EOs such as nanoemulsions containing marjoram, geranium or tea tree EOs [110]. NEs were considered able to increase the bioavailability and have been proved to be an optimal vehicle to deliver active substances. *The need to elucidate* the *mechanisms* for the disruption of biological *membranes* by lipid nanoparticles has been increased to lower the toxicity issues [97, 110].

A NE of *Thymus vulgaris* (thyme oil, TH-EO), *Cymbopogon flexuosus* (lemongrass oil, LG-EO) or *Salvia officinalis* (sage oil, SG-EO) were prepared containing tween 80 (3% v/v), glycerol and alginate dispersions for enhancing encapsulation of EO to obtaining a film [6]. This functional edible film (with high content of thymol) exhibited strong inhibitory effect against inoculated *E. coli* [70]. By hydrophobic interactions, the molecules of thymol were suggested to attach to the microorganisms' membrane proteins and thereby altering the membrane permeability. This may have a role to protect and help in preservation of food products [6]. TH-EO nanoemulsions showed the highest value of whiteness index and SG-EO nanoemulsions showed the lowest value. Strong antimicrobial effect (4.71 Log reductions after 12 h) against *Escherichia coli* was reported for the edible films containing *Thymus vulgaris* essential oil TH-EO [6].

Citral and carvone (Fig. 3) nanoemulsions NEs were reported to enhance antibacterial efficacy through the perturbation of bacterial membrane and reduce oil loss [54]. Another study mentioned the use of ultrasonic emulsification for nanoemulsion droplets obtained in a citral-in-water system [74].

During the forming of an oil/water emulsion, the hydrophilic surfactants have a lesser affinity than the lipophilic surfactant to the dispersed droplets in the emulsion [74]. The mixed surfactants lipophilic-hydrophilic balance values are a critical factor in emulsion droplet formation [74]. These considerations can be based on the required

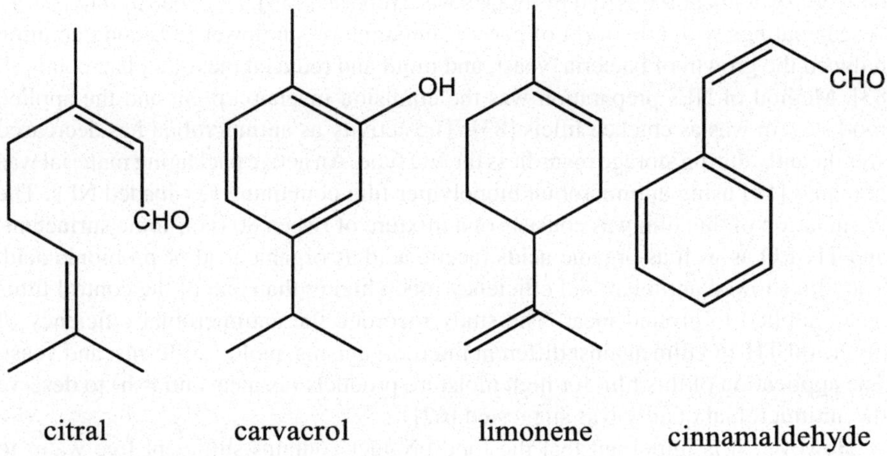

Fig. 3 Compounds with active antimicrobial NEs

function of delivery systems of EOs as antimicrobials in the different industrial fields [74].

BEO is reported to have noteworthy biological activity besides being an expensive and valuable well-known fragrance for cosmetic industry. It is of interest especially as antifungal and antibacterial activity [79]. The prepared NEs of bergamot EO (BEO) and its refined fractions documented enhanced antimicrobial activity with high degree of activity was observed towards *Saccharomyces cerevisiae*, *Lactobacillus delbrueckii* and *Escherichia coli* [79]. The NEs of carvone and citral also showed killing of bacteria on wound dressing cotton gauze bandages [54].

The initial composition for BEO recorded higher antimicrobial activity than its light fraction [79]. The terpenes of γ-terpinene, ß-pinene, and D-limonene were the main constituents of initial BEO and the monoterpenes; linalyl acetate and linalool were the main volatile compounds in the heavy fraction.

However, the role of emulsifiers could not exclude. It might affect the mechanism of action of BEO, by promoting interaction and mass transfer with the microbial cells and exerting an intrinsic antimicrobial activity [145]

The antimicrobial activity of nanoemulsions of mandarin, lemongrass, oregano and thyme EOs and their components on the survival of *Escherichia coli* O157, *Campylobacter jejuni*, *Listeria monocytogenes*, *Staphylococcus aureus* and *Bacillus cereus* in vitro and in food systems was provided by Guerra-Rosas et al. [53].

The *long-term* stability and storage period (*56 days*) of NEs containing EOs were assessed with by tween 80 (non-ionic surfactant) and high methoxyl pectin. Pectin has been added to these EOs as gelling or as a thickening agent to improve their stability [53, 83].

The lemongrass-pectin NEs had higher antimicrobial activity (5.9 log reductions) against *E. coli* and had the smallest droplet size (11 ± 1 nm).

The potential and effectiveness of nanoemulsions NEs to be used as antimicrobial delivery systems in foods and beverages were reported [53]. NEs loaded into pectin-based coatings with NEs form of garlic, cinnamon or sunflower EOs and curcumin reduced the growth of bacteria, yeast, and mold and reduced the total plate count [4, 83]. Method of NEs preparation was the emulsion inversion point and the applied food system was as chicken fillets [83]. The activity as antimicrobial has decreased significantly during storage regardless the EO type. An active packaging material was prepared [14] using antimicrobial biopolymer film containing EO-loaded NEs. The formulation of this film was consists of a mixture of tween 80 (non-ionic surfactant) and TH-EO as well as organic acids (acetic acid as organic acid or propionic acid. The film showed antimicrobial efficiency much higher than that of the control films when applied to ground meat. The study recorded the antimicrobial efficiency of thyme oil TH-EO film against different microorganisms, mold, coliforms, and yeast. The application of this film for high moisture products (as meat and fish) to deserve the antimicrobial quality was suggested [62].

However, it is important that the food product contains sufficient free water to induce spontaneous microemulsion formation when the films come into contact with them [14].

The desirable physicochemical characteristics of the product did not change with the application of NEs into chicken meat. The emulsion inversion point method has been used to form NEs containing a mixture of cinnamon and clove oil and also used to create NEs with antimicrobial properties from a mixture of coconut and cinnamon oils, which were found to act as a good food preservative due to their antimicrobial activity [140]. The activity of NEs was higher against four tested microorganisms of *Bacillus subtilis, S. aureus, E. coli,* and *Salmonella typhimurium,* even at far lower concentrations. Another method (low-energy method) was used for formulation of citral/chitosan oil NEs with potent antimicrobial activity against *Rhizopus stolonifer*, *Aspergillus niger*, and *Erwinia carotovora* [80].

Another work aims at investigating the effect of the nanoemulsion delivery systems on the antimicrobial activity of different EO components (carvacrol, limonene and cinnamaldehyde) (Fig. 3). Different formulations were used for the fabrication of the four different nanoemulsion systems tested for the delivery of the antimicrobial compounds [43].

Representative examples of studies on plant-based antimicrobial NEs are carried out by Donsì et al. [43] and Salvia-Trujillo et al. [110]. The strong antimicrobial activity of nano-emulsified EOs (as plant-based antimicrobials) and other phytochemicals, such as eugenol, curcumin, carvacrol, thymol, cinnamaldehyde, and limonene have been demonstrated in various food systems against a broad spectrum of spoilage pathogens [43, 110]. A main composition of clove essential oil, eugenol has been documented as antimicrobial against several pathogenic bacteria including methicillin-resistant *Staphylococcus aureusand S. epidermidis* [129]. Eugenol *inhibited Candida cells' adhesion capacity as antifungal agent*.

A simple high shear mixer has been used to form cinnamon bark, thymol and eugenol oil NEs stabilized by a mixture of food-grade surfactants (lecithin and lauric arginate), which exhibited good antibacterial effects against *Escherichia coli* O157:H7, *Listeria monocytogenes,* or *Salmonella enterica* [75]. However, MICs of thyme or eugenol oil in microemulsions were much higher than that of free antimicrobials. The methods of high-pressure homogenization and sonication were used to prepare cinnamon oil nanoemulsion-loaded pullulan coatings in the food system of fresh strawberries. The formulation exhibited strong antimicrobial behavior against bacteria and molds during room storage. Enhanced shelf-life was another advantage of NEs and a decreased loss of firmness, fruit mass, and total soluble solids of strawberries were achieved [35].

Intra-nasal formulation of chitosan coated NEs of nanoemulsioned EOs were effective against multi-drug resistant bacterial strains. Methicillin-susceptible *S. aureus* and carbapenem-resistant *Klebsiella pneumoniae* and *Acinetobacter baumannii* were example of Gram-negative multi-drug resistant strains [104].

The antibacterial activity of the EO derived from one of *Citrus* species (*C. medica* L. var. sarcodactylis; fingered citron) and its nanoemulsified was documented against *S. aureus* EO derived from. The reported data exhibited that the nanoemulsified completely inhibit the growth of *S. aureus* within 48 h, while the pure EO exhibited lower inhibition effects. After 48 h, the EO nanoemulsified exhibited higher growth inhibition with 8 Log cfu/g than of those of the pure EO 8.4 Log cfu/g [73].

The nanoemulsion form of the Brazilian *Propolis* ethanolic extract was reported to inhibit *S. aureus*, *S. saprophyticus*, *L. monocytogenes*, and *E. faecalis* higher than the extract itself with the same MIC at 6.2 mg/mL [114].

4.3 Antiviral Activity

Viruses are a small obligate intracellular parasite. They are non-living outside a host cell and spread outside cells via extracellular virions. The interactions with host cell mostly include receptor-ligand interaction [144]. NEs of some natural products have shown as an immense promise for restricting pathogenic virus's entrance into cell and interfacing with their pathogenicity [129]. They may mask viral structures (necessary for entry into host cell or adsorption) or directly inhibit the infection of viruses [18, 19]. Nano-loaded natural products are one of traditional medicines' applications and have also been listed as a major treatment strategy to combat the global pandemic and aid the recovery of patients with COVID-19 [107, 113]. Secondary metabolites and other substances are produced by many plants to protect themselves from invading pathogens, such as viruses. Some NEs of EOs target these type of pathogens and some act as optimal vehicle to deliver active antivirals and vaccines [133, 135, 140, 142]. Curcumin's potential is one example of the secondary metabolites that has investigated to reverse to reverse pathways in COVID-19 infection and brosis-associated pulmonary edema [107]. Nanotechnological carriers such as nanoemulsions have been have developed by researchers to load curcumin, enhance its solubility, modify it to be more easily transported through biofilms, and protect it from metabolic and chemical degradation. In the current COVID-19 pandemic, the form of polymer nanoparticles (NanocurcTM), available on the market, is one of curcumin-based nanotechnology products, which show initial benign therapeutic effect [42, 77, 121]. Many phenolic aromatic compounds obtained from clove oil such as limonene, β-pinene, thymol, and eugenol (Fig. 4) are demonstrated antiviral activity. Eugenol with acyclovir is synergistically inhibiting viruses by limiting viral infection and preventing viral replication [129]. Also, eugenol and its homologues (such as β-caryophyllene and eugenyl acetate) were believed to be the main bioactivity compositions of clove EO against *Burkholderia gladioli* [127]. Eugenol caused a complete remission of HSV lesions when topically applied to the infected area with the virus [76, 127]. Clove oil NEs have shown good inhibitory potential against Potato virus-Y [115]. The use of EOs nanoemulsion needed extensive research before their dominant uses in horticulture and agriculture for effective management of different plant virus diseases [115].

A suggested promising solution for corona therapy was provided using NEs formulations consists of antiviral EO, antiviral substances/drugs and antiviral surfactant/cosurfactant. NEs can be formulated as creams, foams, sprays and liquids administered by intravenous, transdermal, oral, enteric, nasal, and topical routes for treating or preventing the COVID-19 infection [136].

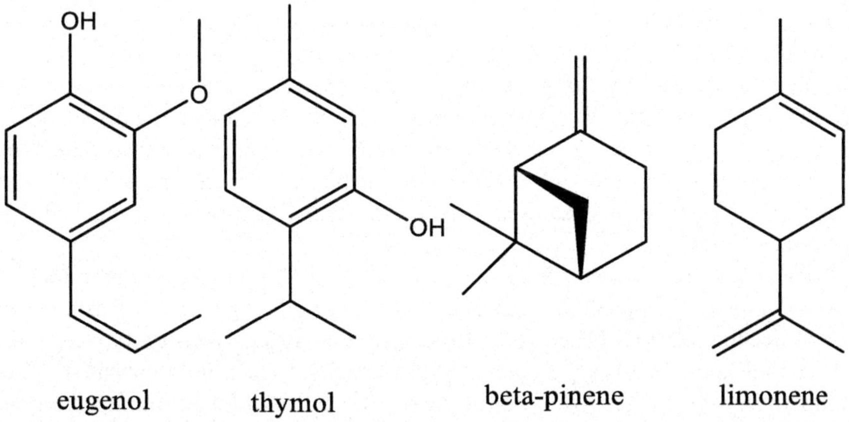

Fig. 4 Compounds with active antiviral NEs

For severe acute respiratory syndrome-corona virus (SARS-CoV), lignoids and terpenoids have been proven as effective antiviral agents [81]. EO's role in vaccine and antiviral drug delivery formulation has been demonstrated with a special focus on SARS-CoV-2 infection by Franklyne et al. [49]. A number of antiviral NEs systems of EOs have been formulated for the proven antiviral EOs [49]. EO has recognized as a potent reservoir with many bioactive compounds and ample antiviral, over the past few decades, and as a micellar structures, these EO bioactive compounds are absorbed in the intestine. The efficacy of the active ingredients in the EOs is significantly improved by the emulsification process [49]. The nanoformulation of EOs is characterized by their kinetic and thermodynamic stability, nonirritant, and nontoxicity property. Moreover the nanoforms with EOs could improve their efficacy and delivery, and availability and also deliver drugs that are needed to be administered (in both water-insoluble and water-soluble forms).

Cinnamaldehyde, pinene, linalool, allyl isothiocyanate, eugenol, thymol, terpineol, and carvacrol, are examples of plant volatiles antipathogenic agents [81, 129]. Further studies by researchers in industry and academy should be focused on eugenol, terpineol, thymol, and EOs from clove, oregano, thyme, and cinnamon, which could be applied in the form of NEs as coatings in many pharmaceutical applications [81]. In the case of human noroviruses such as feline calicivirus (FCV); oregano, thymol and clove EOs have the lower IC_{50}. For murine norovirus, green tea, grape extracts, thymol and clove EOs are effective agents. In the case of hepatitis A virus; cinnamaldehyde, thymol, oregano, zataria oils and carrageenan have high antiviral activity. *Leptospermum scoparium* and *Chamomilla* EOs are the most effective antiviral against the herpes viruses HSV-1 and HSV-2, respectively [81, 129]. NE formulations were prepared for both oleaceous plants EOs (*Jasminum humile* and *J. grandiflorum*), then the NE formula and pure EOs were investigated for the potency of their nanoemulsion preparations as antivirals and to investigate the possible activity enhancement by NE formula preparation [78]. The assay of methylthiazolyl-diphenyl tetrazolium bromide (MTT) was used to evaluate their antiviral activity against hepatitis A and herpes simplex viruses (types 1 and 2). The pure EOs showed weak antiviral activities. However, the prepared NEs of the two EOs (in a concentration-dependent manner) induced significant antiviral activity against the tested viruses. Several constituents in the chemical composition of both EOs were reported to have antiviral activity, such as acyclic alkanes (straight chain hydrocarbons) e.g., nonacosane and tetracosane [29, 34] as well as nerolidol and linalool [33, 106].

Topical administration of NEs has also led to many successful antiviral drug delivery applications [56]. A study recorded the promising results of using penciclovir–lavender oil NEs-based gel as topical antiviral drug. The activity was particularly of anti-herpes virus type against herpes labialis disease and this was suggested due to an improvement of skin permeation has occurred to the protonated surfaces [56]. While curcumin-loaded NEs have exhibited efficacy against the infection of human papillomavirus [26]. Recently, a study by Chelliah et al. [30] reported the purpose of EOs-based NEs in sanitizer formulations was to enhance the efficacy as antiviral. The stability of the NEs based ultrasonication was enhanced, which was applied to promote food safety and illustrate cellular damage of pathogens.

4.4 Anti-inflammatory Activity

Inflammation has an important part in the pathologic process of several chronic illness like diabetes, rheumatoid arthritis, and cancer [31, 143]. If the inflammation is severe, injured nerves cause transmission of pain signals from neurons to brain [58, 116]. Pain is caused by the liberation of inflammatory mediators from injured tissues [105]. Steroidal and non-steroidal drugs, classified as painkillers, act as anti-inflammatory drugs, and are used for relieving the related pains, with minimal effectiveness and side effects [69, 131]. Consequently, efforts to evolve natural nanodrugs acting as major source for innovative therapeutics are recently attracting more attention [47, 124].

EOs are lipotropic compounds which displays little interactivity with water. This concern was regarded as a technological provocation. In pharmaceutical production, lipotropic compounds were encapsulated by colloidal transporters so that they are able to disperse in water [82].

Rosmarinus officinalis L. EO (EORO) was reports to have compounds acting as anti-inflammatory. The study aimed to estimate the nano-emulsions that contain R. officinalis L. EO as anti-inflammatory agents both in vitro and in vivo. This research was fulfilled quantitatively via cellular antioxidant activity (CCA), DPPH and ABTS, cellular viability and anti-inflammatory activity in zebrafish, and verification of nitric oxide production. EORO's were analyzed using gas chromatography coupled with mass spectrometry (GC–MS) analysis. The analysis revealed the presence of 1,8-cineol and camphor in considerable amounts.

NOEROs representing medium size lesser than 200 nm, were obtained via a low-energy technique. The research proved that the nano emulsions were nontoxic towards macrophages, as well; they exhibited antioxidant activity and potency in proliferation of viable fibroblasts. Moreover, NEs enhanced the activity of EOs as anti-inflammatory through applying immunomodulatory activity via hindering pro-inflammatory mediator nitric oxide production. NECHA in zebrafish resulted in the confirmation of the availability of terpenic compounds, camphor 1,8-cineole, and alpha-pinene at the target sites, preventing the inflammatory process in the studied animal species [27].

A pretreatment with *Araucaria heterophylla* resin EO (100, 200 mg/kg and indomethacin) remarkably decreased the induced edema which extended to 32.0%, 30.3%, and 39.0%; respectively from correlated basal value at the second hour, hence proving anti-inflammatory effect of *A. heterophylla* EO and its nano emulsion. At the second hour, *A. heterophylla* EO (100 mg/kg) protective activity only appeared together with indomethacin. At the fourth hour, the edema was reduced by 45.7%, 52.9%, 51.1% and 35.5%, and the rat paw was protected by essential oil of *A. heterophylla* at 100 and 200 mg/kg, nano emulsion 200 mg/kg and indomethacin, respectively. A remarkable decrease in edema was reported as a cause of EO effect at two concentrations (100, 200 mg/kg) or the EO nanoemulsion at concentration (200 mg/kg) comparable with the reference drug (indomethacin) [45].

The EOs of clove (*Syzygium aromaticum*) and Cinnamon (*Cinnamomum zeylanicum*) are both natural substances obtained from plants with great medicinal importance, and a broad biological activities. Moreover, the nano emulsions-based gels (NG) were used extensively to enhance topical drug delivery and efficacy. Their droplets with sizes 28 ± 6 nm and 12 ± 3 nm were developed by using carboxymethylcellulose (3.5% w/v), after that they were gelified and utilized as topical pre-treatment prior to pain and inflammation induction in case of acute and chronic analgesic experimental researches.

The main compounds of Cinnamon EO and Clove EO, cinnamaldehyde and eugenol were reported to have anti-inflammatory activities alike cyclooxygenase (COX) inhibitors, counting indomethacin and celecoxib [39, 71]. As well, they found that cinnamaldehyde stops chondrocyte inflammation induced by lipopolysaccharide [32]. Both clove-NG and cinnamon-NG block the mechanism of peripheral pain, and they are unable to act on central pain via hot plate test. When clove EO is topically applied, and cinnamon EO is intraperitoneally administered, they remarkably reduce acute and chronic pain in formalin tests [40, 41].

Cinnamon-NG stopped the pain induced by formalin in both stages, showing the collaboration of both central and peripheral mediated mechanisms, hence, the results of formalin and paw edema approved the remarkable anti-inflammatory and anti-nociceptive activities of nanogels formulations [48].

Nano emulsions of EOs for both *Deverra triradiata* Hochst and *Deverra tortuosa* DC. showed remarkable anti-inflammatory and antioxidants activities along with remarkable elevation in hydroxyproline levels and growth factors. They also exhibited complete re-epithelialization accompanying activated hair follicles and plentiful collagen fibers, causing boosting wound healing medications used as topical application [60].

Production of Clove EO nano topical delivery systems (CEO), aimed to raising its anti-inflammatory action. Hence, in this research, oil nanoparticulate systems of controlled release, which are called nanofibers (NFs) and nanoemulgel were originated and integrated in the hydrogel matrix of diverse biopolymers—chitosan, gum acacia and guar gum—to prepare nanoemulsion-based nanoemulgel. Moreover, the nanoemulsion-based NFs have been developed by the electrospinning technique utilizing a polymer of polyvinyl alcohol (PVA). Substantial ex vivo, in vitro, and in vivo assessments of the preceding formulae were administered. Twice topical application of CEO-NE-based NEG, and single topical application of CEO-NE-based NFs activated a remarkable anti-inflammatory activity opposing mouse skin inflammation model induced by croton oil alike the control pure CEO. Furthermore, skin irritation test carried in vivo, for the systems applied topically showed their cutaneous safety profile. These motivating preclinical data give an auspicious and safe regimen for topical treatment used as inflammatory drugs alternative to marketed nonsteroidal anti-inflammatory ones, showing side effects [16].

Formulations of EOs-based and coconut oil nano emulgels were tested for their anti-arthritic activity. Nano-emulsions of ginger, coconut oil, lemon, and turmeric have been formulated via spontaneous emulsification, and attained when adding 1% of carboxymethylcellulose and Carbopol.

Assessments of anti-arthritic activity proved that nanoemulgels of lemon and ginger showed elevated and alike percentage inhibition markings between days 7–28. Only nanoemulgel of ginger got to values of basal paw on the 28th day. Investigation of deviation proved statistically remarkable differences ($p < 0.05$) for the 7th day results between Voltaren® and the turmeric, coconut groups, and lemon. On the other hand, on day 21, there were remarkable differences ($p < 0.05$) also between coconut groups and ginger and Voltaren group. The nanoemulgel of ginger was most potent in curing rheumatoid arthritis. The outcome of this research proved the potency of nanoemulgels EOs of ginger and lemon in curing rheumatoid arthritis in animal models anticipating it would be a promising support for additional research investigating its efficacy on humans [93].

Abdelhameed et al. [3] reported that the EOs nanoemulsion extracted from *Araucaria bidiwillii* shoots exhibit remarkable anti-inflammatory activity in case of oral (at concentration of 50 and 100 mg/kg) and topical (5% in soyabean oil) mould of rat paw edema induced by carrageenan. They discovered that this nanoemulsion exhibit a remarkable anti-inflammation by the inflammatory cytokines NO, (IL-1β and IL8), and PGE2 besides the immunohistochemical and histopathological evaluations of MMP-9 and NF-κβ levels in paw tissues.

4.5 Anticancer Activity

Cancer has been considered as one of the most serious challenges conflicting humanity as it bring along several related issues, as lacking the power to prevent diseases, vulnerability of different treatments, deficiency in immunity and high death rate [2]. In their attempts to eradicate this fatal disease, scientists had developed synthetic and semisynthetic treatments to inhibit cell malignant division. Meanwhile, the serious side effects of conventional therapies acted as an important motivator for scientists to investigate alternative natural sources of drugs [10]. EOs are condensed hydrophobic liquid comprising chemical compounds derived from plant source that are prepared by several techniques such as steam distillation and solvent extraction [88].

Anticancer effect of EOs was done in 1960, followed by successive studies against several cancers as melanoma, oral and leukemia [36]. NEs comprise tiny droplet size where they become kinetically consistent colloidal systems. They have improved functional compared with conventional emulsions. Their composition and structure are liable to control for encapsulation and hence effectual delivery of bioactive lipophobic drugs [20].

Cancer prevention, detection, and treatment have all been studied in relation to NEs. Procedures that could target tumors directly and make cancer detection and therapy non-incursive have been proposed and tested. The present-day drug technologies are adjusted to reduce toxicity, and if possible, a tenfold increasing efficacy if the drug is administered without targeting. The potential to target cancer cells alone may manifest to stop therapy side effects such immune system harm and the loss of rapidly multiplying cells. The folate receptor has emerged as a promising target for both cancer diagnosis and treatment. Many different tumour forms have a very high level of folate receptor expression. Through a variety of mechanisms, this expression could be exploited to choose therapeutic medications that target tumour tissues [103].

Because of their variable stability and speedily disorder NEs are prepared to deliver their nonpolar contents by the specific targets. The contents in their main core are habitually expected to cause apoptotic death in several cancer cell lines like hepatocyte, colon, breast, prostate, and lung [64].

A modified MTT test, as well as cellular uptake and nuclear morphological studies, were used to determine the anticancer effects of the *Nigella sativa* L. EO nanoemulsion (NSEO-NE). The treatment significantly reduced the vitality of Michigan Cancer Foundation-7 (MCF-7) breast cancer cells. The nucleo-cytoplasmic morphological properties of NSEO-NE-treated cells covered blebbing of cell membrane, vacuolation of cytoplasmic, chromatin marginalization, and nucleus fragmentation. Hence, the results prove that NSEO-NE induced apoptosis in MCF-7 cells [101].

In his research studies, Alkhatib attempted to overcome the major adverse effects of Docetaxel (DOC), which is used to limit the multiplication of many types of tumour cells. He tested the cytotoxicity of DOC-loaded NE formulated on orange oil (DOC-NEOO). The formulations enhanced the interval of survival times of DOC-NEOO and NEOO groups. On the other hand, the cancer size in their ascetic fluid have been reduced after comparison with the DOC-water group. The formulations that incorporated NEOO improved the amount of high density lipoproteins and decreased the amount of cholesterol and triglyceride in the mice serum. Hence, incorporating the DOC into the NEOO has improved its effectiveness and lowered its effect on the heart [11].

The EO and NE of *Casearia Sylvestris* Sw. (Salicaceae) shown anticancer efficacy against A549 tumour cells, with EC50 values of 4.0 g/mL and 1.0 g/mL, respectively. According to the linear regression test, the essential oil and its NE displayed a dosage dependent pattern ($r = -0.79$, $p = 0.03$), indicating a fourfold better effectiveness of the nano-preparation [100].

Cinnamon oil NE encapsulated with vitamin D and cinnamon oil NE with particle sizes (48.96 nm) and (40.52 nm) displayed anticancer efficacy in human alveolar carcinoma cells, respectively. The formulations induced DNA damage together with comparable elevation in micronucleus frequency, arrested the cell cycle progression in G0/G1 phase, exhibited high expression of Bax, capase-3 and caspase-9 and low expression of BcL2 proteins together with reasonable elevation in apoptotic cell population along with loss of mitochondrial membrane potential [84].

The antitumor activities of frankincense EO (FEO) and ginger EO (GEO) NEs formulated using a high-pressure homogenization method with integrated antineoplastic agent, mitomycin C (MMC), revealed that NE-based EOs enhanced MMC's apoptotic activity on tumour (IC50) values of GEO-MMC and FEO-MCC NEs approximated for HeLa cells were reduced by 44.12 and 29. When compared to MMC solution, those for MCF-7 cells were reduced by 29.29 and 55.3 folds, respectively. Also, FEO-MMC NE resulted in the most breakthroughs on the HeLa cellular morphology, whereas MCF-7 cells had the most damage with segmented nuclei when subjected to high GEO-MMC NE. As a result, combining GEO NE with FEO NE considerably increased its cytotoxicity on MCF-7 and HeLa cells [7].

Mitomycin C synthesized into NEs based on EOs of garlic (GarEO) and chamomile (ChEO) decreased the cell cell viability of HeLa cervical cancer cells by 42- and 20-fold when compared to free MMC, respectively, whereas treatment with GarEO NE or GarEO-MMC NEs exhibited powerful modification in the cell membrane of the HeLa cells when followed by treatment with ChEO or ChEO-MMC NEs.. After staining with 4′,6-diamidino-2-phenyl indole, the nano-emulsion of GarEO and GarEO-MMC damage the cell by attaching to its membrane, while the ChEO or ChEO-MMC migrated within the cell membrane and damaged the nucleus rightly [12].

Nirmala and his colleagues created nanoscale-based emulsions from clove bud essential oil (*Syzygiumaroticum*) in 2019. The cytotoxicity of the formulation was evaluated using the MTT test, colony formation assay, and Annexin V-FITC assay against the thyroid cancer cell line (HTh-7), and all three techniques demonstrated apoptosis and a decrease in cancer cell growth [89].

The MTT assay was used to test the cytotoxicity of *CarumCarvi* oil nano-emulsion (CCONE) on human colon cancer HT-29 cells. The apoptotic properties of the nano-emulsion were evaluated using flow cytometry and real-time qPCR. In compared to Huvec normal cells, the nano-emulsion demonstrated a strong negative interrelationship between HT29 cancer cell viability and C CONE treatment levels (p value 0.001). The IC50 values for HT29 and Huvec were 12.5 g/ml and 50 g/ml, respectively [63]. In a study done by Afshari and other scientists, nano-emulsions prepared using *Anethumgraveolens* (dill) essential oil had dose dependent cytotoxic activity on HepG2 and HUVEC cell lines, with more predominant effect on HepG2 [126].

The work published by AlMotwaa, and his team aimed to evaluate the nano-emulsion prepared from camphor oil (CAM-NE) and Ifosfamide (IFO)-loaded in CAM-NE (CAM-IFO) as anticancer drug against MCF-7 breast and HeLa cervical cancer cell lines. The resulted preparation of CAM-IFO showed significantly remarkable cell growth inhibition and apoptotic activity in vitro after comparison with on-loaded CAM-NE and drug-free IFO. Incorporating IFO with CAM in NE has ameliorated the antineoplastic effect of both of IFO and CAM via enhancing their cellular permeability inside cancer cells [15].

A comparative study was carried by Alkhatib et al. [13], on EONE for lemon (LEM-IFO) and salvia (SAL-IFO), incorporating ifosfamide (IFO) in both formulations to test them against HeLa cervical cancers cells and MCF-7 breast cancer cells. The MTT assay, DAPI strain, and light microscopy were utilised to identify the NE formulations' anticancer activity. The IC_{50} values of SAL-IFO and LEM-IFO, tested against HeLa cells, were 0.141 ± 0.035 and 0.165 ± 0.025 mM, respectively, while the IC_{50} of LEM-IFO and SAL-IFO applied into the MCF-7 cells were 0.200 ± 0.005 mM and 0.270 ± 0.025 mM, respectively. While the IC_{50} of SAL-IFO and LEM-IFO applied into the MCF-7 cells were and 0.270 ± 0.025 mM and 0.200 ± 0.005 mM respectively. According to the study results, SAL-IFO and LEM-IFO have significant apoptotic effect on the HeLa cells and MCF-7, respectively [13].

Moreover, the ultrasonicated-prepared nano-emulsion of celery oil was analyzed to investigate its cytotoxic effects on oral squamous carcinoma cells (SAS), as it showed substantial antiproliferative activities through restraining the growth of cancer cell as well as inducing apoptosis. The optimized nano-emulsion formulation's IC50 value for SAS cells was 1.4 L/mL. where the formulation reduced cell proliferation via restraining anchorage-independent cell growth through inhibiting colony formation and inducing the death of cancer cells (apoptosis) [90].

The cumin EONE has been prepared using Tween 80 non-ionic surfactant employing the technology of ultra-sonication. The cytotoxic activity was assessed via cell viability MTT, antiproliferative activity via clonogenic assay, and apoptosis via Annexin V-FITC assay. MTT assay declared an IC_{50} value at 1.5 μL/mL, where Cumin essential oil nano-emulsion treated tongue carcinoma cell line significantly [91].

As stated by Marchese and his team, the emulsification of BEO enhanced the anticancer activity when compared with free BEO, beyond any measurable variation shown in several formulations evaluated, when used as emulsifiers whey proteins, solely or in combination with modified starch, sugar esters, and a combination of glycerol monooleate with Tween 20. BEO exhibited a significant cytotoxic activity against Caco-2 cells already at a concentration of 100 mg/L [79].

As well, the essential oil base-nano emulsion prepared from *Pinus koraiensis* pinecones (PEO) inhibited tumor growth and enhanced apoptosis. Furthermore, immunohistochemical results proved that PEO nano-emulsion had the ability to inhibit MGC-803 cells proliferation via down-regulating the expression of YAP1/TEAD and its target proteins GLI2, CTGF and AREG, besides managing the HIPPO/YAP signaling pathway and its downstream signaling pathway [141].

The authors Nirmala, Durai, Anusha, and Nagarajan have formulated Mentha oil nano-emulsion (MONE) and tested it as anticancer drug using several cell culture techniques counting MTT, colony formation assay, and Annexin V apoptotic assay. The formulation showed early induction of apoptosis in anaplastic/aggressive HTh-7, suppressing the proliferation of ATC cells, and suggesting its use as potent anticancer drug [88].

Nosrat et al. [92] evaluated the anticancer activity of *Ferula gummosa* EO nano-emulsion (FEGO-NE) against HT-29 human colon cancer cells. He also targeted two major cancer survival strategies counting apoptotic suppression and angiogenesis induction in HT-29 colon cancer cells. The FEGO-NE resulted in inhibition of angiogenesis and an extra effect on the expression of antioxidant genes, where the decrease of tumor volume (69.72% in 14 days) in samples treated with FEGO-NE was reported. These results proved that FEGO-NE by several mechanisms can inhibit tumor cells and decrease effect on induced tumors in the in vivo model [92].

Origanum vulgare EONE (OENE) was synthesized and tested for its toxicity and inhibitory concentration (IC_{50}) of OENE against prostate cancer by MTT analysis. OENE showed an IC_{50} value at 13.82 μg/mL and remarkably produced recognizable morphological changes, counting cell shrinkage, cell shape reduction and cell density. Moreover, OENE remarkably reduced lipid droplet accumulation proved by studying mRNA transcripts of 3-hydroxy-3-methylglutaryl-CoA reductase (HMGCR), sterol regulatory element-binding protein (SREPB1), and fatty acid synthase (FASN). As well, there is a remarkable upregulation BAX (BCL2 associated X) and caspase 3 expressions. Hence, the study proved that OENE could be a therapeutic target for the treatment of prostate cancer and warrants in vivo studies [102].

In another study published by Bashlouei, *Heracleum persicum* oil nano-emulsion (HAE-NE) was formulated and its activity against human breast cancer cells and normal human fibroblasts foreskin was evaluated. Proliferation of cancer cells at $IC_{50} = 2.32$ μg/mL was remarkably inhibited, and cell migration occurred at 1.5 μL/mL. The HAE-NE at 1.5, 2.5 and 3.5 μg/concentration up-regulated caspase 3 and enhanced sub-G1 peak of cell cycle with no cytotoxic effects in the liver, jejunum, and kidney of mice. Hence, HAE-NE was suggested to be used as cytotoxic agent against human breast cancer [22].

Furthermore, the EO-based nano-emulsion preparation derived from *Zingiber ottensii* (ZOEO), was compared with EOs of members from the same family as cytotoxic agents against A549, MCF-7, HeLa, and K562, ZOEO exhibited the most cytotoxicity with IC_{50} of 43.37 ± 6.69, 9.77 ± 1.61, 23.25 ± 7.73, and 60.49 ± 9.41 μg/mL, respectively. ZOEO remarkably enhanced the sub-G1 populations (cell death) in cell cycle analysis as well, it induced cell apoptosis by apoptotic analysis. The synthesized formulation remarkably promoted cytotoxicity of ZOEO, specially against MCF-7 with the IC50 of 3.08 ± 2.58, 0.74 ± 0.45, 2.31 ± 0.91, and 6.45 ± 5.84 μg/mL, respectively [95].

The ultrasound nano emulsification has been used to prepare *Pinus morrisonicola* nano-emulsion (PNEO-NE). Cytotoxic activity of the nano formulation was assessed by cellular and molecular methods. The results showed a great inhibition towards cancer cells than the normal cell (HFF). As well, the noticeable up-regulated genes

(CAT, VEGF/VEGFR, Cas-3, Cas-9, and SOD) and the elevation of subG1 peaks revealed the apoptotic death in HT-29 cells. PNEO-NE suppressed ABTS and DPPH free radicals (IC_{50} 4 and 40 mg/ml). The safe structure of PNEO-NE system exhibited a remarkable effect on cancer cell-sensitive cytotoxic by inducing apoptotic mediated pathways [65].

In the work developed by Alipanah and his team, *Zataria multifora* EONE was evaluated towards breast cancer cell lines (MCF-7 and MDA-MB-468) and human melanoma (A-375). The IC_{50} values of the preparation against both cell lines were 32 (12–84), 46 (32–67), and 105 (85–131) μg/mL. The results also suggested that the ChiNPs containing *Z. multifora* EO might act as a drug supplement to suppress cancer cell proliferation (breast and melanoma cells). The relative study proved that the cytotoxic activity of the formulation was more noticeable in melanoma cells than in breast cancer cells [10].

4.6 Hepatoprotective Activity

The objective of the research carried by Mostafa et al. [86] was to encapsulate cumin seed EO (CEO) in a NE and evaluate it as hepatoprotective remedy. Preparation of S5, comprising of cumin essential oil/oleic acid; oil phase, Smix, distilled water as well as Tween 20/ethanol (2:1), expressed that the most effective pattern in-vivo and in-vitro antioxidant activity, along with significant hepatoprotective potential. After comparing rats that have oxidative stress, the results showed that CEO nano-emulsion exhibited remarkable decrease in plasma ALT and AST. The remarkably higher results of CEO nano-emulsions especially S5 were referred to its high permeation, tiny particles as well as optimum viscosity, therefore giving the noticed fruitful and extended hepatoprotective and antioxidant activities [86].

A study carried out by Al-Otaibi et al. [8], stated that the combination between Mitomycin C (MMC) and Chamomile Oil NE improved the impact of MMC in less-effective concentrations where the half-maximal inhibitory concentration (IC_{50}) has been decreased by 21.94 after comparing with MMC-nano emulsion. In-vivo research proved that mice which administered with CHAM-MMC exhibited a remarkable elevation in the median survival time (MST = 37 days) after comparing the results with the MMC-NS treated group (MST = 29.50 days). Moreover, CHAM-MMC attributed protective effect towards hepatic damage and oxidative stress induced by EAC and MMC treatment [8].

Fennel EONE (FEO-NE) was synthesized using the oil Lauroglycol™ 90 and evaluated for its hepatoprotective activity. Results of the in vivo hepatic dysfunction test in rats showed auspicious remarkable amelioration of liver function shown in ALT, ALP, AST, albumin, bilirubin, ammonia plasma levels, and malondialdehyde. The results manifested the hopeful perspective of FEO NEs in attaining medication for liver toxicity [87].

In 2022, Pascual-Mathey and his team designed nano-emulsion comprising rosemary essential oil, and tested it for its hepatoprotective activity. The prepared NEs administered orally to adult male Wistar rats with thioacetamide (TA)-made hepatic injury. In vivo results cerning the analysis of transaminases and liver illustrated that nano-emulsion formula exhibited hepatoprotective effect on rats' damaged livers. Moreover, the rosemary EONE gave results of ALT and AST activities like those found in control group. It was observed that nano-emulsified essential oil of rosemary has higher bioavailability and enhance the absorption of bioactive constituents [96].

4.7 Antipyretic Activity

The oral administration of the nanoemulsion of the *Araucaria bidiwillii* EO with two concentrations at 50 and 100 mg/kg showed antipyretic potentiality in rats via the inhibition of hyperthermia induced by intramuscular injection of brewer's yeast [3].

4.8 Acetylcholinesterase (AChE) Activity

In vivo studies of the effects of the quercetin (QC, Fig. 5) nanoemulsion (QCNE) were documented via [9]. They found that QC and QCNE attenuate the markers of oxidative stress and inflammatory increased in the Alzheimer's disease (AD). These data were supported via the immunohistochemistry results that confirmed the abilities of QC and QCNE for retracting of the $AlCl_3$ negative impact. QCNE exhibited the more potent effects in $AlCl_3$ impacts mitigating.

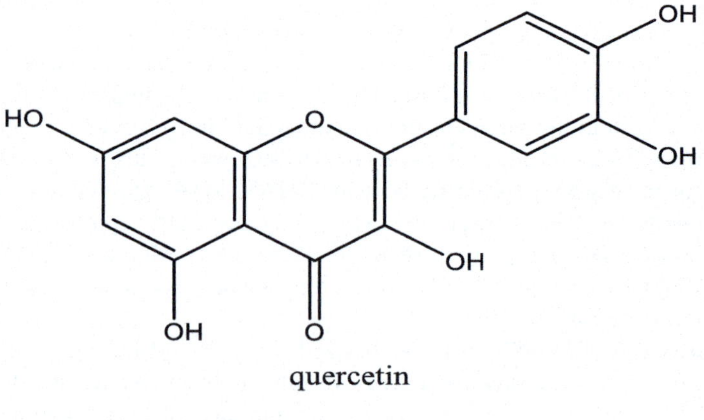

quercetin

Fig. 5 Compounds with active AChE NEs

5 Conclusion

In conclusion, because nanoemulsions have the potential to enhance medication delivery and therapeutic efficacy, they have attracted a lot of attention in the pharmaceutical industry. Nanoemulsions are a promising tool in the creation of novel medication formulations due to their stability and capacity to enhance the absorption and bioavailability of hydrophobic medicines. This chapter has focused on the most current developments and uses of nanoemulsions in pharmaceuticals, including how they can be used to treat a variety of ailments. It is anticipated that nanoemulsions will play a crucial role in the pharmaceutical sector's efforts to create new and better therapeutic formulations as research into them proceeds.

References

1. Abate L, Bachheti A, Bachheti RK, Husen A, Getachew M, Pandey DP (2021) Potential role of forest-based plants in essential oil production: an approach to cosmetic and personal health care applications. In: Non-timber forest products: food, healthcare and industrial applications, pp 1–18
2. Abd El Maksoud AI, Taher RF, Gaara AH, Abdelrazik E, Keshk OS, Elawdan KA, Morsy SE, Salah A, Khalil H (2019) Selective regulation of B-Raf dependent K-Ras/mitogen-activated protein by natural occurring multi-kinase inhibitors in cancer cells. Front Oncol 9:1220. https://doi.org/10.3389/fonc.2019.01220
3. Abdelhameed MF, Asaad GF, Ragab TIM, Ahmed RF, El Gendy AE-NG, Abd El-Rahman SS, Elgamal AM, Elshamy AI (2021) Oral and topical anti-inflammatory and antipyretic potentialities of Araucaria bidiwillii shoot essential oil and its nanoemulsion in relation to chemical composition. Molecules 26:5833. https://doi.org/10.3390/molecules26195833
4. Abdou ES, Galhoum GF, Mohamed EN (2018) Curcumin loaded nanoemulsions/pectin coatings for refrigerated chicken fillets. Food Hydrocoll 83:445–453. https://doi.org/10.1016/j.foodhyd.2018.05.026
5. Abismaıl B, Canselier JP, Wilhelm AM, Delmas H, Gourdon C (2000) Emulsification processes: on-line study by multiple light scattering measurements. Ultrason Sonochem 7:187–192
6. Acevedo-Fani A, Salvia-Trujillo L, Rojas-Graü MA, Martín-Belloso O (2015) Edible films from essential-oil-loaded nanoemulsions: physicochemical characterization and antimicrobial properties. Food Hydrocoll 47:168–177
7. Al-Otaibi WA, Alkhatib MH, Wali AN (2018) Cytotoxicity and apoptosis enhancement in breast and cervical cancer cells upon coadministration of mitomycin C and essential oils in nanoemulsion formulations. Biomed Pharmacother 106:946–955. https://doi.org/10.1016/j.biopha.2018.07.041
8. Al-Otaibi WA, Alkhatib MH, Wali AN (2019) Evaluation of antitumor activity and hepatoprotective effect of mitomycin C solubilized in chamomile oil nanoemulsion. Anticancer Agents Med Chem 19:1232–1242. https://doi.org/10.2174/1871520619666190408114732
9. Alaqeel NK, AlSheikh MH, Al-Hariri MT (2022) Quercetin nanoemulsion ameliorates neuronal dysfunction in experimental Alzheimer's disease model. Antioxidants 11:1986
10. Alipanah H, Yarian F, Rasti F, Safari M, Hatami S, Osanloo M (2022) Cytotoxic effects of chitosan nanoparticles containing Zataria multiflora essential oil against human breast and melanoma cells. Beni-Suef Univ J Basic Appl Sci 11:58. https://doi.org/10.1186/s43088-022-00241-z

11. Alkhatib MH, Alharbi SA, Mahassni SH (2017) In vivo evaluation of the anticancer activity of the docetaxel incorporated into nanoemulsion based on orange oil. Pharmacophore 8:41–47
12. Alkhatib MH, Al-Otaibi WA, Wali AN (2018) Antineoplastic activity of mitomycin C formulated in nanoemulsions-based essential oils on HeLa cervical cancer cells. Chem Biol Interact 291:72–80
13. Alkhatib MH, AlMotwaa SM, Alkreathy HM (2019) Incorporation of ifosfamide into various essential oils-based nanoemulsions ameliorates its apoptotic effect in the cancer cells. Sci Rep 9:1–10
14. Almasi L, Radi M, Amiri S, McClements DJ (2021) Fabrication and characterization of antimicrobial biopolymer films containing essential oil-loaded microemulsions or nanoemulsions. Food Hydrocoll 117:106733
15. AlMotwaa SM, Alkhatib MH, Alkreathy HM (2019) Nanoemulsion-based camphor oil carrying ifosfamide: preparation, characterization, and in vitro evaluation in cancer cells. Int J Pharm Sci Res 10:2018–2026
16. Aman RM, Abu Hashim II, Meshali MM (2020) Novel clove essential oil nanoemulgel tailored by Taguchi's model and Scaffold-based nanofibers: phytopharmaceuticals with promising potential as cyclooxygenase-2 inhibitors in external inflammation. Int J Nanomedicine 15:2171–2195. https://doi.org/10.2147/IJN.S246601
17. Ammar NM, Hassan HA, Ahmed RFG, El-Gendy AE-N, Abd-ElGawad AM, Farrag ARH, Farag MA, Elshamy AI, Afifi SM (2022) Gastro-protective effect of Artemisia sieberi essential oil against ethanol-induced ulcer in rats as revealed via biochemical, histopathological and metabolomics analysis. Biomarkers 27:247–257
18. Asaad GF, Abdelhameed MF, El Raey MA, Roshdy WH, Elgamal AM, Moemen YS (2022) Citrus clementine peels essential oil exhibited anti-SARS-CoV-2 and its modulatory effect against cytokine storm: evidence from in vitro and in silico studies. Egypt J Chem 65:419–427. https://doi.org/10.21608/EJCHEM.2022.116657.5271
19. Astani A, Schnitzler P (2014) Antiviral activity of monoterpenes beta-pinene and limonene against herpes simplex virus in vitro. Iran J Microbiol 6:149
20. Aswathanarayan JB, Vittal RR (2019) Nanoemulsions and their potential applications in food industry. Front Sustain Food Syst 3. https://doi.org/10.3389/fsufs.2019.00095
21. Bachheti RK, Abate L, Hunde Y, Tadesse MG, Bachheti A, Pandey DP, Sharma A, Husen A (2023) Essential oils from medicinal plants and their role in nanoparticles synthesis, characterization, and applications. In: Secondary metabolites from medicinal plants. CRC Press, pp 177–198
22. Bashlouei SG, Karimi E, Zareian M, Oskoueian E, Shakeri M (2022) Heracleum persicum essential oil nanoemulsion: a nanocarrier system for the delivery of promising anticancer and antioxidant bioactive agents. Antioxidants (Basel, Switzerland) 11. https://doi.org/10.3390/antiox11050831
23. Baudron V, Gurikov P, Smirnova I (2019) A continuous approach to the emulsion gelation method for the production of aerogel micro-particle. Colloids Surf A Physicochem Eng Asp 566:58–69
24. Bhavaniramya S, Vishnupriya S, Al-Aboody MS, Vijayakumar R, Baskaran D (2019) Role of essential oils in food safety: antimicrobial and antioxidant applications. Grain Oil Sci Technol 2:49–55
25. Bilia AR, Guccione C, Isacchi B, Righeschi C Firenzuoli F, Bergonzi MC (2014) Essential oils loaded in nanosystems: a developing strategy for a successful therapeutic approach. Evidence-based complement. Altern Med
26. Bonfim CMD, Monteleoni LF, Calmon MDF, Cândido NM, Provazzi PJS, Lino VDS, Rabachini T, Sichero L, Villa LL, Quintana SM, Melli PPDS (2020) Antiviral activity of curcumin-nanoemulsion associated with photodynamic therapy in vulvar cell lines transducing different variants of HPV-16. Artif Cells Nanomed Biotechnol 48(1):515–524
27. Borges RS, Keita H, Ortiz BLS, dos Santos Sampaio TI, Ferreira IM, Lima ES, de Jesus Amazonas da Silva M, Fernandes CP, de Faria Mota Oliveira AE, da Conceição EC (2018) Anti-inflammatory activity of nanoemulsions of essential oil from Rosmarinus officinalis L.: in vitro and in zebrafish studies. Inflammopharmacology 26:1057–1080

28. Chang Y, McLandsborough L, McClements DJ (2012) Physical properties and antimicrobial efficacy of thyme oil nanoemulsions: influence of ripening inhibitors. J Agric Food Chem 60:12056–12063. https://doi.org/10.1021/jf304045a
29. Chathuranga K, Weerawardhana A, Dodantenna N, Ranathunga L, Cho W-K, Ma JY, Lee J-S (2021) Inhibitory effect of Sargassum fusiforme and its components on replication of respiratory syncytial virus in vitro and in vivo. Viruses 13:548
30. Chelliah R, Jo KH, Yan P, Chen X, Jo HY, Madar IH, Sultan G, Oh DH (2023) Unravelling the sanitization potential of slightly acidic electrolyzed water combined Thymus vulgaris based nanoemulsion against foodborne pathogens and its safety assessment. Food Control 146:109527
31. Chen L, Deng H, Cui H, Fang J, Zuo Z, Deng J, Li Y, Wang X, Zhao L (2018) Inflammatory responses and inflammation-associated diseases in organs. Oncotarget 9:7204–7218. https://doi.org/10.18632/oncotarget.23208
32. Chen P, Ruan A, Zhou J, Huang L, Zhang X, Ma Y, Wang Q (2020) Cinnamic aldehyde inhibits lipopolysaccharide-induced chondrocyte inflammation and reduces cartilage degeneration by blocking the nuclear factor-kappa B signaling pathway. Front Pharmacol 11:949
33. Chiang L, Ng L, Cheng P, Chiang W, Lin C (2005) Antiviral activities of extracts and selected pure constituents of Ocimum basilicum. Clin Exp Pharmacol Physiol 32:811–816
34. Chidewe C, Castillo UF, Sem DS (2017) Structural analysis and antimicrobial activity of chromatographically separated fractions of leaves of Sesamum angustifolium (Oliv.) Engl. J Biol Act Prod Nat 7:463–474
35. Chu Y, Gao C, Liu X, Zhang N, Xu T, Feng X, Yang Y, Shen X, Tang X (2020) Improvement of storage quality of strawberries by pullulan coatings incorporated with cinnamon essential oil nanoemulsion. LWT-Food Sci Technol 122:109054
36. Cosentino S, Tuberoso CIG, Pisano B, Satta ML, Mascia V, Arzedi E, Palmas F (1999) In-vitro antimicrobial activity and chemical composition of Sardinian thymus essential oils. Lett Appl Microbiol 29:130–135
37. Costa AM, Bueno KTL, da Rosa APC, Costa JAV (2019) The antioxidant activity of nanoemulsions based on lipids and peptides from Spirulina sp. LEB18. Lwt 99:173–178
38. Courrier HM, Vandamme TF, Krafft MP (2004) Reverse water-in-fluorocarbon emulsions and microemulsions obtained with a fluorinated surfactant. Colloids Surf A Physicochem Eng Asp 244:141–148
39. Daniel AN, Sartoretto SM, Schmidt G, Caparroz-Assef SM, Bersani-Amado CA, Cuman RKN (2009) Anti-inflammatory and antinociceptive activities A of eugenol essential oil in experimental animal models. Rev Bras Farmacogn 19:212–217
40. Daroogari Sh, Parandin R, Yousofvand N, Shakibaie D (2017) The analgesic effect of topical clove oil using formalin test in male mice. J Ardabil Univ Med Sci 17:252–260
41. Dashti-Rahmatabadi MH, Vahidi Merjardi AR, Pilavaran AA, Farzan F (2009) Antinociceptive effect of cinnamon extract on formalin induced pain in rat. SSU_J 17:190–199
42. de M Ribeiro LN, Fonseca BB (2020) The role of pharmaceutical nanotechnology in the time of COVID-19 pandemic. Future Microbiol 15:1571–1582
43. Donsì F, Annunziata M, Vincensi M, Ferrari G (2012) Design of nanoemulsion-based delivery systems of natural antimicrobials: effect of the emulsifier. J Biotechnol 159:342–350
44. El-Newary SA, Afifi SM, Aly MS, Ahmed RF, El Gendy AE-NG, Abd-ElGawad AM, Farag MA, Elgamal AM, Elshamy AI (2021) Chemical profile of Launaea nudicaulis ethanolic extract and its antidiabetic effect in streptozotocin-induced rats. Molecules 26:1000
45. Elshamy AI, Ammar NM, Hassan HA, Al-Rowaily SL, Ragab TI, El Gendy AE-NG, Abd-ElGawad AM (2020) Essential oil and its nanoemulsion of Araucaria heterophylla resin: chemical characterization, anti-inflammatory, and antipyretic activities. Ind Crops Prod 148:112272. https://doi.org/10.1016/j.indcrop.2020.112272
46. Enayatifard R, Akbari J, Babaei A, Rostamkalaei SS, Hashemi SM, Hassan, Habibi E (2021) Anti-microbial potential of nano-emulsion form of essential oil obtained from aerial parts of Origanum vulgare L. as food additive. Adv Pharm Bull 11(2):327–334

47. Esmaeili F, Rajabnejhad S, Partoazar AR, Mehr SE, Faridi-Majidi R, Sahebgharani M, Syedmoradi L, Rajabnejhad MR, Amani A (2016) Anti-inflammatory effects of eugenol nanoemulsion as a topical delivery system. Pharm Dev Technol 21:887–893
48. Esmaeili F, Zahmatkeshan M, Yousefpoor Y, Alipanah H, Safari E, Osanloo M (2022) Anti-inflammatory and anti-nociceptive effects of cinnamon and clove essential oils nanogels: an in vivo study. BMC Complement Med Ther 22:143. https://doi.org/10.1186/s12906-022-03619-9
49. Franklyne JS, Gopinath PM, Mukherjee A, Chandrasekaran N (2021) Nanoemulsions: the rising star of antiviral therapeutics and nanodelivery system—current status and prospects. Curr Opin Colloid Interface Sci 54:101458. https://doi.org/10.1016/j.cocis.2021.101458
50. Godeto YG, Ayele A, Ahmed IN, Husen A, Bachheti RK (2023) Medicinal plant-based metabolites in nanoparticles synthesis and their cutting-edge applications: an overview. Secondary metabolites from medicinal plants, pp 1–34
51. Gonfa YH, Tessema FB, Gelagle AA, Getnet SD, Tadesse MG, Bachheti A, Bachheti RK (2022) Chemical compositions of essential oil from aerial parts of Cyclospermum leptophyllum and its application as antibacterial activity against some food spoilage bacteria. J Chem 2022.https://doi.org/10.1155/2022/5426050
52. Gonfa YH, Tessema FB, Tadesse MG, Bachheti A, Bachheti RK (2023) Medicinally important plant roots and their role in nanoparticles synthesis and applications. In: Secondary metabolites from medicinal plants. CRC Press, pp 227–242
53. Guerra-Rosas MI, Morales-Castro J, Cubero-Márquez MA, Salvia-Trujillo L, Martín-Belloso O (2017) Antimicrobial activity of nanoemulsions containing essential oils and high methoxyl pectin during long-term storage. Food Control 77:131–138
54. Guliani A, Verma M, Kumari A, Acharya A (2021) Retaining the 'essence' of essential oil: nanoemulsions of citral and carvone reduced oil loss and enhanced antibacterial efficacy via bacterial membrane perturbation. J Drug Deliv Sci Technol 61:102243
55. Hardt S, Schönfeld F (2007) Microfluidic technologies for miniaturized analysis systems. Springer Science & Business Media
56. Hosny KM, Sindi AM, Alkhalidi HM, Kurakula M, Alruwaili NK, Alhakamy NA, Abualsunun WA, Bakhaidar RB, Bahmdan RH, Rizg WY, Ali SA (2021) Oral gel loaded with penciclovir–lavender oil nanoemulsion to enhance bioavailability and alleviate pain associated with herpes labialis. Drug Deliv 28(1):1043–1054
57. Infante V, Darvin M, Silke L, Schleusener J, Schanzer S, Lademann J, Meinke M (2022) Cosmetic formulations with *Melaleuca alternifolia* essential oil for the improvement of photoaged skin: a double-blind, randomized, placebo-controlled clinical study. Photochem Photobiol 99:176–183
58. Jancalek R (2011) Signaling mechanisms in mirror image pain pathogenesis. Ann Neurosci 18:123–127
59. Jordens J, Gielen B, Xiouras C, Hussain MN, Stefanidis GD, Thomassen LC, Braeken L, Van Gerven T (2019) Sonocrystallisation: observations, theories and guidelines. Chem Eng Process: Process Intensif 139:130–154
60. Kamel RA, Teiama MS, El-Hagrassi AM, Elgayed SH, Khattab MA, El-Sayed EK, Ibrahim MT, Mady MS, Moharram FA (2022) Appraisal on the wound healing potential of Deverra tortuosa DC. and Deverra triradiata Hochst essential oil nanoemulsion topical preparation. Front Pharmacol 13
61. Karami Z, Zanjani MRS, Hamidi M (2019) Nanoemulsions in CNS drug delivery: recent developments, impacts and challenges. Drug Discov Today 24:1104–1115
62. Khan MR, Sadiq MB, Mehmood Z (2020) Development of edible gelatin composite films enriched with polyphenol loaded nanoemulsions as chicken meat packaging material. CYTA J Food 18:137–146
63. Khatamian N, Homayouni Tabrizi M, Ardalan P, Yadamani S, Darchini Maragheh A (2019) Synthesis of Carum carvi essential oil nanoemulsion, the cytotoxic effect, and expression of caspase 3 gene. J Food Biochem 43:e12956. https://doi.org/10.1111/jfbc.12956

64. Khatamian N, Soltani M, Shadan B, Neamati A, Tabrizi MH, Hormozi B (2021) Pinus morrisonicola needles essential oil nanoemulsions as a novel strong antioxidant and anticancer agent. Inorg Nano-Metal Chem 1–9
65. Khatamian N, Soltani M, Shadan B, Neamati A, Tabrizi MH, Hormozi B (2022) Pinus morrisonicola needles essential oil nanoemulsions as a novel strong antioxidant and anticancer agent. Inorg Nano-Metal Chem 52:253–261
66. Koroleva MY, Yurtov EV (2012) Nanoemulsions: the properties, methods of preparation and promising applications. Russ Chem Rev 81:21
67. Kumar V (2020) Ultrasonic-assisted de-agglomeration and power draw characterization of silica nanoparticles. Ultrason Sonochem 65:105061
68. Kumar M, Bishnoi RS, Shukla AK, Jain CP (2019) Techniques for formulation of nanoemulsion drug delivery system: a review. Prev Nutr Food Sci 24:225
69. Lazzaroni M, Bianchi Porro G (2004) Gastrointestinal side-effects of traditional non-steroidal anti-inflammatory drugs and new formulations. Aliment Pharmacol Ther 20:48–58
70. Li S, Sun J, Yan J, Zhang S, Shi C, McClements DJ, Liu X, Liu F (2021) Development of antibacterial nanoemulsions incorporating thyme oil: layer-by-layer self-assembly of whey protein isolate and chitosan hydrochloride. Food Chem 339:128016
71. Liao J-C, Deng J-S, Chiu C-S, Hou W-C, Huang S-S, Shie P-H, Huang G-J (2012) Anti-inflammatory activities of Cinnamomum cassia constituents in vitro and in vivo. Evidence-based complement. Altern Med
72. López-Montilla JC, Herrera-Morales PE, Pandey S, Shah DO (2002) Spontaneous emulsification: mechanisms, physicochemical aspects, modeling, and applications. J Dispers Sci Technol 23:219–268
73. Lou Z, Chen J, Yu F, Wang H, Kou X, Ma C, Zhu S (2017) The antioxidant, antibacterial, antibiofilm activity of essential oil from Citrus medica L. var. sarcodactylis and its nanoemulsion. LWT 80:371–377
74. Lu W-C, Huang D-W, Wang C-C, Yeh C-H, Tsai J-C, Huang Y-T, Li P-H (2018) Preparation, characterization, and antimicrobial activity of nanoemulsions incorporating citral essential oil. J food drug Anal 26:82–89
75. Ma Q, Davidson PM, Zhong Q (2016) Antimicrobial properties of microemulsions formulated with essential oils, soybean oil, and Tween 80. Int J Food Microbiol 226:20–25
76. Mancini DAP, Torres RP, Pinto JR, Mancini-Filho J (2009) Inhibition of DNA virus: Herpes-1 (HSV-1) in cellular culture replication, through an antioxidant treatment extracted from rosemary spice. Braz J Pharm Sci 45:127–133
77. Manoharan Y, Haridas V, Vasanthakumar KC, Muthu S, Thavoorullah FF, Shetty P (2020) Curcumin: a wonder drug as a preventive measure for COVID19 management. Indian J Clin Biochem 35:373–375
78. Mansour KA, El-Neketi M, Lahloub M-F, Elbermawi A (2022) Nanoemulsions of Jasminum humile L. and Jasminum grandiflorum L. Essential oils: an approach to enhance their cytotoxic and antiviral effects. Molecules 27:3639. https://doi.org/10.3390/molecules27113639
79. Marchese E, D'onofrio N, Balestrieri ML, Castaldo D, Ferrari G, Donsì F (2020) Bergamot essential oil nanoemulsions: antimicrobial and cytotoxic activity. Z Naturforsch C 75:279–290. https://doi.org/10.1515/znc-2019-0229
80. Marei GIK, Rabea EI, Badawy ME (2018) Preparation and characterizations of chitosan/citral nanoemulsions and their antimicrobial activity. Appl Food Biotechnol 5:69–78. https://doi.org/10.22037/afb.v5i2.19005
81. Marquez R, Zwilling J, Zambrano F, Tolosa L, Marquez ME, Venditti R, Jameel H, Gonzalez R (2022) Nanoparticles and essential oils with antiviral activity on packaging and surfaces: an overview of their selection and application. J Surfactants Deterg
82. McClements DJ, Rao J (2011) Food-grade nanoemulsions: formulation, fabrication, properties, performance, biological fate, and potential toxicity. Crit Rev Food Sci Nutr 51:285–330
83. McClements DJ, Das AK, Dhar P, Nanda PK, Chatterjee N (2021) Nanoemulsion-based technologies for delivering natural plant-based antimicrobials in foods. Front Sustain Food Syst 5:643208. https://doi.org/10.3389/fsufs.2021.643208

84. Meghani N, Patel P, Kansara K, Ranjan S, Dasgupta N, Ramalingam C, Kumar A (2018) Formulation of vitamin D encapsulated cinnamon oil nanoemulsion: its potential anti-cancerous activity in human alveolar carcinoma cells. Colloids Surf B Biointerfaces 166:349–357. https://doi.org/10.1016/j.colsurfb.2018.03.041
85. Modarres-Gheisari SMM, Gavagsaz-Ghoachani R, Malaki M, Safarpour P, Zandi M (2019) Ultrasonic nano-emulsification—a review. Ultrason Sonochem 52:88–105
86. Mostafa DM, Kassem AA, Asfour MH, Al Okbi SY, Mohamed DA, Hamed TE-S (2015) Transdermal cumin essential oil nanoemulsions with potent antioxidant and hepatoprotective activities: in-vitro and in-vivo evaluation. J Mol Liq 212:6–15
87. Mostafa DM, Abd El-Alim SH, Asfour MH, Al-Okbi SY, Mohamed DA, Hamed TE-S, Awad G (2019) Transdermal fennel essential oil nanoemulsions with promising hepatic dysfunction healing effect: in vitro and in vivo study. Pharm Dev Technol 24:729–738. https://doi.org/10.1080/10837450.2019.1584633
88. Nirmala MJ, Durai L, Anusha GS, Nagarajan R (2021) Nanoemulsion of Mentha arvensis essential oil as an anticancer agent in anaplastic thyroid cancer cells and as an antibacterial agent in Staphylococcus aureus. Bionanoscience 11:1017–1029. https://doi.org/10.1007/s12668-021-00893-8
89. Nirmala MJ, Durai L, Gopakumar V, Nagarajan R (2019) Anticancer and antibacterial effects of a clove bud essential oil-based nanoscale emulsion system. Int J Nanomed 14:6439–6450. https://doi.org/10.2147/IJN.S211047
90. Nirmala MJ, Durai L, Gopakumar V, Nagarajan R (2020) Preparation of celery essential oil-based nanoemulsion by ultrasonication and evaluation of its potential anticancer and antibacterial activity. Int J Nanomed 15:7651–7666. https://doi.org/10.2147/IJN.S252640
91. Nirmala MJ, Durai L, Rao KA, Nagarajan R (2020) Ultrasonic nanoemulsification of cuminum cyminum essential oil and its applications in medicine. Int J Nanomed 15:795–807. https://doi.org/10.2147/IJN.S230893
92. Nosrat T, Tabrizi MH, Etminan A, Irani M, Zarei B, Rahmati A (2022) In vitro and in vivo anticancer activity of Ferula gummosa essential oil nanoemulsions (FGEO-NE) for the colon cancer treatment. J Polym Environ 30:4166–4177. https://doi.org/10.1007/s10924-022-02495-1
93. Okpalaku O, Uronnachi E, Okoye E, Umeyor C, Nwakile C, Okeke T, Attama A (2022) Evaluating some essential oils-based and coconut oil nanoemulgels for the management of rheumatoid arthritis
94. Pagar KR, Darekar AB (2019) Nanoemulsion: a new concept of delivery system. Asian J Res Pharm Sci 9:39–46
95. Panyajai P, Chueahongthong F, Viriyaadhammaa N, Nirachonkul W, Tima S, Chiampanichayakul S, Anuchapreeda S, Okonogi S (2022) Anticancer activity of Zingiber ottensii essential oil and its nanoformulations. PLoS ONE 17:e0262335. https://doi.org/10.1371/journal.pone.0262335
96. Pascual-Mathey LI, Briones-Concha JA, Jiménez M, Beristain CI, Pascual-Pineda LA (2022) Elaboration of essential oil nanoemulsions of Rosemary (Rosmarinus officinalis L.) and its effect on liver injury prevention. Food Bioprod Process 134:46–55
97. Pavoni L, Perinelli DR, Bonacucina G, Cespi M, Palmieri GF (2020) An overview of micro- and nanoemulsions as vehicles for essential oils: formulation, preparation and stability. Nanomaterials 10(1):135
98. Pathania R, Khan H, Kaushik R, Khan MA (2018) Essential oil nanoemulsions and their antimicrobial and food applications. Curr Res Nutr Food Sci J 6:626–643. https://doi.org/10.12944/CRNFSJ.6.3.05
99. Perazzo A, Preziosi V, Guido S (2015) Phase inversion emulsification: current understanding and applications. Adv Colloid Interface Sci 222:581–599
100. Pereira FG, Marquete R, Oliveira-Cruz L, Quintanilha-Falcão D, Mansur E, de Lima Moreira D (2017) Cytotoxic effects of the essential oil from leaves of Casearia sylvestris Sw.(Salicaceae) and its nanoemulsion on A549 tumor cell line. Boletín Latinoam y del Caribe Plantas Med y Aromáticas 16:506–512

101. Periasamy VS, Athinarayanan J, Alshatwi AA (2016) Anticancer activity of an ultrasonic nanoemulsion formulation of Nigella sativa L. essential oil on human breast cancer cells. Ultrason Sonochem 31:449–455. https://doi.org/10.1016/j.ultsonch.2016.01.035
102. Perumalsamy H, Shanmugam R, Kim J-R, Anandapadmanaban G, Huq MA, Dua K, Chellappan DK, Yoon TH, Balusamy SR (2022) Nanoemulsion and encapsulation strategy of hydrophobic oregano essential oil increased human prostate cancer cell death via apoptosis by attenuating lipid metabolism. Bioinorg Chem Appl 2022:9569226. https://doi.org/10.1155/2022/9569226
103. Praveen Kumar G, Divya A (2015) Nanoemulsion based targeting in cancer therapeutics. Med Chem (Los Angeles) 5. https://doi.org/10.4172/2161-0444.1000275
104. Rinaldi F, Oliva A, Sabatino M, Imbriano A, Hanieh PN, Garzoli S, Mastroianni CM, De Angelis M, Miele MC, Arnaut M, Di Timoteo F, Marianecci C, Ragno R, Carafa M (2020) Antimicrobial essential oil formulation: chitosan coated nanoemulsions for nose to brain delivery. Pharmaceutics 12:678. https://doi.org/10.3390/pharmaceutics12070678
105. Ronchetti S, Migliorati G, Delfino DV (2017) Association of inflammatory mediators with pain perception. Biomed Pharmacother 96:1445–1452
106. Ryabchenko B, Tulupova E, Schmidt E, Wlcek K, Buchbauer G, Jirovetz L (2008) Investigation of anticancer and antiviral properties of selected aroma samples. Nat Prod Commun 3:1934578X0800300710
107. Saber-Moghaddam N, Salari S, Hejazi S, Amini M, Taherzadeh Z, Eslami S, Rezayat SM, Jaafari MR, Elyasi S (2021) Oral nano-curcumin formulation efficacy in management of mild to moderate hospitalized coronavirus disease-19 patients: an open label nonrandomized clinical trial. Phytother Res 35(5):2616–2623
108. Salager J-L, Forgiarini A, Marquez L, Pena A, Pizzino A, Rodriguez MP, Rondon-Gonzalez M (2004) Using emulsion inversion in industrial processes. Adv Colloid Interface Sci 108:259–272
109. Saleh I, Abd-ElGawad A, El Gendy AE-N, Abd El Aty A, Mohamed T, Kassem H, Aldosri F, Elshamy A, Hegazy M-EF (2020) Phytotoxic and antimicrobial activities of Teucrium polium and Thymus decussatus essential oils extracted using hydrodistillation and microwave-assisted techniques. Plants 9:716
110. Salvia-Trujillo L, Rojas-Graü A, Soliva-Fortuny R, Martín-Belloso O (2015) Physicochemical characterization and antimicrobial activity of food-grade emulsions and nanoemulsions incorporating essential oils. Food Hydrocoll 43:547–556
111. Sattari A, Hanafizadeh P, Hoorfar M (2020) Multiphase flow in microfluidics: from droplets and bubbles to the encapsulated structures. Adv Colloid Interface Sci 282:102208
112. Savun-Hekimoğlu B (2020) A review on sonochemistry and its environmental applications. Acoustics 2:766–775
113. Scheel D, Parker JE (1990) Elicitor recognition and signal transduction in plant defense gene activation. Zeitschrift für Naturforsch C 45:569–575
114. Seibert JB, Bautista-Silva JP, Amparo TR, Petit A, Pervier P, dos Santos Almeida JC, Azevedo MC, Silveira BM, Brandão GC, de Souza GHB (2019) Development of propolis nanoemulsion with antioxidant and antimicrobial activity for use as a potential natural preservative. Food Chem 287:61–67
115. Shafie R (2017) Antiviral activity of clove oil nanoemulsion against potato virus-Y (PVY). Egypt J Phytopathol 45:15–31
116. Shin J, Cho H, Hwang SW, Jung J, Shin CY, Lee S-Y, Kim SH, Lee MG, Choi YH, Kim J (2002) Bradykinin-12-lipoxygenase-VR1 signaling pathway for inflammatory hyperalgesia. Proc Natl Acad Sci 99:10150–10155
117. Shinoda K, Friberg S (1986) Emulsions and solubilization. Wiley-Interscience
118. Shinoda K, Saito H (1968) The effect of temperature on the phase equilibria and the types of dispersions of the ternary system composed of water, cyclohexane, and nonionic surfactant. J Colloid Interface Sci 26:70–74
119. Shinoda K, Hanrin M, Kunieda H, Saito H (1981) Principles of attaining ultra-low interfacial tension: the role of hydrophile—lipophile-balance of surfactant at oil/water interface. Colloids Surf 2:301–314

120. Singh Y, Meher JG, Raval K, Khan FA, Chaurasia M, Jain NK, Chourasia MK (2017) Nanoemulsion: concepts, development and applications in drug delivery. J Control release 252:28–49
121. Sivasankarapillai VS, Pillai AM, Rahdar A, Sobha AP, Das SS, Mitropoulos AC, Mokarrar MH, Kyzas GZ (2020) On facing the SARS-CoV-2 (COVID-19) with combination of nanomaterials and medicine: possible strategies and first challenges. Nanomaterials 10:852
122. Sohrabi S, Moraveji MK (2020) Droplet microfluidics: fundamentals and its advanced applications. RSC Adv 10:27560–27574
123. Stone HA, Stroock AD, Ajdari A (2004) Engineering flows in small devices: microfluidics toward a lab-on-a-chip. Annu Rev Fluid Mech 36:381–411
124. Subramanian B, Kuo F, Ada E, Kotyla T, Wilson T, Yoganathan S, Nicolosi R (2008) Enhancement of anti-inflammatory property of aspirin in mice by a nano-emulsion preparation. Int Immunopharmacol 8:1533–1539
125. Suzuki K, Shuto I, Hagura Y (1996) Characteristics of the membrane emulsification method combined with preliminary emulsification for preparing corn oil-in-water emulsions. Food Sci Technol Int Tokyo 2:43–47
126. Tavakkol Afshari HS, Homayouni Tabrizi M, Ardalan T (2019) Evaluation of antioxidant and anticancer effects of nanoemulsions prepared using dill essential oil. J Arak Univ Med Sci 22:40–51
127. Tragoolpua Y, Jatisatienr A (2007) Anti-herpes simplex virus activities of Eugenia caryophyllus (Spreng.) Bullock & SG Harrison and essential oil, eugenol. Phyther Res Int J Devot Pharmacol Toxicol Eval Nat Prod Deriv 21:1153–1158
128. ud Din F, Aman W, Ullah I, Qureshi OS, Mustapha O, Shafique S, Zeb A (2017) Effective use of nanocarriers as drug delivery systems for the treatment of selected tumors. Int J Nanomed 12:7291
129. Ulanowska M, Olas B (2021) Biological properties and prospects for the application of eugenol—a review. Int J Mol Sci 22(7):3671
130. Vladisavljević GT, Williams RA (2005) Recent developments in manufacturing emulsions and particulate products using membranes. Adv Colloid Interface Sci 113:1–20
131. Vongtau HO, Abbah J, Ngazal IE, Kunle OF, Chindo BA, Otsapa PB, Gamaniel KS (2004) Anti-nociceptive and anti-inflammatory activities of the methanolic extract of Parinari polyandra stem bark in rats and mice. J Ethnopharmacol 90:115–121
132. Wang Y (2014) Preparation of nano-and microemulsions using phase inversion and emulsion titration methods: a thesis presented in partial fulfilment of the requirements for the degree of Master of Food Technology at Massey University, Auckland, New Zealand
133. Wang SH, Chen J, Smith D, Cao Z, Acosta H, Fan Y, Ciotti S, Fattom A, Baker J (2020) A novel combination of intramuscular vaccine adjuvants, nanoemulsion and CpG produces an effective immune response against influenza A virus. Vaccine 38:3537–3544. https://doi.org/10.1016/j.vaccine.2020.03.026
134. Williams RA (2001) Making the perfect particle. Ingenia 1
135. Wong PT, Wang SH, Ciotti S, Makidon PE, Smith DM, Fan Y, Schuler CF IV, Baker JR Jr (2014) Formulation and characterization of nanoemulsion intranasal adjuvants: effects of surfactant composition on mucoadhesion and immunogenicity. Mol Pharm 11:531–544
136. Yan R, Zhang Y, Li Y, Xia L, Guo Y, Zhou Q (2020) Structural basis for the recognition of SARS-CoV-2 by full-length human ACE2. Science (80–)367:1444–1448
137. Yao C, Zhao S, Liu L, Liu Z, Chen G (2022) Ultrasonic emulsification: basic characteristics, cavitation, mechanism, devices and application. Front Chem Sci Eng 16:1560–1583
138. Yener FG, Acar C, Çelik BÖ, Kara EM (2022) Development of innovative cosmetic formulations to help fungal treatment and testing the efficiency of formulations. Fabad J Pharm Sci 47:347–354
139. Yukuyama MN, Ghisleni DDM, de Pinto TJA, Bou-Chacra NA (2016) Nanoemulsion: process selection and application in cosmetics—a review. Int J Cosmet Sci 38:13–24
140. Zhang J, Miao J, Han X, Lu Y, Deng B, Lv F, Zhao Y, Ding C, Hou J (2018) Development of a novel oil-in-water emulsion and evaluation of its potential adjuvant function in a swine influenza vaccine in mice. BMC Vet Res 14:415. https://doi.org/10.1186/s12917-018-1719-2

141. Zhang Y, Xin C, Cheng C, Wang Z (2020) Antitumor activity of nanoemulsion based on essential oil of Pinus koraiensis pinecones in MGC-803 tumor-bearing nude mice. Arab J Chem 13:8226–8238
142. Zhao L, Zhu Z, Ma L, Li Y (2020) O/W nanoemulsion as an adjuvant for an inactivated h3n2 influenza vaccine: based on particle properties and mode of carrying. Int J Nanomed 15:2071–2083. https://doi.org/10.2147/IJN.S232677
143. Zhong J, Shi G (2019) Regulation of inflammation in chronic disease. Front Immunol
144. Zhong P, Agosto LM, Munro JB, Mothes W (2013) Cell-to-cell transmission of viruses. Curr Opin Virol 3:44–50
145. Ziani K, Chang Y, McLandsborough L, McClements DJ (2011) Influence of surfactant charge on antimicrobial efficacy of surfactant-stabilized thyme oil nanoemulsions. J Agric Food Chem 59:6247–6255. https://doi.org/10.1021/jf200450m

Safety Regulation of Nanoemulsion in the Food, Agriculture and Biomedical Sector

Upendra Singh, Atreyi Pramanik, Aashna Sinha, Anish Kumar Pal, Pranchal Rajput, Kundan Kumar Chaubey, Sapna Yadav, Chetan Shrivastava, Sujata Hariharan, Rakesh Kumar Bachheti, and Archana Bachheti

Abstract One of the most crucial instruments in contemporary agriculture is nanotechnology and agri-food nanotechnology is predicted to become a major economic force. The preservation and sustainability of agriculturally produced foods, including both crops for human consumption and those used as animal feed, are particularly stressed by agri-food themes. The use of pesticides is expected to decline thanks to the introduction of novel agrochemical agents and delivery systems made possible by nanotechnology. In medicine, improved medication bioavailability is made possible, along with fewer negative side effects, less non-specific absorption, and precise targeting of particular target cells. Meanwhile, it plays a significant role in food preservation and the bio-fortification of food with beneficial compounds in the food sector. New generations of agrochemicals are produced in large quantities using nanotechnology. Natural components can be found in bio-based nanoemulsions (NEs) that are members of the lipid nanocarrier class. These components include the plant oils used for the oil phase, emulsifiers, biosurfactants, cosurfactants, targeting ligands on the surface of the NEs (like folate), or encapsulating active substances. In addition to protecting encapsulated chemicals from deterioration, ensuring their

Upendra Singh and Atreyi Pramanik are equally contributed.

U. Singh · A. Pramanik · A. Sinha · A. K. Pal · P. Rajput · K. K. Chaubey (✉) · S. Yadav · C. Shrivastava · S. Hariharan
Division of Research and Innovation, School of Applied and Life Sciences, Uttaranchal University, Arcadia Grant, P.O. Chandanwari, Premnagar, Dehradun 248007, Uttarakhand, India
e-mail: kundan2006chaubey@gmail.com

K. K. Chaubey
School of Basic and Applied Sciences, Sanskriti University, Mathura, Uttar Pradesh-281401, India

R. K. Bachheti
Department of Industrial Chemistry, Centre of Excellence in Nanotechnology, Addis Ababa Sciences and Technology University, P.O. Box 16417, Addis Ababa, Ethiopia

A. Bachheti
Department of Environment Science, Graphic Era University, Dehradun 248002, Uttarakhand, India

A. Husen et al. (eds.), *Current Trends in Green Nano-emulsions*, Smart Nanomaterials Technology, https://doi.org/10.1007/978-981-99-5398-1_16

lasting discharge, and lowering the quantity of an active component required for the desired effect, the bio components in these NEs are not hazardous to living things.

Keywords Nanoemulsion · Biomedical · Agriculture · Composition · Safety

1 Introduction

Severe health-related concerns contribute to the international dissemination of healthier, safer, and cost-effective food items. Functional foods were also developed as a tool to give food an additional purpose. This can be accomplished by producing more already physiologically active molecules or including new bioactive components. As a result, food products typically have qualities that promote health or prevent disease in addition to their nutritional value. Nevertheless, it has become obvious that many health-promoting drugs' limited bioavailability or ineffective long-term stability may not retain their effects. Nanoemulsions (NEs) are one of the most fascinating delivery technologies in the food industry. Food stability is increased, and the bioavailability of the encapsulated bioactive components is improved by NEs-based delivery methods [36]. One of the most popular nanocarriers is NEs, particularly in the food business. The droplet size of NEs ranges from 20 to 200 nm, and it is a biphasic heterogeneous, thermodynamically unstable, kinetically stable, transparent, colloidal system. It is made up of surfactant, cosurfactant, aqueous, and oil. Due to their compositional flexibility, NEs are widely employed in food science. They also considerably affect food systems' texture, rheology, and organoleptic aspects [18]. The dispersion of two immiscible liquids is known as an emulsion, with the spherical droplets constituting the dispersed phase and the surrounding liquid the continuous phase. Water and oil are the two most frequently used liquids to create emulsions [26]. Large oil droplets with water droplets scattered in an aqueous phase make up the w/o/w emulsions.

In contrast, water droplets, including oil droplets, are dispersed in an oil phase in an o/w/o emulsion system. Continuous NEs have inter-distributed oil and water microdomains in the system [3]. Applications for NEs include medicine delivery, pharmaceuticals, cosmetics, and food (Fig. 1). NEs are one of the most fascinating delivery technologies in the food industry. Food stability is increased, and the bioavailability of the encapsulated bioactive components is improved by NEs-based delivery methods [36]. It is crucial to ensure they are safe to use and have no negative impacts on human health. In order to better understand the potential toxicity of swallowed nanoparticles (NPs) like NEs, significant efforts are being made [1]. The potential to encapsulate bioactive compounds with improved solubility, bioavailability, protection, and targeted delivery makes the NEs system a compelling choice. Regarding stability against creaming, flocculation, and sedimentation, the nano-based system performs better than micro- and macroemulsions [25]. This chapter discusses the contribution of NEs in the agri-food industry and biomedical sector.

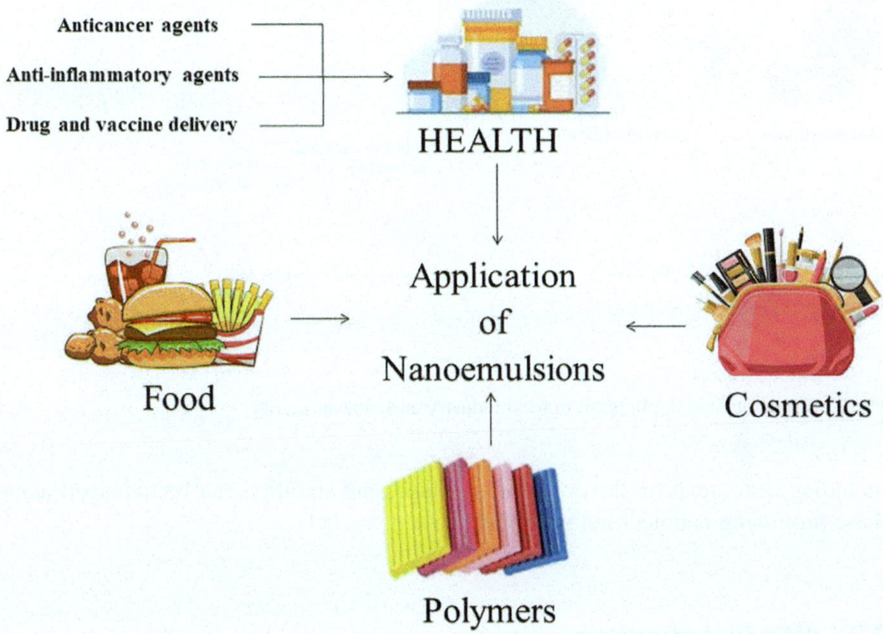

Fig. 1 NEs in food, agriculture and cosmetics

2 NEs Security

2.1 NEs in Food

Several industries, including medicine delivery, pharmaceuticals, cosmetics, and food, use NEs. This section focuses on the uses of NEs in the food business. As a suitable form, NEs have been utilized to enhance food digestion, the bioavailability of active ingredients, the pharmacological actions of certain chemicals, and the solubilization of medications. NEs have been employed as an effective form to increase food digestibility, the bioavailability of active ingredients, the pharmacological actions of certain chemicals, and the solubilization of medications (Fig. 2). The amount of food absorbed by the gastrointestinal tract and enters the bloodstream is referred to as food digestibility. In order to enhance the characteristics of food and natural extracts' digestion, NEs have been used as a suitable form [36].

Depending on the purpose, NE-based delivery systems may include nutrients, flavorings, preservatives, colors, or preservatives. Compared to samples packaged using traditional methods, this technology can stop microbe development, changes in food color and appearance, weight loss, moisture content reduction, unwanted flavor and taste, and slow the rate of oxidation and browning. Several food qualities,

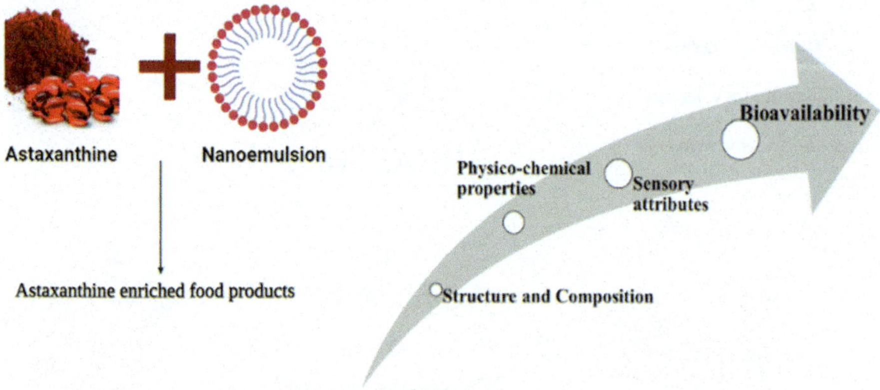

Fig. 2 Nanoemulsion application in food industry and bioavailability

including taste, texture, flavor, color, spoilage, and stability, can be managed using these promising nanotechnologies [1].

2.2 *NEs in Agriculture*

In order to increase crop yield while ensuring food safety and nutritional value, nanotechnology is used in agriculture [11, 14]. Using nanomaterials, which enable the delayed release and transport of these nanoagrochemicals, is necessary to boost crop output by controlling the release of fertilizers, insecticides, herbicides, and plant growth regulators/promoters [21–23, 30, 38]. A farming technique called precision agriculture is used to maximize agricultural productivity without harming the soil and water. Another way nanotechnology is used in agriculture for industrial purposes involves growing plants in certain soils to extract NPs from the soil, a practice known as particle farming (Ali et al. 2018). Another method used to reduce the dosage of nanoagrochemicals and improve the environment is nanoencapsulation. Insect-resistant plants have been developed through gene or DNA transfer using NPs. Certain NPs themselves behave as insecticides with great toxicity and sensitivity. Due to their inherent properties, metal oxide NPs like Zinc oxide (ZnO), Cupric oxide (CuO), and Titanium dioxide (TiO_2) can be used to protect plants from infections by plant pathogens (Nikolova and Chavali 2020). ZnO NPs' antifungal and antibacterial properties prevent the growth of microorganisms. By improving crop growth and quality, nanofertilizers are aimed to reduce nutrient waste while increasing nutritional value, which increases economic gains (Nikolova and Chavali 2020) [9–13, 15]. Commercialized Nano fertilizers are sold as micronutrients with nanoscale of Manganese (Mn), Copper (Cu), Iron (Fe), Zinc (Zn), Molybdenum (Mo), Nitrogen (N), and Boron (B). Using wireless Nano sensors dispersed across the fields, nanotechnology is used in crop disease surveillance, crop growth, plant

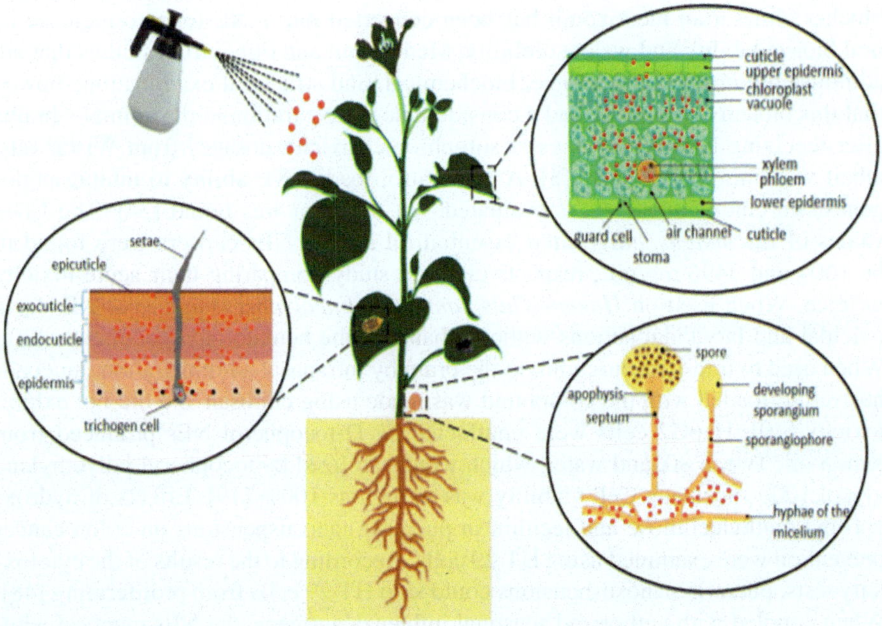

Fig. 3 Diagrammatic representation of NEs-based agricultural chemicals penetrating into whole plant

health, and environmental factors (Shang et al. 2019). Nanosensors in the food and agricultural systems can identify Nano agrochemicals and diseases even in extremely low quantities. According to a recent study, copper-doped montmorillonite is used for online monitoring of the aquatic environment's exposure to the fungicide propineb (in both fresh and saltwater). Aquaculture can benefit from graphene NPs since they can detect pathogens in effluent and remove them. For real-time monitoring in agriculture, nanosensors made of gold, silver, copper, and CNT materials are also used [38] (Shaw and Honeychurch 2022) (Fig. 3).

2.3 NEs in Biomedical Sector

How NEs interact with biological systems determines their toxicity and safety. In the past several years, numerous comprehensive studies have been conducted on the interactions of inorganic NPs with biological systems. However, very little study has been conducted on the interactions of NEs droplets with biological systems. NEs containing vitamin E altered the expression of numerous genes, which altered how ketone bodies were formed (β-hydroxybutyrate and palmitoleic acid), how energy and xenobiotic were metabolized (*CYP1A1* and glutathione S-transferase), and how much stearoyl-CoA desaturase was produced [33]. Angiotensin II receptor blocker olmesartan medoxomil has been converted into a NE form to increase its oral bioavailability and water solubility. Medoxomil and olmesartan studies that are administered orally, hematologic, biochemical, and structural examination showed that this bioactive molecule had a considerable concentration in the animal's brains after receiving NEs for 28 days at sub-chronic toxicity dosages from Wistar rats, albeit with minimal toxicity [8]. A resveratrol-loaded NE ability to inhibit angiogenesis in chick embryos was evaluated, but no harm was found [34]. The LD_{50} values of *Eucalyptus staigeriana* essential oil and its NEs carriers were found to be 1604 and 3496 mg/mL, respectively, in a study comparing their acute toxicity in mice. When used on *Haemonchus contortus*, *Eucalyptus staigeriana* NEs have ovicidal and larvicidal actions without changing the hematologic parameters [20]. When used to transport curcumin to the brain by intranasal administration, mucoadhesive NE loaded with the compound was made using chitosan and did not exhibit toxicity [40]. HepG2 cells were unaffected by D-tocopherol NEs produced from lemon oil, Tween 80, and water, which were stabilized by tocopheryl polyethylene glycol 1000 succinate. Cell viability was as high as 100% [19]. Effects of hydroxypropyl methylcellulose and lecithin in puerarin nanosuspensions on colon cancer prevention were examined using HT-29 cells. According to the results of the cytotoxicity tests, puerarin nanosuspensions could stop HT-29 cells from proliferating [48]. When coupled with authorized seasonal influenza antigens, the NEs mucosal adjuvant W805EC demonstrated no toxicity, and healthy adult human subjects tolerated it well [41]. In a 28-day subchronic toxicity trial, paclitaxel NEs (6.5 and 3 mg/kg oral dosages), an anticancer medication, exhibited no significant hematologic, biochemical, or structural abnormalities. However, at a dose of 12.8, mg/kg body weight, a noticeable drop in RBCs, hemoglobin, and neutrophils were seen [5]. When used in large doses, pomegranate seed oil NEs had a hemolytic impact [31]. In an oral administration of 18.63 mL/kg of garlic oil NEs to female rats as part of an acute toxicity investigation, all rats died immediately. However, 0.46 mL/kg of garlic oil NEs and 0.5 mL/kg of Tween 80 did not cause any hematologic or histological alterations [35]. It has been demonstrated that quercetin NEs have good skin biocompatibility and cause little eye discomfort [6]. A recent study found that curcumin-loaded NEs could induce the death of colon cancer cells while having no cytotoxic effects on healthy fibroblasts and having strong anti-inflammatory characteristics [45]. Most recent research on the toxicity and safety of NEs in biological systems points to the safety of their use. Only a few studies have found NEs to be harmful, likely due to the extremely high concentrations used. Thus, it is possible to conclude that most NEs are safe when utilized in low quantities (Fig. 4).

3 Safety Regulation in NEs

It is difficult to determine the safety of nanoscale products in general and of NEs in particular. Numerous research studies have been conducted to determine the possible toxicity of inorganic NEs. However, relatively few researches have been done to

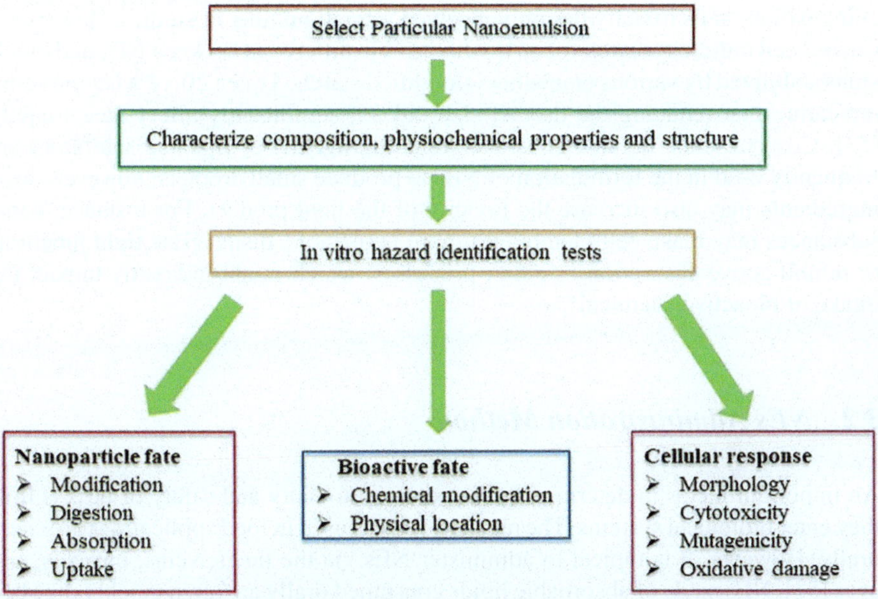

Fig. 4 Proposed protocol for the evaluation of potential toxicity of food-grade NEs

assess the toxicity and safety of NEs. So, much of what we now know about the study, NEs may harm other nanoscale materials. The possible toxicological effects of NEs on biological systems are potentially influenced by a wide range of variables, which can be grouped into many major categories:

- Composition of NEs
- NEs administration method

3.1 Composition of NEs

A multitude of organic or synthetic ingredients, such as oils, surfactants, cosurfactants, weighing agents, ripening inhibitors, thickeners, and gelling agents, may be used to create NEs [2, 16, 28] (Wani et al. 2016). Some of these ingredients are either not used in traditional emulsions or are used far less frequently than in NEs [7, 17, 29]. Since the components used to create NEs differ from those used to create emulsions, they are likely to have toxicity profiles that are distinct from emulsions. Given their widespread use at high concentrations in other products, many components used to create NEs, such as protein-emulsifiers with phospholipid and polysaccharide bases, would not be expected to enhance toxicity. However, several synthetic surfactants, including anionic surfactants (such as lactic acid esters), nonionic surfactants, and cationic surfactants (such as lauricarginate), may increase itchiness or high amounts of usage can be dangerous (e.g., Tweens and Spans) [26]. The effect of the NEs's composition on its toxicity has only received a small number of studies. However, a Caco-2 cell culture research found no discernible difference between NEs and emulsions stabilized by various emulsifiers (modified starch, Tween 20, or whey protein), indicating that reducing the droplet size did not significantly affect their toxicity [53]. Cosurfactants, organic solvents, weighing agents, or ripening inhibitors are frequently used in the formulation of NEs to produce small droplets; however, these ingredients may also increase the toxicity of the final product. For instance, some substances may make cell membranes more permeable, disintegrate tight junctions or inhibit active absorption or efflux processes, which might indirectly impact the intake of bioactive chemicals.

3.2 NEs Administration Method

An important factor in determining the potential toxicity and safety of NEs is how they enter biological systems. The majority of NEs used in food applications are taken orally. However, it is typical to administer NEs via the lungs, veins, muscles, and eyes [53]. NEs made of absorbable lipids consumed orally are often quickly digested in the digestive system, resulting in mixed micellar that contain monoacylglycerides and free fatty acids, which are then absorbed. As a result, even if they are created at a faster rate, As a result of digesting NEs, the final products may be the same as those produced by traditional emulsions, which may change the levels of blood serum [27]. Therefore, only a very small part of the swallowed NEs droplets will typically be able to pass through the epithelial cells will absorb the substance without being harmed by the mucous layer [27] (Müller et al. 2011). A small percentage of NPs are absorbed in the gut without being digested; it is unknown how harmful these particles may be.

When inedible lipids are utilized to create NEs, a far more significant proportion of NPs may be directly absorbed. The properties of the NEs droplets may then impact their absorption, distribution, metabolism, and excretion (such as particle size, charge, and composition) [4]. There is certainly a great need for more research in the future because our current grasp of this crucial subject needs to be improved. Due to its potential to amplify the permeability and retention of bioactive elements, NEs may also be administered intravenously in medicinal applications [48]. The acceptable range for the mean particle diameter of NEs designed for these applications is between 100 and 300 nm [42]. The droplets may obstruct the blood vessels if they are sufficiently big; nevertheless, if they are too tiny, they could be absorbed rapidly and unevenly. One of the key determinants of the component that was injected's toxicity and safety is the dosage of the injection. Another method for continuously releasing bioactive components in biological systems is an intramuscular injection [51]. To manage medication pharmacokinetics, it is crucial to regulate the size of the droplets in NEs designed for this kind of application. It has been demonstrated that administering curcumin and curcumin didecanoatenano suspensions intra muscularly has no adverse effects on the spleen, heart, lung, liver, or kidney [51].

Consequently, there is room for research into NEs designed for intravenous and intramuscular injection. Since it has been shown that NEs can improve their permeability profiles, risk relating to the skin penetration pattern of active substances should be considered for cosmetic-loaded NEs. Cosmetic components can typically penetrate the top layer of human skin, but they should not go much deeper than that. As a result, it is essential to examine the potential that employing NEs could improve its penetration profile as a key component of ensuring consumer safety [32]. The risk analysis of cosmetic products based on NEss suggests that the regulatory framework may need reevaluation (Table 1).

NEs offer very promising multisectoral usage in the healthcare, food, polymer manufacturing, and cosmetics industries based on their physicochemical and functional characteristics. Consequently, they have attracted considerable attention from the scientific community. When hydrophobic bioactive substances have to be encapsulated and released, NEs is one of the most straightforward and efficient delivery strategies. Depending on the kind of substances utilized and the manufacturing processes employed to create them, they have different compositions and architectures. Depending on their intended use, they may also be given via various delivery methods (including the mouth, skin, nose, eyes, or intravenously). Because the system composition of NEs has changed from traditional NEs, the additional components or higher ingredient levels frequently needed to form the very small droplets may

Table 1 Safety regulations used in different publication of nano-emulsion in the food, agriculture and biomedical sector

Nano-emulsions	Impact of nano-emulsion	Key references
Agriculture		
Neem oil (0.5–3 w/v%)	*Aspergillus flavus* and *Penicillium citrinum*-specific antifungal activities	Silva et al. [39]
Clove oil (1 w/v%)	Activity against *Glomerella cingulate* that is antifungal and cytotoxic	Silva et al. [39]
	Effect of genotoxicity and ecotoxicityon *Folsomia candida*	
Eugenol oil (3.3 v/v%)	Action against *Glomerella cingulata* that is antifungal	Velho et al. [46]
Food		
Ripening retardant	Sodium chloride	Schuch et al. (2014)
Citral	Flavoured substance	Yang et al. [52]
β-Carotene	Coloured substance	Tan and Nakajima [43]
Biomedical sector		
Insulin	Increasing the effectiveness and oral absorption of insulin	Li et al. [24]
Aspirin	Minimise any negative consequences	Tang et al. [44]
Docetaxel	To provide a reliable, efficient, and secure substitute	Verma et al. [47]

impact their toxicity. By altering the solubility, stability, or absorption of encapsulated bioactive compounds, the tiny size of the droplets in NEs may also impact their biological fate. However, this issue can frequently be solved by creating NEs solely from substances proven to be secure when utilized at the necessary concentrations. Applications for bioactive substances and functional food components in NEs in the food sectors are quite promising. The functions of food can be improved, as well as their quality and shelf life, with the help of emulsion-based delivery methods and NEs edible coatings. However, using food-grade NEs will be common only if their manufacturing cost is commercially viable and exceeds the industry's safety standards. The bioactivity of the encapsulated components must be maximized due to increased production. More research should concentrate on the biological effects and dangers of using delivery systems based on NEs in food products and packaging applications to assure consumer safety.

Nanotechnology makes potential new applications for agriculture, food, and medicine possible. With its scope including complete industrial operations, starting with plant cultivation and animal nutrition, then moving on to food manufacturing, monitoring, and packaging, and possibly beyond, which encompasses the entire chain from farm to fork, nanomaterials have sparked a tremendously positive change. By using methods including targeted delivery, slow/controlled release, conditional release of active substances in response to environmental cues, accurately matching biological demands and other methods, nanotechnology has introduced nanostructure formulations in agriculture, revolutionizing the industry. Applications of nanotechnology already in use include eradicating the harmful effects of traditional. The effects of agrochemicals and food business practices on human health and the environment and moving toward a more modern approach that makes use of this technology's potential and high reliability.

4 Conclusion and Perspectives

One of the most straightforward and efficient delivery techniques for encapsulating and releasing hydrophobic bioactive substances is NEs. Depending on the type of ingredients used and the manufacturing techniques, they may have various compositions and architectures. Depending on their intended use, they may also be given in other ways (such as the mouth, skin, nose, eyes, or intravenously). Due to the different system composition from typical NEs, the additional components or higher ingredient concentrations frequently needed to form the very small droplets in NEs may impact their toxicity. However, this issue can frequently be solved by creating NEs solely from substances proven to be secure when utilized at the necessary concentrations. By changing the stability, solubility, or absorption of encapsulated bioactive substances, the tiny size of the droplets in NEs may also impact their biological fate. It is often preferable to boost the bioactivity of bioactive substances, which increases their bioavailability. However, in certain circumstances, this can be undesirable because the bioactive agent is harmful by nature (like pesticides) or because the bioactive agent is toxic at higher levels, even while it is useful when absorbed at relatively low to intermediate levels (such as some oil-soluble vitamins). The physicochemical underpinnings of the possible toxicity of NEs need to be better understood. Additional in vitro and in vivo research is needed in this area. There is no global regulatory body in this field, and the regulation of nanoscale items (including NEs) is still in its early phases. It is still urgent for regulatory bodies worldwide to create reliable, consistent processes for evaluating the safety of nanoscale items. The creation of these laws will ensure that the advantages of nanotechnology may be reaped without negatively impacting consumer and environmental health.

References

1. Ahari H, Naeimabadi M (2021) Employing nanoemulsions in food packaging: shelf life enhancement. Food Eng Rev 13(4):858–883
2. Assadpour E, Jafari SM (2017) Spray drying of folic acid within nano-emulsions: optimization by Taguchi approach. Drying Technol 35(9):1152–1160
3. Aswathanarayan JB, Vittal RR (2019) Nanoemulsions and their potential applications in food industry. Front Sustain Food Syst 3:95
4. Bouwmeester H, Dekkers S, Noordam MY, Hagens WI, Bulder AS, De Heer C, Sips AJ et al (2009) Review of health safety aspects of nanotechnologies in food production. Regul Toxicol Pharmacol 53(1):52–62
5. Choudhury H, Gorain B, Chatterjee B, Mandal UK, Sengupta P, Tekade RK (2017) Pharmacokinetic and pharmacodynamic features of nanoemulsion following oral, intravenous, topical and nasal route. Curr Pharm Des 23(17). https://doi.org/10.2174/13816128226661612011 43600
6. Dario MF, Santos MSC, Viana AS, Areas EP, Bou-Chacra NA, Oliveira MC, Velasco MVR et al (2016) A high loaded cationic nanoemulsion for quercetin delivery obtained by sub-PIT method. Colloids Surf A: Physicochem Eng Aspects 489:256–264
7. Esfanjani AF, Assadpour E, Jafari SM (2018) Improving the bioavailability of phenolic compounds by loading them within lipid-based nanocarriers. Trends Food Sci Technol 76:56–66
8. Gorain B, Choudhury H, Tekade RK, Karan S, Jaisankar P, Pal TK (2016) Comparative biodistribution and safety profiling of olmesartan medoxomil oil-in-water oral nanoemulsion. Regul Toxicol Pharmacol 82:20–31
9. Husen A (2020) Interactions of metal and metal-oxide nanomaterials with agricultural crops: an overview. In: Husen A, Jawaid M (eds) Nanomaterials for agriculture and forestry applications. Elsevier Inc., Cambridge, pp 167–197. https://doi.org/10.1016/B978-0-12-817852-2.00007-X
10. Husen A (2020) Carbon-based nanomaterials and their interactions with agricultural crops. In: Husen A, Jawaid M (eds). Elsevier Inc., Cambridge, pp 199–218. https://doi.org/10.1016/B978-0-12-817852-2.00008-1
11. Husen A (2022) Engineered nanomaterials for sustainable agricultural production, soil improvement and stress management. Elsevier Inc., Cambridge
12. Husen A (2023) Nanomaterials and nanocomposites exposures to plants (Response, interaction, phytotoxicity and defense mechanisms). Springer Nature Singapore Pte Ltd., Singapore
13. Husen A, Iqbal M (2019) Nanomaterials and plant potential: an overview. In: Husen A, Iqbal M (eds) Nanomaterials and plant potential. Springer International Publishing AG, Cham, pp 3–29. https://doi.org/10.1007/978-3-030-05569-1_1
14. Husen A, Jawaid M (2020) Nanomaterials for agriculture and forestry applications. Elsevier Inc., Cambridge. https://doi.org/10.1016/C2018-0-02349-X. ISBN 9780128178522

15. Husen A, Siddiqi KS (2023) Advances in smart nanomaterials and their applications. Elsevier Inc., Cambridge
16. Jafari SM, Paximada P, Mandala I, Assadpour E, Mehrnia MA (2017) Encapsulation by nanoemulsions. In: Nanoencapsulation technologies for the food and nutraceutical industries. Academic Press, pp 36–73
17. Jafari SM, Fathi M, Mandala I (2015) Emerging product formation. In: Food waste recovery: processing technologies and industrial techniques
18. Jena GK, Parhi R, Sahoo SK (2022) Nanoemulsions in food industry. In: Application of nanotechnology in food science, processing and packaging. Springer International Publishing, Cham, pp 73–91
19. Kaur A, Katiyar SS, Kushwah V, Jain S (2017) Nanoemulsion loaded gel for topical co-delivery of clobitasol propionate and calcipotriol in psoriasis. Nanomed: Nanotechnol Biol Med 13(4):1473–1482
20. Klojdová I, Milota T, Smetanová J, Stathopoulos C (2023) Encapsulation: a strategy to deliver therapeutics and bioactive compounds? Pharmaceuticals 16(3):362
21. Kumar A, Choudhary A, Kaur H, Guha S, Mehta S, Husen A (2022) Potential applications of engineered nanoparticles in plant disease management: a critical update. Chemosphere 295:133798. https://doi.org/10.1016/j.chemosphere.2022.133798
22. Kumar A, Choudhary A, Kaur H, Mehta S, Husen A (2021) Smart nanomaterial and nanocomposite with advanced agrochemical activities. Nanoscale Res Lett 16(156):1–26. https://doi.org/10.1186/s11671-021-03612-0
23. Kumar S, Sharma AK, Lalhlenmawia H, Kumar D (2021) Natural compounds targeting major signaling pathways in lung cancer. Targeting cellular signalling pathways in lung diseases, pp 821–846. https://doi.org/10.1007/978-981-33-6827-9_37
24. Li M, Ma Y, Cui J (2014) Whey-protein-stabilized nanoemulsions as a potential delivery system for water-insoluble curcumin. LWT-Food Sci Technol 59(1):49–58
25. Madni A, Jameel F, Bajwa SZ, Rehman A, Khan WS (2022) Toxicity, biological fate, and bioavailability of nanoemulsion formulations. In: Bio-based nanoemulsions for agri-food applications. Elsevier, pp 91–104
26. McClements DJ, Rao J (2011) Food-grade nanoemulsions: formulation, fabrication, properties, performance, biological fate, and potential toxicity. Crit Rev Food Sci Nutr 51(4):285–330
27. McClements DJ (2013) Edible lipid nanoparticles: Digestion, absorption, and potential toxicity. Progr Lipid Res 52(4):409–423
28. Mehrnia MA, Jafari SM, Makhmal-Zadeh BS, Maghsoudlou Y (2017) Rheological and release properties of double nano-emulsions containing crocin prepared with Angum gum, Arabic gum and whey protein. Food Hydrocolloids 66:259–267
29. Mohammadi A, Jafari SM, Esfanjani AF, Akhavan S (2016) Application of nano-encapsulated olive leaf extract in controlling the oxidative stability of soybean oil. Food Chem 190:513–519
30. Mondal R, Dam P, Chakraborty J, Paret ML, Katı A, Altuntas S, Sarkar R, Ghorai S, Mandal AK, Husen A (2022) Potential of nanobiosensor in sustainable agriculture: the state-of-art. Heliyon. https://doi.org/10.2139/ssrn.4134873
31. Mota Ferreira L, Gehrcke M, Ferrari Cervi V, Eliete Rodrigues Bitencourt P, Ferreira da Silveira E, Hofstatter Azambuja J, Cruz L et al (2016) Pomegranate seed oil nanoemulsions with selective antiglioma activity: optimization and evaluation of cytotoxicity, genotoxicity and oxidative effects on mononuclear cells. Pharm Biol 54(12):2968–2977
32. Musazzi UM, Marini V, Casiraghi A, Minghetti P (2017) Is the European regulatory framework sufficient to assure the safety of citizens using health products containing nanomaterials? Drug Discovery Today 22(6):870–882
33. Park J, Fong PM, Lu J, Russell KS, Booth CJ, Saltzman WM, Fahmy TM (2017) PEGylated PLGA nanoparticles for the improved delivery of doxorubicin. In: Nanomedicine in cancer. Jenny Stanford Publishing, pp 575–596
34. Pund S, Thakur R, More U, Joshi A (2014) Lipid based nanoemulsifying resveratrol for improved physicochemical characteristics, in vitro cytotoxicity and in vivo antiangiogenic efficacy. Colloids Surf B: Biointerfaces 120:110–117

35. Ragavan G, Muralidaran Y, Sridharan B, Ganesh RN, Viswanathan P (2017) Evaluation of garlic oil in nano-emulsified form: Optimization and its efficacy in high-fat diet induced dyslipidemia in Wistar rats. Food Chem Toxicol 105:203–213
36. Salem AH, Mietzel T, Brunstermann R, Widmann R (2018) Two-stage anaerobic fermentation process for bio-hydrogen and bio-methane production from pre-treated organic wastes. Bioresour Technol 265:399–406
37. Salem MA, Ezzat SM (2019) Nanoemulsions in food industry. In: Some new aspects of colloidal systems in foods, vol 2, pp 238–267
38. Sharma P, Pandey V, Sharma MMM, Patra A, Singh B, Mehta S, Husen A (2021) A review on biosensors and nanosensors application in agroecosystems. Nano Res Lett 16:136. https://doi.org/10.1186/s11671-021-03593-0
39. Silva HD, Beldíková E, Poejo J, Abrunhosa L, Serra AT, Duarte CM, Vicente AA et al (2019) Evaluating the effect of chitosan layer on bioaccessibility and cellular uptake of curcumin nanoemulsions. J Food Eng 243:89–100
40. Sood S, Jain K, Gowthamarajan K (2014) Optimization of curcumin nanoemulsion for intranasal delivery using design of experiment and its toxicity assessment. Colloids Surf B: Biointerfaces 113:330–337
41. Stanberry LR, Simon JK, Johnson C, Robinson PL, Morry J, Flack MR, Baker Jr JR et al (2012) Safety and immunogenicity of a novel nanoemulsion mucosal adjuvant W805EC combined with approved seasonal influenza antigens. Vaccine 30(2):307–316
42. Sun B, Yeo Y (2012) Nanocrystals for the parenteral delivery of poorly water-soluble drugs. Curr Opin Solid State Mater Sci 16(6):295–301
43. Tan CP, Nakajima M (2005) Effect of polyglycerol esters of fatty acids on physicochemical properties and stability of β-carotene nanodispersions prepared by emulsification/evaporation method. J Sci Food Agric 85(1):121–126
44. Tang SY, Manickam S, Wei TK, Nashiru B (2012) Formulation development and optimization of a novel Cremophore EL-based nanoemulsion using ultrasound cavitation. Ultrason Sonochem 19(2):330–345
45. Vecchione R, Quagliariello V, Giustetto P, Calabria D, Sathya A, Marotta R, Netti PA et al (2017) Oil/water nano-emulsion loaded with cobalt ferrite oxide nanocubes for photo-acoustic and magnetic resonance dual imaging in cancer: In vitro and preclinical studies. Nanomed: Nanotechnol Biol Med 13(1):275–286
46. Velho MC, de Oliveira DA, da Silva Gündel S, Favarin FR, Santos RCV, Ourique AF et al (2019) Nanoemulsions containing mancozeb and eugenol: development, characterization, and antifungal activity against Glomerella cingulata. Appl Nanosci 9:233–241
47. Verma P, Meher JG, Asthana S, Pawar VK, Chaurasia M, Chourasia MK (2016) Perspectives of nanoemulsion assisted oral delivery of docetaxel for improved chemotherapy of cancer. Drug Deliv 23(2):479-488
48. Wang YG, Kim H, Mun S, Kim D, Choi Y (2013) Indocyanine green-loaded perfluorocarbon nanoemulsions for bimodal 19F-magnetic resonance/nearinfrared fluorescence imaging and subsequent phototherapy. Quant Imaging Med Surg 3(3):132
49. Wang L, Zhang Y (2017) Eugenol nanoemulsion stabilized with zein and sodium caseinate by self-assembly. J Agric Food Chem 65(14):2990–2998
50. Wani TA, Masoodi FA, Jafari SM, McClements DJ (2018) Safety of nanoemulsions and their regulatory status. In Nanoemulsions. Academic Press, pp 613–628
51. Wei W, Lv PP, Chen XM, Yue ZG, Fu Q, Liu SY, Ma GH et al (2013) Codelivery of mTERT siRNA and paclitaxel by chitosan-based nanoparticles promoted synergistic tumor suppression. Biomaterials 34(15):3912–3923
52. Yang X, Tian H, Ho CT, Huang Q (2011) Inhibition of citral degradation by oil-in-water nanoemulsions combined with antioxidants. J Agric Food Chem 59(11):6113–6119
53. Yu J, Huang DY, Yousaf MZ, Hou YL, Gao S (2013) Magnetic nanoparticle-based cancer therapy. Chin Phys B 22(2):027506

The manufacturer's authorised representative in the EU is Springer Nature Customer Service Centre GmbH, Europaplatz 3, 69115 Heidelberg, Germany. If you have any concerns regarding our products, please contact ProductSafety@springernature.com

Printed and bound by CPI Group (UK) Ltd, Croydon, CR0 4YY
01/10/2026
02225867-0013